RENEWABLE RESOURCES AND RENEWABLE ENERGY

A GLOBAL CHALLENGE

RENEWABLE RESOURCES AND RENEWABLE ENERGY

A GLOBAL CHALLENGE

Edited by

Mauro Graziani
Paolo Fornasiero

CRC Press
Taylor & Francis Group
Boca Raton   London   New York

CRC Press is an imprint of the
Taylor & Francis Group, an informa business

CRC Press
Taylor & Francis Group
6000 Broken Sound Parkway NW, Suite 300
Boca Raton, FL 33487-2742

© 2007 by Taylor & Francis Group, LLC
CRC Press is an imprint of Taylor & Francis Group, an Informa business

No claim to original U.S. Government works
Printed in the United States of America on acid-free paper
10 9 8 7 6 5 4 3 2 1

International Standard Book Number-10: 0-8493-9689-1 (Hardcover)
International Standard Book Number-13: 978-0-8493-9689-2 (Hardcover)

Library of Congress Cataloging-in-Publication Data

Renewable resources and renewable energy : a global challenge / Mauro Graziani and Paolo Fornasiero, editors.
        p. cm.
    Includes bibliographical references and index.
    ISBN-13: 978-0-8493-9689-2 (alk. paper)
    1. Renewable energy sources. I. Graziani, Mauro. II. Fornasiero, Paolo, 1968- III. Title.

    TJ808.R458 2007
    621.042--dc22   2006020515

Visit the Taylor & Francis Web site at
http://www.taylorandfrancis.com

and the CRC Press Web site at
http://www.crcpress.com

Preface

The continuous increase of the world population, especially in developing countries, is exponentially escalating the requirements for energy and raw materials. These goods represent the fundamental needs to ensure sustainable human development. The limited availability of resources makes urgent the adoption of suitable strategies in the raw materials and energy sectors in order to prevent an economic and social emergency that, in the absence of adequate strategies — not even considering pessimistic scenarios, will arrive sooner or later.

The fulfillment of basic human needs — food, health, and acceptable environmental quality — requires an enormous amount of energy. It is urgent to act immediately, before it is too late to bridge the gap between developed and developing countries. The lack of adequate actions could lead to worldwide social perturbations that can easily be foreseen on the basis of today's situation.

The large-scale use of renewable resources is becoming urgent. It is generally accepted that the renewable energy produced today is well below the world potential and contributes only marginally to human needs. This is due not only to economic reasons related to the cost of traditional fossil resources and to political choices, but also to technological limitations. The real problem is to know when science and technology will be able to answer the open questions regarding renewable materials and energy, allowing the adoption on a global scale of well-considered decisions before the depletion of traditional energy sources becomes an unsolvable problem. Science and technological development are the key tools for achieving these objectives. Therefore, we must recognize that the progress of scientific research and its applications represents the primary way to solve the major human problems. This "scientific optimism" cannot solve the problem by itself, especially in the developing countries. In fact, in those countries, the lack of infrastructure and human resources with adequate scientific knowledge often precludes the possibility of carrying out innovative research or of adopting technologies derived from the developed world.

There are many proposed sustainable solutions in the energy sector, but so far, only a few are competitive with the use of fossil hydrocarbons. These, besides being our primary energy source, are important building blocks for the synthesis of most of the chemical products that we commonly use every day. Therefore, the end of oil will lead not only to an energy crisis, it will also affect the availability of most of the products that satisfy our elementary needs. Indeed, 90% of the organic substances derive, through chemical transformations/reactions, from seven oil derivates: ethylene, propylene, butenes, benzene, toluene, xylene, and methane. Approximately 16% of oil is transformed into chemicals. Notably, the extraction of oil is also becoming increasingly expensive. In 1920, the energy of a single barrel of oil was sufficient for the extraction and refinement of 50 barrels of oil. Today, with the same energy input, it is possible to get only 5 barrels.

The replacement of oil with new energy or raw feedstocks should not be our primary goal. Rather, our reasonable objective must be the maximization of source diversification. Some general aspects must be constantly kept in mind. First, each geographic area has its own characteristics and resources. Then, the distance between the sources of energy and raw materials and the cities where the electricity and the products are used requires huge investments for transportation and distribution.

Besides the investments for production, we must also consider those related to the exploitation of energy resources. The discontinuous availability of some energy/feedstock sources can be also a serious problem. However, the richness in biodiversity, even considering the sensitivity to climate changes, and the exploitation of agro-overproduction represent a great opportunity for some developing countries. The possible use of agro-food waste for countries lacking in fossil resources, the needs of diversification of resources, and the development of national/local capacities are additional challenges.

Great attention is dedicated nowadays to the so-called hydrogen economy. It must be immediately clarified that hydrogen is not an energy source; rather, it is an energy vector, and it can be considered renewable. Its utilization in fuel cells leads to water as a final product, and hydrogen can be produced from water using solar energy. However, despite some promising experimental successes, hydrogen today is still mainly produced from methane. New, clean, and efficient methods for hydrogen production are necessary to really move to the "hydrogen economy." In any case, science and technology, with their limitations, are the only way to the solution of the problem.

The importance of the topics related to the exploitation of renewable resources and renewable energy is reflected in several international institutions and programs on a global level. The European Union and the United States have been coordinating their strategic plans, setting up working groups and cooperative programs. The UNIDO (United Nations Industrial Development Organization) programs on renewable energy are further evidence of global efforts to address these issues, especially in developing countries.

The technologies of exploiting renewable feedstocks has received particular attention from the International Centre for Science and High Technology of UNIDO. A series of awareness and capacity-building programs as well as the promotion of pilot projects in developing countries have been organized by ICS-UNIDO. In fact, their effort has inspired the present initiative, resulting in the preparation of a comprehensive survey in this field. We appreciate the collaboration of Professor Stanislav Miertus in this initiative.

Mauro Graziani and Paolo Fornasiero

Editors

Mauro Graziani has been a full professor of inorganic chemistry at the University of Trieste since 1975. His scientific interests have moved from organometallic chemistry, to homogeneous catalysis, to catalyst heterogenization on various types of supports, and finally, to heterogeneous catalysis using transition metals supported on different oxides. With regard to these catalytic systems, the topics of hydrogen production, the water–gas shift, and the corresponding NO + CO reactions have been of particular interest. He is coauthor of more than 190 publications and four patents and has been invited to present lectures at the most prestigious congresses in the field. Among his professional experiences, he has been a visiting research associate at Ohio State University and M.I.T.; a visiting professor at the Universities of Cambridge, Campinas, and Zaragoza; a UNESCO scientific advisor; prorector of the University of Trieste, dean of the Faculty of Science, vice-president of the Synchrotron Elettra, and vice-president of the Area Science Park of Trieste. He is an associate fellow of the Third World Academy of Science.

Paolo Fornasiero obtained a PhD in heterogeneous catalysis in 1997. A postdoctoral fellow at the Catalysis Research Center of the University of Reading (U.K.), he became an assistant professor in inorganic chemistry at the University of Trieste in 1998 and an associate professor in 2006. His scientific interests are in the technological application of material science and heterogeneous catalysis to the solution of environmental problems such as the design of innovative materials for catalytic converters, the development of catalysts for the reduction of nitrogen oxides under oxidizing conditions, the combustion of diesel particulate, and the design of new catalysts for the production and purification of hydrogen to be used in fuel cells. He is coauthor of 80 publications in international journals and books, four patents, and a number of communications to national and international meetings, in many cases as an invited lecturer. He was awarded the Stampacchia Prize in 1994 for his first publication, and he received the Nasini Gold Medal in 2005, awarded by the Italian Chemical Society, for his contribution to research in the field of inorganic chemistry.

Contributors

Simone Albertazzi
Department of Industrial Chemistry and
Materials Science
University of Bologna
Italy

Carlos R. Apesteguía
Institut of Research in Catalysis and
Petrochemistry
Nacional University of
Litoral–CONICET
Santa Fe, Argentina

Francesco Basile
Department of Industrial Chemistry and
Materials Science
University of Bologna
Italy

Yusof Basiron
Malaysian Palm Oil Board
Selangor, Malaysia

Andrea Bodor
Department of Chemical Technology
and Environmental Chemistry
Eötvös University
Budapest, Hungary

Rodolfo Bona
Institute of Biotechnology &
Biochemical Engineering
Graz University of Technology
Austria

Gerhart Braunegg
Institute of Biotechnology &
Biochemical Engineering
Graz University of Technology
Austria

Olivia L. Castillo
UN Sec. Gen. Kofi Annan Water &
Sanitation
Advisory Board Member

Kien Yo Cheah
Malaysian Palm Oil Board
Selangor, Malaysia

Yuen May Choo
Malaysian Palm Oil Board
Selangor, Malaysia

Emo Chiellini
Department of Chemistry and Industrial
Chemistry
University of Pisa
Italy

Patrizia Cinelli
Department of Chemistry and Industrial
Chemistry
University of Pisa
Italy

Andrea Corti
Department of Chemistry and Industrial
Chemistry
University of Pisa
Italy

Paolo Fornasiero
Chemistry Department
University of Trieste
Italy

Mauro Graziani
Chemistry Department
University of Trieste
Italy

Gajanana Hegde
Department of Electrical and Computer
Engineering
Curtin University of Technology
Perth, Australia

Peter Heidebrecht
Max-Planck-Institut for Dynamics of
Complex Technical Systems
Magdeburg, Germany

Carmen Hermann
Institute of Biotechnology &
Biochemical Engineering
Graz University of Technology
Austria

Paula Hesse
Institute of Biotechnology &
Biochemical Engineering
Graz University of Technology
Austria

Neal Hickey
Chemistry Department
University of Trieste
Italy

Stanko Hočevar
Laboratory of Catalysis and Chemical
Reaction Engineering
National Institute of Chemistry
Ljubljana, Slovenia

Predrag Horvat
Institute of Biotechnology &
Biochemical Engineering
Graz University of Technology
Austria

Istvan T. Horvath
Department of Chemical Technology
and Environmental Chemistry
Eötvös University
Hungary

A. Inci Işli
Department of Environmental
Engineering
Fatih University
Istanbul, Turkey

Martin Koller
Institute of Biotechnology &
Biochemical Engineering
Graz University of Technology
Austria

Christoph Kutschera
Institute of Biotechnology &
Biochemical Engineering
Graz University of Technology
Austria

Ah Ngan Ma
Malaysian Palm Oil Board
Ministry of Plantation Industries and
Commodities,
Selangor, Malaysia

Hasan Mehdi
Department of Chemical Technology
and Environmental Chemistry
Eötvös University
Hungary

László T. Mika
*Department of Chemical Technology
and Environmental Chemistry
Eötvös University
Budapest, Hungary*

Ramani Narayan
*Department of Chemical Engineering
and Materials Science
Michigan State University
East Lansing, Michigan
United States*

Chem V. Nayar
*Department of Electrical and Computer
Engineering
Curtin University of Technology
Perth, Australia*

José Neto
*Institute of Biotechnology &
Biochemical Engineering
Graz University of Technology
Austria*

Martin Patel
*Department of Science, Technology, and
Society
Utrecht University
The Netherlands*

Luis Pereira
*Institute of Biotechnology &
Biochemical Engineering
Graz University of Technology
Austria*

Kai Sundmacher
*Max-Planck-Institute for Dynamics of
Complex Technical Systems
and
Process Systems Engineering
Otto-von-Guericke University
Magdeburg, Germany*

Kornél Torkos
*Department of Chemical Technology
and Environmental Chemistry
Eötvös University
Budapest, Hungary*

Ferruccio Trifirò
*Department of Industrial Chemistry and
Materials Science
University of Bologna
Italy*

Robert Tuba
*Department of Chemical Technology
and Environmental Chemistry
Eötvös University
Hungary*

Herman van Bekkum
*Ceramic Membrane Centre "The Pore"
Delft University of Technology
The Netherlands*

T. Nejat Veziroglu
*UNIDO-ICHET
Istanbul, Turkey*

Martin Weiss
*Department of Science, Technology, and
Society
Utrecht University
The Netherlands*

Bernard Witholt
*Institute of Biotechnology
ETH
Zurich, Switzerland*

Contents

PART III Technologies for Renewable Energy...... 155

PART IV Trends, Needs, and Opportunities in Developing Countries 289

Part I

Technologies for Application and Utilization of Renewable Resources

1 Rationale, Drivers, Standards, and Technology for Biobased Materials

Ramani Narayan

CONTENTS

Annually renewable resources should be used for the production of materials/products, especially plastics and chemicals, because of the abundant availability of biomass and agricultural feedstocks, their role in managing our carbon emissions in a sustainable and environmentally responsible manner, their potential to create a positive environmental footprint, and the value they add to agriculture. The use of renewable resources would contribute to a country's economic growth, especially in developing countries, many of which have abundant biomass and agricultural resources that provide the potential for achieving self-sufficiency in materials. Furthermore, new environmental regulations, societal concerns, and a growing environmental awareness throughout the world are triggering a paradigm shift toward producing plastics and other materials from inherently biodegradable and annually renewable biomass/agricultural feedstocks. This chapter reviews these drivers and the rationale for designing and engineering biobased materials/products from renewable feedstock resources. Standards for identifying and quantifying biobased content are presented, and the role of biodegradability is assessed. The platform for biobased products technology is discussed, and examples of emerging biobased products are presented.

1.1 INTRODUCTION

1.1.1 BIOMASS/RENEWABLE RESOURCES DRIVERS

There is an abundance of natural, renewable biomass resources. Indeed, the primary production of biomass, estimated in energy equivalents, is 6.9×10^{17} kcal/year [1] (Figure 1.1).

Mankind utilizes only 7% of this amount, i.e., 4.7×10^{16} kcal/year. In terms of mass units, the net photosynthetic productivity of the biosphere is estimated to be 155 billion tons/year [2], or over 30 tons per capita, and this is the case under the current conditions of nonintensive cultivation of biomass. Forests and croplands contribute 42 and 6%, respectively, of that 155 billion tons/year. The world's plant biomass is about 2×10^{12} tons, and the renewable resources amount to about 10^{11} tons/year of carbon, of which starch provided by grains exceeds 10^9 tons (half of which comes from wheat and rice), and sucrose accounts for about 10^8 tons. Another estimate of the net productivity of the dry biomass gives 172 billion tons/year, of which 117.5 and 55 billion tons/year are obtained from terrestrial and aquatic sources, respectively [3].

Fortunately, we are growing trees faster than they are being consumed, although sometimes the quality of the harvested trees is superior to those being planted. Agriculture represents only a small fraction of these vast biomass resources, and this acreage does not include idle croplands and pastures. Again, these figures clearly illustrate the potential for biomass utilization.

AVAILABILITY OF BIOMASS RESOURCES

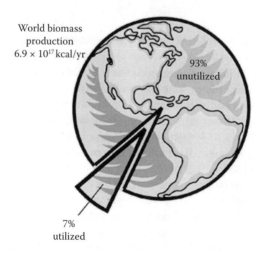

World biomass
production
6.9×10^{17} kcal/yr

93%
unutilized

7%
utilized

FIGURE 1.1 Biomass resources: production and utilization.

In the United States, it is estimated that agriculture accounts directly and indirectly for about 20% of the gross national product (GNP) by contributing $750 billion to the economy through the production of foods and fiber, the manufacture of farm equipment, the transportation of agricultural products, etc. It is also interesting that while agricultural products contribute to the U.S. economy with $40 billion of exports, with each $1 billion of export creating 31,600 jobs (1982 figures), foreign oil imports drain the economy and make up 23% of the U.S. trade deficit (U.S. Department of Commerce, 1987 estimate).

Given these scenarios of abundance of biomass feedstocks and the value added to a country's economy, it seems logical to pursue the use of agricultural and biomass feedstocks for production of materials, chemicals, and fuels [4].

1.1.2 GOVERNMENT POLICY DRIVERS

Biobased materials/products are nonfood, nonfeed agricultural products used in a variety of commercial and industrial applications, thereby harnessing the energy of the sun to provide raw materials. Biobased products include fuels, energy, chemicals, construction materials, lubricants, oils, automotive supplies, and a host of other products. The U.S. government has set the goal of tripling U.S. use of bioenergy and biobased products by the year 2010. Meeting this goal could create an additional $15 to 20 billion a year in new income for farmers and rural America while reducing annual greenhouse gas emission by an amount equal to as much as 100 million metric tons of carbon [5]. Europe and several Asian countries rich in renewable resources are also putting in place policies and goals for the use of their annually renewable resources.

Biomass-derived materials are being produced at substantial levels. For example, paper and paperboard production from forest products was around 139 billion lb in 1988 [6], and production of biomass-derived textiles was around 2.4 billion lb [7]. About 3.5 billion pounds of starch from corn is used in paper and paperboard applications, primarily as adhesives [8]. However, biomass use in production of plastics, coatings, resins, composites, and other articles of commerce is negligible. These areas are dominated by synthetics derived from oil and represent the industrial materials of today.

The development of biobased materials/products — including new commercial (nonfood, nonfeed) markets for agricultural products, the use of agricultural by-products, and the development of new crops — will directly help farmers by providing direct new streams of income and opportunities to become involved through ownership in technology-based value-added enterprises. The farmer also receives benefits through enhanced crop production of current row crops through the inclusion of new crops. New crops help build soil health, reduce perennial weed cycles, and help build resistance to insects and other predators.

1.1.3 ENVIRONMENTAL CONSIDERATIONS

New environmental regulations, societal concerns, and a growing environmental awareness throughout the world have triggered the search for new products and processes that are compatible with the environment. Sustainability, industrial ecology, ecoefficiency, and green chemistry are the new principles that are guiding the development of the next generation of products and processes. Thus, new products have to be designed and engineered "from conception to reincarnation," incorporating a holistic life-cycle-thinking approach. The ecological impact of raw material resources used in the manufacture of a product and the ultimate fate (disposal) of the product when it enters the waste stream has to be factored into the design of the product. The use of annually renewable resources and the biodegradability or recyclability of the product is becoming an important design criterion. This has opened up new market opportunities for developing biodegradable and biobased products as the next generation of sustainable materials that meet ecological and economic requirements — ecoefficient products [9–12].

Currently, most products are designed with limited consideration of their ecological footprint, especially as it relates to their ultimate disposability. Of particular concern are plastics used in single-use disposable packaging and consumer goods. Designing these materials to be biodegradable and ensuring that they end up in an appropriate disposal system is environmentally and ecologically sound. For example, by composting our biodegradable plastic and paper waste along with other "organic" compostable materials like yard, food, and agricultural wastes, we can generate much-needed carbon-rich compost (humic material). Compost-amended soil has beneficial effects by increasing soil organic carbon, increasing water and nutrient retention, reducing chemical inputs, and suppressing plant disease. Composting is increasingly a critical element for maintaining the sustainability of our agriculture system. Figure 1.2 shows a conceptual schematic for the closed-loop use of corn feedstock to prepare starch and protein and then process them into biodegradable, single-use, disposable packaging and plasticware for use in fast-food restaurants. The food wastes, along with other biowastes, are separately collected and composted to generate a valuable soil amendment that goes back on the farmland to reinitiate the carbon cycle [13, 14].

Polymer materials have been designed in the past to resist degradation. The challenge is to design polymers that have the necessary functionality during use but that self-destruct under the stimulus of an environmental trigger after use. The trigger could be microbial, hydrolytically or oxidatively susceptible linkage built into the backbone of the polymer, or additives that catalyze breakdown of the polymer chains in specific environments. More importantly, the breakdown products should not be toxic or persist in the environment, and they should be completely utilized by soil microorganisms within a defined time frame. To ensure market acceptance of biodegradable products, the ultimate biodegradability of these materials in appropriate waste-management infrastructures within reasonable time frames needs to be demonstrated beyond doubt.

1.2 RATIONALE AND DRIVERS FOR BIOBASED PRODUCTS

Manufacturers traditionally have not concerned themselves with the impact on the environment of using various feedstocks. They have also not worried about the

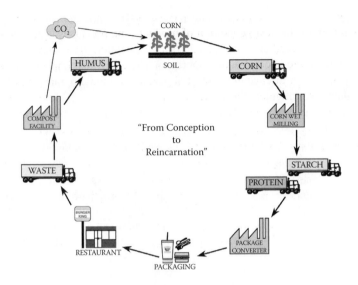

FIGURE 1.2 Biomass resources: production and utilization.

ultimate disposability of their products. The products of the future must be designed "from conception to reincarnation" or "cradle to cradle" using holistic life-cycle concepts. The use of annually renewable biomass — corn, cellulosics, soy, or other vegetable oils, as opposed to petrochemicals (oil or natural gas) — as the feedstocks for the production of polymers, chemicals, and fuel needs to be understood from a global carbon-cycle basis.

1.2.1 CARBON MANAGEMENT

Carbon is the major basic element that is the building block of polymeric materials: biobased products, petroleum-based products, biotechnology products, fuels, even life itself. To summarize:

Carbon is what we, as well as all of the other plants and animals on earth, are made of; it is 50% of our dry weight. Fossil carbon is in the form of $(CH_2)_n$, and biocarbon is in the form of $(CH_2O)_x$.

Carbon is of interest because carbon, in the form of carbon dioxide (CO_2), is the major greenhouse gas released to the atmosphere as a result of human activities. Excessive levels of greenhouse gases can:

- Raise the temperature of the Earth
- Disrupt the climates that we and our agricultural systems depend on
- Raise the sea level

Concentration of CO_2 in the atmosphere has already increased by about 30% since the start of the Industrial Revolution.

Most of the increase in atmospheric CO_2 concentrations has come from and will continue to come from the use of fossil fuels (coal, oil, and natural gas) for energy.

Much (20 to 25%) of the increase over the last 150 years has come from changes in land use, e.g., the clearing of forests and the cultivation of soils for food production.

Therefore, discussions on sustainability, sustainable development, and environmental responsibility center on the issue of managing carbon-based materials in a sustainable and environmentally responsible manner. Natural ecosystems manage carbon through the biological carbon cycle, and so it makes sense to review how carbon-based polymeric materials fit into nature's carbon cycle and address any issues that may arise.

Carbon is present in the atmosphere as CO_2. Photoautotrophs like plants, algae, and some bacteria fix this inorganic carbon to organic carbon (carbohydrates) using sunlight for energy:

$$CO_2 + H_2O \xrightarrow{\text{Sunlight Energy}} (CH_2O)_x + O_2 . \qquad (1.1)$$

Over geological time frames ($>10^6$ years), this organic matter (plant materials) is fossilized to provide our petroleum, natural gas, and coal. Clearly, petrochemical feedstocks are therefore also "natural." We consume these fossil resources to make our polymers, chemicals, and fuel, thereby releasing the carbon back into the atmosphere as CO_2 within a short time frame of 1 to 10 years (see Figure 1.3). Thus, the rate at which biomass is converted to fossil resources is in total imbalance with the rate at which they are consumed and liberated ($>10^6$ years vs. 1 to 10 years). The release of more CO_2 than we sequester as fossil resources creates a kinetics problem. Clearly, this is not sustainable, and we are not managing carbon in a sustainable and environmentally responsible manner.

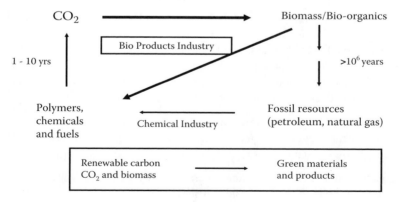

FIGURE 1.3 Global carbon cycling: sustainability driver.

However, if we use annually renewable crops or biomass as the feedstocks for manufacturing our carbon-based polymers, chemicals, and fuels, the rate at which CO_2 is fixed equals the rate at which it is consumed and liberated. This is sustainable, and the use of annually renewable crops/biomass would allow us to manage carbon in a sustainable manner. Furthermore, if we manage our biomass resources effectively by making sure that we plant more biomass (trees, crops) than we utilize, we can begin to start reversing the CO_2 rate equation and move toward a net balance between CO_2 fixation/sequestration and release due to consumption.

Thus, using annually renewable biomass feedstocks to manufacture carbon-based materials and products allows for:

- Sustainable development of carbon-based polymer materials and products
- Control and even reduction of CO_2 emissions, thus helping to meet the global CO_2 emissions standards established by the Kyoto Protocol
- An improved environmental profile

1.2.2 STANDARDS FOR BIOBASED MATERIALS/PRODUCTS

Implementing the use of biobased products requires that the following questions be addressed to prevent confusion and misrepresentation in the marketplace:

- How do you differentiate between biobased (renewable) and nonbiobased materials, i.e., fossil, inorganics?
- How do you define and quantify biobased content? This is an important issue, since most products would continue to have some fossil carbon content to obtain adequate performance and cost benefits.
- How do you evaluate and report on the environmental profile/footprint?
  - Quantitatively show improved environmental profile of biobased products
  - Demonstrate cost and performance benefits

To address these questions, ASTM (American Society for Testing and Materials) International Committee D20.96 [15] developed standards for identifying and quantifying biobased materials and products.

Based on earlier discussions and Equation 1.1, one defines biobased and organic materials as follows:

Biobased materials: organic materials in which the carbon comes from contemporary (nonfossil) biological sources
Organic materials: materials containing carbon-based compounds in which the carbon is attached to other carbon atoms, hydrogen, oxygen, or other elements in a chain, ring, or three-dimensional structure (IUPAC [International Union of Pure and Applied Chemistry] nomenclature)

Therefore, to be classified as biobased, the materials must be organic and contain recently fixed (new) carbon from biological sources. A 100% biobased material

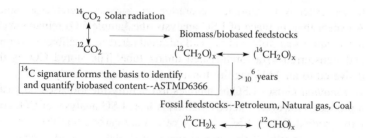

FIGURE 1.4 Carbon-14 method to identify and quantify biobased content.

would be an ideal scenario, but realistically, in practice, most products would contain some nonbiobased materials (inorganic fillers, fossil-based materials) to satisfy performance and cost requirements. Therefore, quantifying the biobased content of a material/product is of paramount importance. ASTM has developed an elegant and absolute method to identify and determine the biobased content of biobased materials using carbon-14 radioactive signatures associated with biocarbon.

As shown in Figure 1.4, the ^{14}C signature forms the basis for identifying and quantifying biobased content. The CO_2 in the atmosphere is in equilibrium with radioactive $^{14}CO_2$. Radioactive carbon is formed in the upper atmosphere through the effect of cosmic-ray neutrons on ^{14}N. It is rapidly oxidized to radioactive $^{14}CO_2$ and enters the Earth's plant and animal lifeways through photosynthesis and the food chain. Plants and animals that utilize carbon in biological food chains take up ^{14}C during their lifetimes. They exist in equilibrium with the ^{14}C concentration of the atmosphere, that is, the numbers of ^{14}C atoms and nonradioactive carbon atoms stay approximately the same over time. As soon as a plant or animal dies, it ceases the metabolic function of carbon uptake; there is no replenishment of radioactive carbon, only decay. Since the half-life of carbon is around 5730 years, the fossil feedstocks formed over millions of years will have no ^{14}C signature. Thus, by using this methodology, one can identify and quantify biobased content. ASTM Subcommittee D20.96 developed a test method (D 6866) to quantify biobased content using this approach [16].

Test method D 6866 involves combusting the test material in the presence of oxygen to produce CO_2 gas. The gas is analyzed to provide a measure of the products. The $^{14}C/^{12}C$ content is determined relative to the modern carbon-based oxalic acid radiocarbon standard reference material (SRM) 4990c, referred to as HOxII. Three different methods can be used for the analysis. The methods are:

- *Test method A* utilizes liquid scintillation counting (LSC) radiocarbon (^{14}C) techniques by collecting the CO_2 in a suitable absorbing solution to quantify the biobased content. The method has an error of 5 to 10%, depending on the LSC equipment used.
- *Test method B* utilizes accelerator mass spectrometry (AMS) and isotope ratio mass spectrometry (IRMS) techniques to quantify the biobased content of a given product with possible uncertainties of 1 to 2% and 0.1 to

0.5%, respectively. Sample preparation methods are identical to method A except that, in place of LSC analysis, the sample CO_2 remains within the vacuum manifold and is distilled, quantified in a calibrated volume, and transferred to a torch-sealed quartz tube. The stored CO_2 is then delivered to an AMS facility for final processing and analysis.

- *Test method C* uses LSC techniques to quantify the biobased content of a product. However, whereas method A uses LSC analysis of CO_2 cocktails, method C uses LSC analysis of sample carbon that has been converted to benzene. This method determines the biobased content of a sample with a maximum total error of ±2% (absolute).

Although test methods A and C are less sensitive than that of method B, using AMS/IRMS, they have two distinct advantages:

- Lower costs per evaluation
- Much greater instrument availability worldwide

The nuclear testing programs of the 1950s resulted in a considerable enrichment of ^{14}C in the atmosphere. Although it continues to decrease by a small amount each year, the current ^{14}C activity in the atmosphere has not reached the pre-1950 level. Because all ^{14}C sample activities are referenced to a "prebomb" standard, and because nearly all new biobased products are produced in a postbomb environment, all values (after correction for isotopic fractionation) must be multiplied by 0.93 (as of the writing of this standard) to better reflect the true biobased content of the sample.

Thus, the biobased content of a material is based on the amount of biobased carbon present and is defined as follows:

Biobased content or gross biobased content: the amount of biobased *carbon* in the material or product as fraction weight (mass) or percent weight (mass) of the total organic carbon in the material or product (ASTM D 6866).

1.2.3 Examples of Biobased Content Calculations

The following examples illustrate biobased content determinations on a theoretical basis.

Let us say that a fiber-reinforced composite with the composition 30% biofiber (cellulose fiber) + 70% PLA (poly[lactic acid], a biobased material) is formulated. The biobased content of this composition would be 100% since all of the carbon is biocarbon.

If the fiber-reinforced composite is formulated with the composition 30% glass fiber (inorganic material) + 70% PLA (biobased material), the biobased content is still 100%, and not 70%. This is because the biobased content is on the basis of carbon, and glass fiber has no carbon associated with it; therefore, all carbon in this product is biocarbon. This would be the case if one were to formulate a product with water or other inorganic fillers. As discussed in previous sections, the rationale for using biobased products is to manage carbon in a sustainable and efficient manner as part of the natural carbon cycle; therefore, it makes sense to use carbon as the basis for determining biobased content. It is also fortuitous that an absolute method using ^{14}C is available to measure the biobased carbon present in a material. Clearly, there may be positive or negative environmental impacts on the use of the noncarbon materials, and this needs to be addressed separately. In any case, one must define biobased content and organic content. Thus, the biobased content of the glass-fiber-reinforced composite is 100%, but the organic content is 70%, implying that the balance 30% is inorganic material. In the earlier example, the biobased content is 100% and the organic content is 100%. This distinction allows end-users/customers to clearly differentiate between two 100% biobased products and make their choice on additional criteria, e.g., by looking at the LCA (life-cycle assessment) profile of the two products (using ASTM D 7075).

As another example, let us say that a fiber-reinforced composite with the composition 30% biofiber (cellulose) + 70% polypropylene (petroleum-based organic) is formulated. The biobased content of this formulation is 18.17% and not 30%. Again, biobased content is based not on weight (mass), but on a carbon basis, i.e., the amount of biobased *carbon* as fraction weight (mass) or percent weight (mass) of the total organic carbon. Therefore, the biobased content for this product is

$$\frac{0.3 \times 44.4 \text{ (percent biocarbon in cellulose)}}{0.7 \times 85.7 \text{ (percent carbon in polypropylene)} + 0.3 \times 44.4 \text{ (percent total biocarbon)}} \times 100,$$

which computes to 18.17%.

The theoretical calculations presented here have been validated in experimental observations using ASTM D 6866 and are in agreement within ±2%.

The U.S. Congress passed the Farm Security and Rural Investment Act of 2002 (P.O. 107–171), requiring the purchase of biobased products by the federal government. The U.S. Department of Agriculture (USDA) was charged with developing guidelines for designating biobased products and publishing a list of designated biobased product classes for mandated federal purchase [17]. In its rule making, the USDA adopted the previously described methodology for identifying and quantifying biobased content and required the use of ASTM D 6866 to establish biobased content of products.

1.3 MATERIAL DESIGN PRINCIPLES FOR THE ENVIRONMENT

So far, we have discussed the use of biobased (as opposed to petroleum) feedstocks to manage our carbon-based products in a sustainable and environmentally responsible manner. However, that is only part of the equation; environmental responsibility requires us to look at the entire product cycle, from feedstock to ultimate disposal, from a holistic point of view. Therefore, material design principles for the environment also require that we address the issue of what happens to the material/product after use by the customer, when it enters the waste stream, i.e., the ultimate disposal

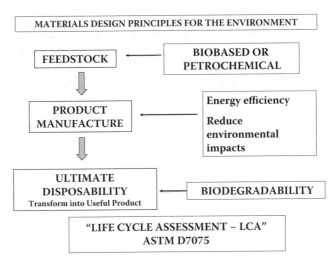

FIGURE 1.5 Material design principles for the environment using LCA.

of a product by using biodegradability, recyclability, or other recovery options. It is equally important that the product-manufacturing step involve energy efficiency and reduced environmental emissions as part of this overall goal toward sustainability and environmental responsibility. Figure 1.5 schematically shows the material design principles for the environment.

Figure 1.5 illustrates the needs to conduct life-cycle assessment (LCA) over the product's entire life cycle to establish its environmental footprint. A recently published ASTM standard (D 7075) [18] provides guidance on evaluating and reporting the environmental performance of biobased products using LCA methodology. Chapter 7 provides a detailed discussion of LCA.

1.3.1 BIODEGRADABILITY CRITERION

Biobased polymers are synthesized by many types of living matter — plants, animals, and bacteria — and are an integral part of ecosystem function. Because they are synthesized by living matter, biopolymers are generally capable of being utilized by living matter (after being biodegraded) and thus can be recycled in safe and ecologically sound ways through such disposal processes (waste management) as composting, soil application, and biological wastewater treatment. Therefore, for single-use, short-life, disposable materials applications such as packaging and consumer articles, biobased materials can and should be engineered to retain a biodegradable functionality. For durable, long-life articles, biobased materials need to be engineered for long life and performance, and biodegradability may not be an essential criterion. Currently, most products are designed with limited consideration to their ecological footprint, especially as it relates to its ultimate disposability. Designing these materials to be biodegradable and ensuring that they end up in an appropriate disposal system is environmentally and ecologically sound. For example, by composting our biodegradable plastic and paper waste along with other "organic" compostable materials such as yard, food, and agricultural wastes, we can generate much-needed carbon-rich compost (humic material). Compost-amended soil has beneficial effects by increasing soil organic carbon, increasing water and nutrient retention, reducing chemical inputs, and suppressing plant disease. Composting is increasingly a critical element for maintaining the sustainability of our agriculture system. The food wastes, along with other biowastes, are separately collected and composted to generate a valuable soil amendment that is returned to the farmland to reinitiate the carbon cycle [19, 20].

Designing products to be only partially biodegradable causes irreparable harm to the environment. Partially degraded products may be invisible to the naked eye, but out of sight does not make the problem go away. Indeed, the toxic effects of partially biodegradable products has been documented [21, 22]. One must ensure complete biodegradability within a short defined time frame (determined by the disposal infrastructure). Typical time frames would be up to one growing season or 1 year.

1.3.2 BIOBASED MATERIALS TECHNOLOGY

Polymer materials based on annually renewable agricultural and biomass feedstocks can form the basis for a portfolio of sustainable, environmentally preferable alternatives to current materials based exclusively on petroleum feedstocks. Two basic routes are possible.

Direct extraction from biomass yields a series of natural polymer materials (cellulose, starch, proteins), fibers, and vegetable oils that provide a platform on which polymer materials and products can be developed, as shown in Figure 1.6. (The bolded items in Figure 1.6 represent our work in this area.)

Alternatively, the biomass feedstock (annually renewable resources) can be converted to biomonomers by fermentation or hydrolysis. The biomonomers can be further modified by a biological or chemical route. As shown in Figure 1.7, the biomonomers can be fermented to give succinic acid, adipic acid, and 1,3-propanediol, which are precursor chemicals for the manufacture of polyesters; for example, DuPont, in a joint venture with Tatel & Lyle PLC, produces Sorona® polyester from a biologically obtained 1,3-propanediol. Biomonomers can also be fermented to lactic acid, which is then converted into poly(lactic acid), a process currently being commercialized by NatureWorks LLC in a 300-MM-lb manufacturing plant in Blair, NE [23]. Biomonomers can also be microbially transformed to biopolymers such as the polyhydroxyalkanoates (PHAs). Chapter 6 provides a thorough discussion of PHA polymers.

In contrast to microbial fermentative processes, chemical conversion of biomonomers yields intermediate chemicals such as ethylene and propylene glycols. Vegetable oils offer a platform to make a portfolio of polyols, lubricants, polyesters, and polyamides. We have reported on a new ozone-mediated transformation of vegetable oils to polyols, urethane foams, polyesters, and polyamides, including biodiesel fuel [24–26].

Surfactants, detergents, adhesives, and water-soluble polymers can be engineered from biomass feedstocks. As discussed previously, biobased materials targeted for

BIOBASED FEEDSTOCKS

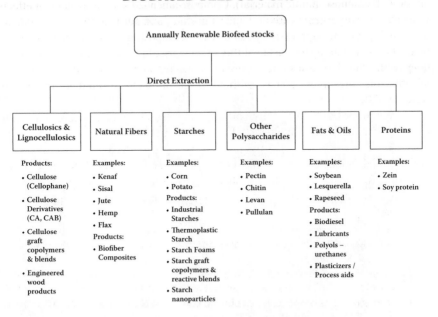

FIGURE 1.6 Direct extraction of biomass to provide biopolymers for use in manufacture of biobased products.

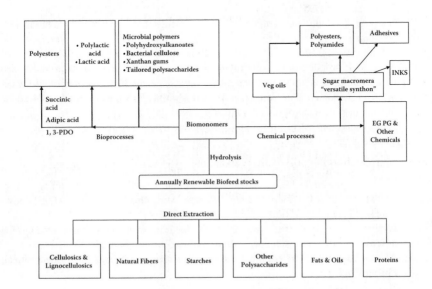

FIGURE 1.7 Conversion to biomonomers and other chemicals and polymers.

short-life, single-use, disposable packaging materials and consumer products can and should be engineered to retain inherent biodegradability properties, thereby offering an environmentally responsible disposal option for such products.

REFERENCES

1. Lieth, H. and Whittaker, H.R., Eds., *Primary Productivity of the Biosphere*, Springer-Verlag, Heidelberg, 1975.
2. Institute of Gas Technology, *Symposium on Clean Fuels from Biomass and Wastes*, Orlando, FL, 1977.
3. Szmant, H.H., *Industrial Utilization of Renewable Resources*, Technomic, Lancaster, PA, 1986.
4. Rowell, R.M., Schultz, T.P., and Narayan, R., Eds., Emerging technologies for materials and chemicals from biomass, in ACS Symp. Ser. 476, ACS, Washington, DC, 1991.
5. DOE, The technology roadmap for plant/crop-based renewable resources 2020, DOE/GO-10099-706, Feb. 1999; available on-line at www.oit.doe.gov/agriculture.
6. Cavaney, R., *Pulp Pap. Int.*, 31, 37, 1989.
7. Chum, H.L. and Power, A.J., in ACS Symp. Ser. 476, ACS, Washington, DC, 1991, Vol. 976, p. 28.
8. Doane, W.M., Swanson, C.L., and Fanta, G.F., in ACS Symp. Ser. 476, ACS, Washington, DC, 1991, Vol. 97, p. 197.
9. Narayan, R., Environmentally degradable plastics, *Kunststoffe*, 79, 1022, 1989.
10. Narayan, R., Biomass (renewable) resources for production of materials, chemicals, and fuels: a paradigm shift, in ACS Symp. Ser. 476, ACS, Washington, DC, 1992, p . 1.
11. Narayan, R., Polymeric materials from agricultural feedstocks, in *Polymers from Agricultural Coproducts*, Fishman, M.L., Friedman, R.B., and Huang, S.J., Eds., ACS Symp. Ser. 575, ACS, Washington, DC, 1994, p. 2.
12. Narayan, R., Commercialization technology: a case study of starch-based biodegradable plastics, in *Paradigm for Successful Utilization of Renewable Resources*, Sessa, D.J. and Willett, J.L., Eds., AOCS Press, Champaign, IL, 1998, p. 78.
13. Narayan, R., Biodegradation of polymeric materials (anthropogenic macromolecules) during composting, in *Science and Engineering of Composting: Design, Environmental, Microbiological and Utilization Aspects*, Hoitink, H.A.J. and Keener, H.M., Eds., Renaissance Publications, Columbus, OH, 1993, p. 339.
14. Narayan, R., Impact of governmental policies, regulations, and standards activities on an emerging biodegradable plastics industry, in *Biodegradable Plastics and Polymers*, Doi, Y. and Fukuda, K., Eds., Elsevier, New York, 1994, p. 261.
15. ASTM International, Committee D20 on Plastics, Subcommittee D20.96, Biobased and Environmentally Degradable Plastics; available on-line at www.astm.org.
16. ASTM, Standard D 6866-04, *Annual Book of Standards*, Vol. 8.03, ASTM International, Philadelphia, 2004.
17. Federal Biobased Preferred Product Procurement Program (FB4P); available on-line at www.biobased.oce.usda.gov.
18. ASTM, Standard D 7075, *Annual Book of Standards*, Vol. 8.03, ASTM International, Philadelphia, 2004.

19. Narayan, R., in *Science and Engineering of Composting: Design, Environmental, Microbiological and Utilization Aspects*, Hoitink, H.A.J. and Keener, H.M., Eds., Renaissance Publications, Columbus, OH, 1993, pp. 339.
20. Narayan, R., in *Biodegradable Plastics and Polymers*, Doi, Y. and Fukuda, K., Eds., Elsevier, New York, 1994, p. 261.
21. Anonymous, *Science*, 304, 838, 2004.
22. Mato, Y., Isobe, T., Takada, H., Kahnehiro, H., Ohtake, C., and Kaminuma, T., *Environ . Sci. Technol.*, 35, 318–324, 2001.
23. Nature works, PLA producer; available on-line at www.natureworks.com.
24. Narayan, R. and Graiver, D., *Lipid Technol.*, 16, 2, 2006.
25. Tran, P., Graiver, D., and Narayan, R., *J. Am. Oil Chemists Soc.*, Vol. 82, pp. 183–196, 2005.
26. Baber, T.M., Graiver, D., Lira, C.T., and Narayan, R., *Biomacromolecules*, 6, 1334–1344, 2005.

2 Biobased Key Molecules as Chemical Feedstocks

Herman van Bekkum

CONTENTS

2.1 INTRODUCTION

World biomass production amounts to 120,000 million metric tons per annum [1], of which some 20% per annum is being cultivated, harvested, and used (food, feed, and nonfood). Human consumption ranges from 6% (South America) to 80% (South Central Asia).

Biobased chemicals and materials can be approached in various ways:

Nature already produces the desired structures [2], and isolation of these components requires only physical methods

Examples: polysaccharides (cellulose, starch, alginate, pectin, agar, chitin, inulin, etc.), disaccharides (sucrose and lactose [animal origin]), triglycerides, lecithin, natural rubber, gelatin (animal origin), flavors and fragrances, quinine (flavor as well as pharmaceutical), etc. Some present-day production volumes are sucrose, at 145×10^6 metric tons/year (t/yr); triglycerides, 130×10^6 t/yr; and natural rubber, 6.0×10^6 t/yr. Cotton, the natural cellulose fiber, is produced in a volume of over 20×10^6 t/yr.

One-step (bio)chemical modification of naturally produced structures

    Examples: cellulose and starch derivatives, glucose and fructose, glycerol, and fatty acids. Nature offers various starting materials for pharmaceuticals. Thus morphine is converted by one methylation step into the antitussive and analgesic codeine (200 t/yr), whereas one (generally illegal) acetylation step leads to heroin. Fermentation is a one-step process in which the enzymes of a micro-organism catalyze the conversions in a multistep synthesis without isolation of intermediates. Bulk chemical examples include ethanol 30×10^6 t/yr, citric acid 10^6 t/yr, sodium glutamate 10^6 t/yr, and lactic acid 250×10^3 t/yr, starting from starch or sucrose.

Multistep derivation of organic chemicals and organic materials from natural products

    Examples: vitamin C in several steps from glucose; (S)-β-hydroxybutyro-lactone in two steps from lactose; the fragrance linalool in four steps from α-pinene, (–)-menthol in six steps from β-pinene [3]; fatty alcohols and amines from triglycerides; alkyl polyglucosides from glucose and fatty alcohols, etc.

2.2 TOTAL CROP USE

Total crop use — in which coproduction of bulk and fine chemicals takes place, while waste organic materials are liquefied or gasified or are fermented to methane (biogas) — is coming to the fore. Direct burning is also an option that is executed worldwide with bagasse, delivering the energy required in sugar cane refineries.

As an example of total crop use, we mention the soybean crop, which is first split into beans and waste biomass. The beans are rolled and divided, by hydrocarbon extraction, into soybean meal (goes mainly to animal feed but also provides an isoflavones concentrate, a health-enhancing food supplement) and crude soybean oil. The crude oil undergoes several purification steps in which, in the modern technology, valuable side products like lecithin and steroids (e.g., β-sitosterol) are isolated and refined soybean oil is obtained.

The steroids can serve as starting compounds for various steroid-type pharmaceuticals [4] and, after transesterification with sunflower oil, find application in cholesterol-level-lowering margarines. Tall-oil, a by-product of wood pulp manufacture, is a source of related steroids, serving the same health purpose [5].

It is often the case that waste material volumes are substantial and sometimes much larger (e.g., sugar cane) than the volumes of the target compounds or materials. It has been estimated [6] that the caloric value of agricultural waste streams is more than 50% of the world's annual oil consumption.

2.3 BIOMASS TO TRANSPORTATION FUELS

There is a growing interest in processes that convert biomass to (green) transportation fuels or to fuel boosters. Some options are given in Figure 2.1. On a large and rapidly increasing scale, hydrolysis of the polysaccharide starch and the disaccharide sucrose to the monosaccharides is carried out, followed by fermentation to ethanol. Two ethanol and two CO_2 molecules are formed from one C6-monosaccharide (glucose). The largest ethanol producers are Brazil [7, 8] and the United States. [9]. Zeolites play a role (adsorption [KA] or membrane techniques) in the dewatering of ethanol, which is required for mixing with gasoline. Forthcoming are processes in which cellulosics, as present in agricultural waste streams (wheat straw, corn stalks, bagasse), are also hydrolyzed (enzymatically) and converted to ethanol [10] by a yeast.

When the ethanol is not blended but used as such fuel, hydrous ethanol (e.g., 95%) can be applied. In several other chemical conversions of ethanol, there is no need to remove all the water. Two examples follow.

2.3.1 Example 1

In a cascade-type continuous setup [11], a partial evaporation of the fermentation liquid was carried out, and the ethanol-water mixture was passed over an H-ZSM5 zeolite catalyst (350°C, 1 atm). As in the MTG (methanol to gasoline) [12] process, a mixture of alkanes and aromatics was obtained. In this approach, the ethanol-water

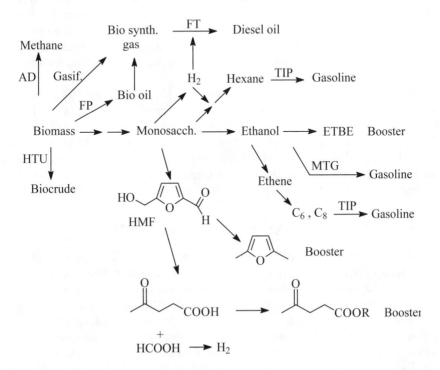

FIGURE 2.1 Carbohydrates to transportation fuels.

separation is avoided, as the hydrocarbons and water are nonmiscible and separate by gravity. Though its octane number is good, there is no future in MTG-gasoline because of the trend to lower the aromatics content.

When passing the ethanol-water mixture over H-ZSM5 at lower temperature (200°C) or over zeolite H-Y, only dehydration occurs and ethene is obtained [13]. The track to gasoline is then oligomerization to C_6-C_8 alkene and hydrogenation/isomerization, e.g., over the TIP-catalyst Pt-H-mordenite.

Recently, Dumesic [14] disclosed a route from glucose to n-hexane (together with some pentane and butane) consisting of hydrogenation to sorbitol followed by stepwise hydrogenolysis over a Pt-catalyst in acid medium (pH 2). The hydrogen required is obtained [15] by aqueous-phase reforming of sorbitol or glycerol over a relatively inexpensive Raney Ni-Sn catalyst. Roughly 1.6 mol of glucose are required per mol hexane.

By gasification of biomass, and with supplementation of hydrogen, the Fischer-Tropsch synthesis can be used to produce clean diesel oil.

2.3.2 EXAMPLE 2

Figure 2.1 also shows routes via hydroxymethylfurfural (HMF), a compound that can be selectively made from fructose using a dealuminated H-mordenite or a niobium-based catalyst. HMF might well develop to become a new key chemical; its chemistry (and that of furfural) has recently been reviewed by Moreau et al. [16]. When preparing HMF, it is advantageous to apply a cascade reaction by using a fructose precursor. Thus the hydrolysis of the fructan inulin (glucose [fructose]$_n$) or of the glucose-fructose combination sucrose is coupled with the dehydration to HMF. In the case of sucrose as a starting compound, HMF and the remaining glucose can be easily separated.

HMF can be hydrogenated over a Pd-catalyst [17] to 2,5-dimethylfuran, a compound with the very high blending research octane number (BRON) of 215 [18].

Another interesting HMF-derived compound is levulinic acid formed together with formic acid by solid acid catalysis. A one-pot cascade route from, for example, inulin seems feasible. Manzer et al. [19] give examples in which an esterification step with an alkene is coupled as well. The levulinic esters exhibit good octane numbers. The by-product formic acid might also be esterified or used as a hydrogen source.

2.4 ETHANOL AS A KEY CHEMICAL

Some further conversions of ethanol are shown in Figure 2.2. We have already mentioned preparation of today's number one organic chemical, ethene. The decrease of mass in the conversion of carbohydrates via ethanol to ethene is large (65%), and the annual production of ethene exceeds 100×10^6 t, so the total world sugar production (145×10^6 t) would not be nearly enough to cover that amount. In India, where cheap sugar streams are available, over 400,000 t of ethanol were used in 1997 [20] to make "alco-chemicals" with acetic acid as the main product. In China and India, aqueous ethanol is directly applied in aromatic ethylation (ethylbenzene, 1,4-diethylbenzene, and 4-ethyltoluene).

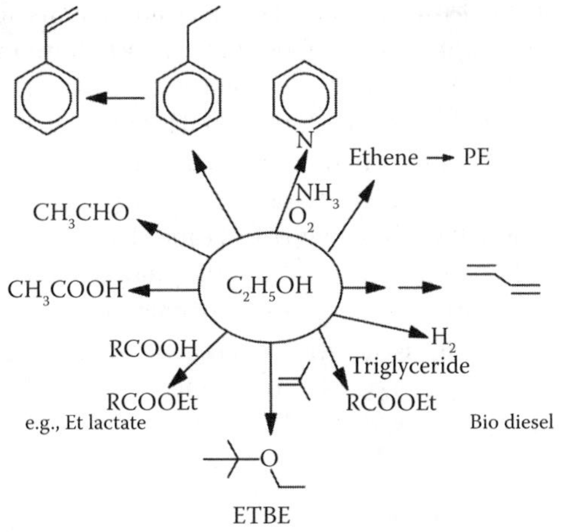

FIGURE 2.2 Ethanol as key chemical.

The reaction of ethanol with isobutene affords ethyl t-butyl ether (ETBE), a partly green compound, which can develop as a successor of methyl t-butyl ether (MTBE) as a gasoline booster. Biodiesel becomes fully renewable based on when ethanol is applied instead of methanol in the transesterification of triglycerides.

The oxidative dehydrocyclization of ethanol and ammonia toward pyridine and 4-methylpyridine was studied at Delft University of Technology [21]. The medium-pore zeolites H-ZSM-5 and H-TON were shown to be the catalysts of choice.

Finally, aqueous ethanol can be considered as a future hydrogen carrier. Up to 6 mol of hydrogen can be obtained from 1 mol of ethanol.

2.5 THE MONOTERPENES

Limonene and α- and β-pinene can be regarded as natural key molecules for the monoterpene (C_{10}) sector of the fragrance industry. Limonene is a component of orange and lemon peels (different enantiomers present) and is a cheap by-product of the citrus industry. The two pinenes are major components of crude sulfate turpentine (CST), a by-product of wood processing, and are relatively inexpensive starting compounds.

Catalytic conversions in the monoterpene field have recently been reviewed [22]. For instance, limonene is commercially converted to alkoxylated systems by solid-acid-catalyzed addition of lower alcohols [23]. Quite another limonene conversion is dehydrogenation toward a "green" aromatic p-cymene compound that can be converted by oxidation to the hydroperoxide and rearranged to p-cresol.

The networks around α- and β-pinene are versatile. By way of example, Figure 2.3 shows how the important fragrance linalool is approached industrially from both α- and β-pinene. These semisynthetic linalool products are in competition with synthetic linalool, made by building up the C_{10}-system starting from isobutene.

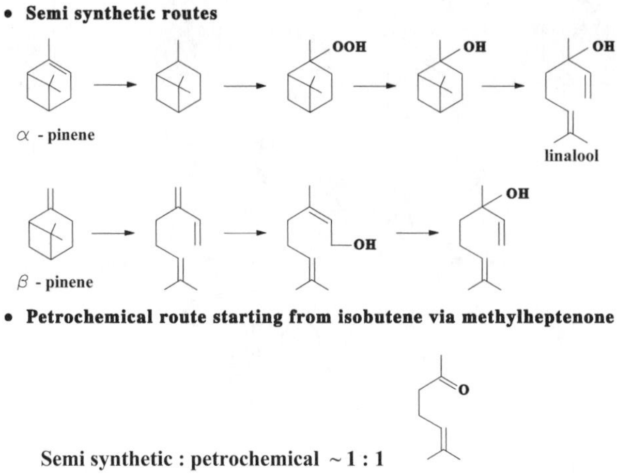

- **Semi synthetic routes**

α - pinene

linalool

β - pinene

- **Petrochemical route starting from isobutene via methylheptenone**

Semi synthetic : petrochemical ~ 1 : 1

FIGURE 2.3 Linalool syntheses.

Another example is the synthesis of campholenic aldehyde by epoxidation of α-pinene followed by isomerization [24]. Campholenic aldehyde is the starting compound for several sandalwood fragrances.

2.6 TRIGLYCERIDES, GLYCEROL

Oils and fats (triglycerides) serve primarily food applications. Some 15% of the annual production of 130×10^6 t (2004 estimate) goes to oleochemicals, such as fatty acids and soaps, fatty amines, long-chain alcohols, and biodiesel (fatty acid methyl esters). In all cases, glycerol is a by-product. Especially because of the growth of biodiesel (2004 estimate 2.5×10^6 t, with Germany as the leading producer [25]), world glycerol production is now approaching 1 million t/yr. Less than 10% of that amount is made by the conventional petrochemical process starting with chlorination of propene. The natural way is a clear winner here in terms of both economy and ecofriendliness.

Glycerol is applied as such in pharmaceutical and cosmetic formulations as well as in food, tobacco, and cellophane. Some chemical conversions are given in Figure 2.4.

A major reaction application of glycerol is as triol in alkyd resin formulations, moreover it is used to grow polyetherpolyols to be used in polyurethanes and in the manufacture of triacetin (glycerol triacetate). New opportunities include selective oxidation over Au-catalysts [26] toward glyceric acid and synthesis of glycidol [27] via the carbonate. In view of the growing glycerol production, further new applications would be welcomed.

2.7 TRIGLYCERIDES, FATTY ACIDS

The fatty acid composition of triglycerides depends strongly on the crop that provides them. Thus, coconut oil and palm-kernel oil contain mostly relatively short (C_{10},

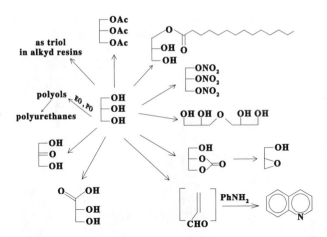

FIGURE 2.4 Glycerol conversions.

C_{12}, C_{14}, C_{16}) saturated fatty acids, whereas soybean, rapeseed, and sunflower oil contain mainly the C_{18} unsaturated acids oleic and linoleic acid, with one and two double bonds, respectively. In palm oil, the second in production volume after soybean oil, the dominant fatty acid components are palmitic acid (C_{16} saturated) and oleic acid (C_{18} monounsaturated).

Insight into health aspects has been subject to fluctuation. Linolenic acid (C_{18}, three double bonds), belonging to the so-called ω-3 acids, was earlier removed from margarine triglycerides by selective hydrogenation, but nowadays it is treasured. Moreover, it is agreed now that trans double bonds, formed upon partial hydrogenation of natural *cis* double bonds (catalytic hardening), are unhealthy.

Commercial products obtained from saturated fatty acids include a broad spectrum of esters, linear alcohols, primary and secondary linear amines, amides, and various metal salts. Relatively new are the direct hydrogenation of carboxylic acids to aldehydes over chromia [28] and the coupling of fatty acids (by an amide bond) to amino acids, leading to a new class of surfactants of all natural origin (Ajinomoto). Biocatalytic λ- and ω-hydroxylation of fatty acids is a challenge.

In the unsaturated oleic acid molecule, the double bond is an extra reaction site. As shown in Figure 2.5, metathesis [29] especially opens up some interesting conversions, although these are not yet operated industrially. Thus self-metathesis of oleic acid (as methyl ester) gives a C_{18} dicarboxylic acid (together with a C_{18} alkene), whereas metathesis with ethene leads to 1-decene and 9-decenoic acid, a precursor of nylon-10. By oxidative ozonization, oleic acid is industrially converted in the C_9 dicarboxylic acid azelaic acid and nonanoic acid.

Some recent patents deal with isomerization (branching) of oleic and stearic acid, with the aim of lowering the melting point. For instance, oleic acid is isomerized [30] at 250°C over an acidic zeolite, followed by hydrogenation over Pd. The product obtained is similar to the "isostearic acid" found as a by-product in the dimerization of oleic acid over an acid clay.

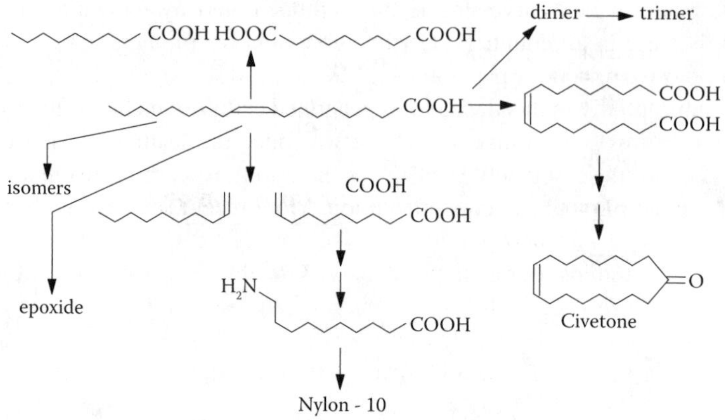

FIGURE 2.5 Oleic acid as a key chemical.

2.8 CARBOHYDRATES

Carbohydrates are by far the most abundant class of renewables. The big three among the carbohydrates (Figure 2.6) are the glucose polymers (glucans), cellulose and starch, and the disaccharide sucrose. Chitin is also widespread, but its actual production — from waste material of the seafood industry — is small. Its monomer, glucosamine, is receiving much attention as a health supplement.

Some natural polysaccharides find use [31] as thickening and stabilizing agents (hydrocolloids) [32] in food and beverages. Sources can be seaweed (agar, alginate, carrageenan), seeds (guar gum, locust bean gun), fruits (pectin), or bacteria (xanthan gum). A newcomer in the polysaccharide field is the fructan inulin.

2.9 CELLULOSE

Wood harvested annually for energy, construction, and paper, cardboard, and hygiene products amounts to over 3 billion m³ [33]. In many countries, forestry is in a

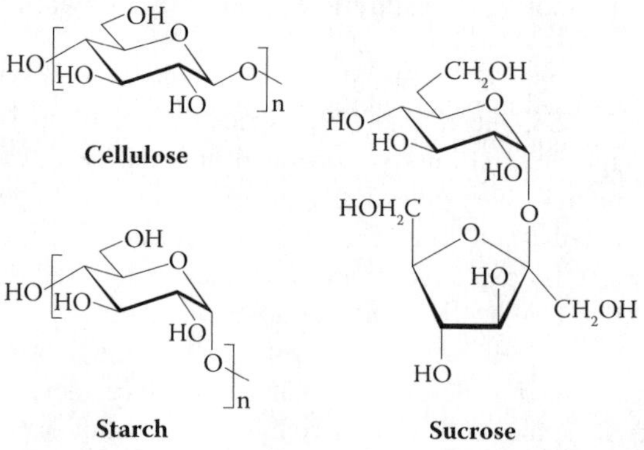

FIGURE 2.6 Carbohydrates, the big three.

sustainable balance, i.e., harvesting is fully compensated by replanting; however, this is not the case in all countries. Certification is an instrument here.

The cellulose demand for paper amounts to some 200×10^6 t/yr. Recycle streams contribute substantially to this demand. Regenerated cellulose includes fibers (rayon, mainly used in tires) and films (cellophane was once the leading clear packaging film). Classic cellulose solvents used in regeneration are carbon disulfide/sodium hydroxide and an ammoniacal copper solution. More recent solvents are N-methyl-morpholine-N-oxide and phosphoric acid. The most recently developed cellulose solvents are ionic liquids. For example, 1-butyl-3-methyl-imidazolium chloride dissolves 100 g/l of cellulose at 100°C [34]. Improved solubility of carbohydrates in ionic liquids is observed when dicyanamide is applied as the anion [35].

Cellulose derivatives [36] can be divided into nonionic and anionic materials. The degree of substitution (up to three) is always an important variable. The nonionic derivatives comprise ethers (hydroxyethyl and hydroxypropyl cellulose, methyl cellulose) and esters (cellulose acetate, cellulose nitrate). The anionic carboxymethyl cellulose is produced in the greatest volume. The cellulose derivatives serve a broad spectrum of applications. For instance, cellulose acetate is applied in cigarette filters, as membranes, as fibers, etc. The anionic class can be extended by 6-carboxycellulose [37] and by 2,3-dicarboxycellulose; both materials show promise but are still in the research and development stage. A challenge is to arrive at cellulose-based superabsorbing materials.

2.10 STARCH

Starch is a mixture of a linear α-1,4-glucan (amylose, see Figure 2.6) and a branched glucan (amylopectin), containing also 1,4,6-bonded glucose units. Generally, the weight ratio for amylopectin:amylose is about 75:25, but high-amylopectin starches can be obtained by genetic modification of corn or potato.

The big starch-containing grains, wheat, rice, and corn, are each annually produced at amounts of over 600×10^6 t. Some 40×10^6 t of starch is industrially isolated. The dominant raw material (almost 80%) is corn.

The starch serves food and nonfood applications. In Europe the ratio is about 1:1. The largest nonfood application in Europe as well as in the United States is in the paper and board area. Both native and modified starches are applied here. Of the starch derivatives used in papermaking, cationized starch is of particular importance. Here, starch is equipped with C_3-chains carrying a quaternary ammonium group. The reagent is made by reacting epichlorohydrine with trimethylamine.

Figure 2.7 lists the major starch-derived chemicals and materials. The conversion pathways are marked with either a circle or a square, indicating whether the conversion step is industrially biocatalyzed or chemocatalyzed, respectively. For instance, the starch-to-vitamin C route involves consecutively enzymatic steps (hydrolysis), a metal-catalyzed hydrogenation, a biocatalytic bacterial regioselective oxidation to L-sorbose, and chemocatalytic protection/oxidation/deprotection/ring-closure steps. North China Pharmaceutical Corp. possesses fermentation technology for the direct conversion of glucose to vitamin C. With an annual production of over 100,000 t, vitamin C is becoming a bulk chemical.

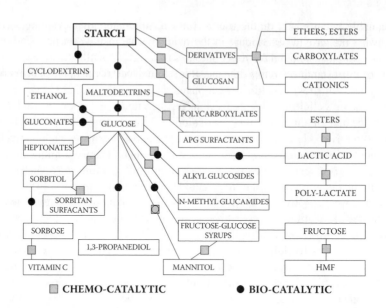

FIGURE 2.7 Starch network.

Fermentative processes for the polyester monomer 1,3-propanediol and for the diabetic-friendly sweetener mannitol have been developed by DuPont and zuChem [38], respectively. In chemocatalysis, sometimes steps can be combined; thus starch can be directly converted to sorbitol by applying a bifunctional Ru-HUSY zeolitic catalyst [39]. The outer surface of the zeolite provides the Bronsted acidity required for the starch hydrolysis. The Ru component catalyzes the hydrogenation of glucose. This process has recently come into industrial practice.

Note that there are also some commercial green surfactants shown in Figure 2.7: the sorbitan esters, known for a long time, and the more recently developed (Henkel) alkyl polyglucosides (APG surfactants) and the N-methyl glucamides (Procter and Gamble/Hoechst). Together with dicarboxylate-polysaccharides as Ca-complexing materials, peracetylated sugar polyols as peracetate precursors, and carboxymethyl cellulose as antiredeposition agent, it is not difficult to imagine production of fully green detergent formulations.

Lactic acid, made by fermentation of glucose or sucrose in an enantiomeric pure form, is used traditionally in meat preservation, but it is increasingly being developed as a key chemical. Major outlets are in esters (biodegradable solvents and bread improvers) and especially in the green polyester poly-L-lactate. The volume of lactic acid (racemate) produced using the chemical process, starting with acetaldehyde, is much less than that produced by fermentation.

Anionic starches can be obtained by sulfatation, by carboxymethylation, or by oxidation. Starch oxidation skills have improved substantially. In particular, the TEMPO-catalyzed oxidation [40] displays an amazing selectivity. Indeed, potato starch has been oxidized to 6-carboxy starch with a selectivity of >98% at 98% conversion. Salt-free enzymatic TEMPO oxidations (O_2/laccase/TEMPO) have also been patented recently. Here, 6-carboxylate as well as 6-aldehyde groups are introduced. Gallezot et al. reported recently [41] on the use of iron tetrasulfonatophthalocyanine as catalyst and hydrogen peroxide as oxidant in the oxidation of several starches. Carboxyl as well as carbonyl functions are introduced, leading to hydrophilic materials.

For further starch derivatives, the reader is referred to a recently published book [42].

2.11 SUCROSE

With its present-day world market price of about 26 cents/kg, the disaccharide sucrose [43] is probably the cheapest chiral compound. The three largest producers are Brazil and India (sugar cane) and the European Union (EU) (sugar beet). The top four exporters are Brazil >> Thailand > EU > Australia.

Some industrial sucrose conversions are shown in Figure 2.8. Part of the conversions, e.g., alcohol manufacture, are executed with molasses, the mother liquor of the sugar crystallization. Potential side products of beet sugar manufacture are pectin (from the pulp) and betaine and raffinose (from the molasses).

Sucrose can be transesterified with fatty acid methyl esters toward mono- and di-esters, applied as emulsifiers or to the sucrosepolyesters (SPEs), which materials have been proposed as fat replacers. The fully esterified sucroseoctaacetate is known for its bitterness.

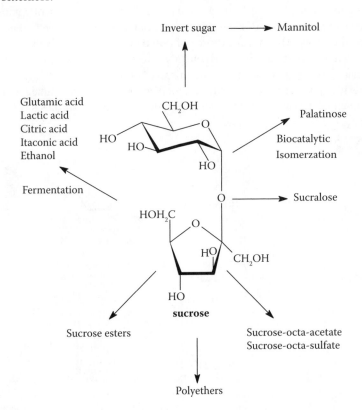

FIGURE 2.8 Sucrose as a key chemical.

Full use of the hydroxyl groups of sucrose is also made in the reaction with ethene oxide or propene oxide, leading to polyether polyols that are used in polyurethane manufacture.

The high-intensity sweetener sucralose was recently admitted to the American market. In sucralose, three hydroxyl groups of sucrose have been replaced by chlorine. This pertains to the 1- and 6-positions in the fructose part and to the 4-position in the glucose part of sucrose. Sucralose has a similar taste profile to sucrose [44]; it is nontoxic, nonnutritive, and 60 times more stable to acid hydrolysis than sucrose. Sucralose is 650 times as sweet as sucrose.

Sucrose is biocatalytically isomerized (Südzucker), whereby the $1 \rightarrow 2$ bond between glucose and fructose changes to a $1 \rightarrow 6$ bond. The disaccharide obtained is named Palatinose, which upon hydrogenation gives a 1:1 mixture of two C_{12} systems under the name Palatinit. Both Palatinose and Palatinit are commercial sweetening compounds. Due to the low rate of hydrolysis compared with sucrose, both systems are suitable for diabetics and are mild to the teeth.

2.12 INULIN AND FRUCTOSE

A relative newcomer in the carbohydrate field is inulin, a fructan (Figure 2.9) consisting of β-linked $(2 \rightarrow 1)$ fructofuranose units with an α-glucopyranose unit at the reducing end. Inulin (GF_n) is obtained in Belgium and The Netherlands from the roots of chicory. The average degree of polymerization is relatively low, $n \approx 12$ [45]. Small amounts of purefructan F_m, are also present in unulin. Another interesting inulin-containing crop is the Jerusalem artichoke [46].

Inulin is applied as a health additive in food applications; it is claimed to improve the intestinal bacterial flora. Moreover, it is a direct source (hydrolysis) of fructose

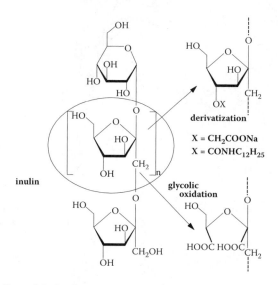

FIGURE 2.9 Inulin and derivatives.

that, on a weight basis, is 1.5 times as sweet as sucrose. Fructose is also the precursor of the versatile compound hydroxymethylfurfural (HMF) from which several furan-2,5-disubstituted "biomonomers" (diol, dialdehyde, dicarboxylic acid) can be derived. Furthermore, HMF is a precursor of levulinic acid (cf. Figure 2.1).

Worldwide derivatization studies on inulin are ongoing. For a review, see Stevens et al. [47]. A first successful example has been carboxymethylation leading to the new material carboxymethyl inulin (CMI). In the reaction with chloroacetate (Figure 2.9), the 4-position of the fructose units turned out to be the most reactive [48], but positions 3 and 6 contribute, too. CMI appeared to be an excellent low-viscosity inhibitor of calcium carbonate crystallization [49] and is industrially manufactured now.

Another new commercial inulin derivative is inulin lauryl carbamate [50], an efficient stabilizer of oil droplets and hydrophobic particles against coalescence or flocculation. The material is used in personal-care applications.

Other promising inulin derivatives are 3,4-dicarboxyinulin, an excellent calcium-complexing material [51] and alkoxylated inulins [52].

2.13 LACTOSE

The major source of the disaccharide lactose (a galactose-glucose combination) is cheese whey, which contains 4.4 wt.% lactose (together with 0.8% of protein and 0.8% of minerals). Lactose can be crystallized in its α- or β-form. Common grades are edible and pharmaceutical [53]. Altogether, it is admirable that the processes of collecting, transporting, concentrating, demineralization, and crystallization allows such a modest price for lactose.

Lactose is used as such in the food industry, in infant nutrition, and in the pharmaceutical industry (matrix material of pills). In view of the growing chemical network around it, lactose is developing into a key molecule (see Figure 2.10).

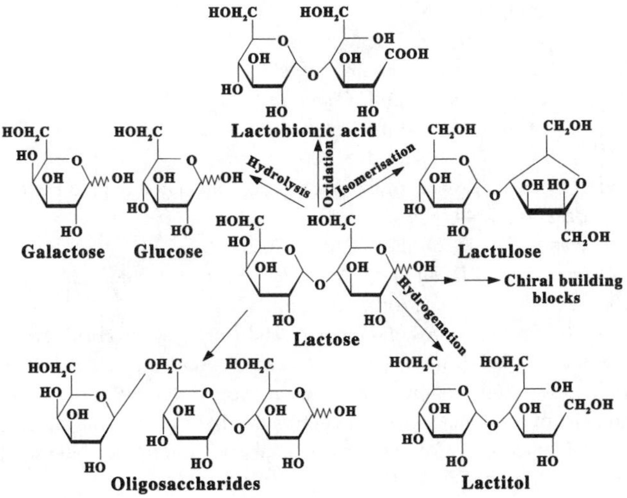

FIGURE 2.10 Lactose as a key chemical.

Industrial conversions include hydrogenation to the low-caloric sweetener lactitol, isomerization to the laxative lactulose (galactose-fructose combination), oligomerization to health supplements, and oxidation to the metal-complexing lactobionic acid. Moreover lactose serves to make chiral compounds such as (S)-3-hydroxy-γ-butyrolactone. Part of the lactose molecule is sacrificed in this synthesis. Via its monomers, new vitamin-C-type antioxidants can be made from lactose [54].

2.14 CONCLUSION

In conclusion, it can be stated that renewables offer a broad spectrum of structures that are applied as such or form starting points for further upgrading. The networks around the renewable key molecules are steadily expanding. In the future, we will increasingly rely on green feedstocks.

REFERENCES

1. Imhoff, M.L. et al., Global patterns in human consumption of net primary production, *Nature*, 429, 870, 2004.
2. *Römpp Encyclopedia of Natural Products*, Steglich, E., Fugmann, B., and Lang-Fugmann, S., Eds., Thieme, Stuttgart, 2000.
3. Sheldon, R.A., *Chirotechnology*, Marcel Dekker, New York, 1993, p. 304.
4. Kleemann, A. and Engel, J., *Pharmaceutical Substances*, 3rd ed., Thieme, Stuttgart, 1999.
5. De Guzman, D., Sterol Investments on the Rise, *Chem. Mark. Rep.*, April 11, 2005, p. 25.
6. Groeneveld, M.J., Research for a sustainable shell, in *Conference Report Gratama Workshop*, Osaka, Japan, 2000, p. 44.
7. Knight, P., No sign of a halt to Brazil's sugar output: madness or foresight? *Int. Sugar J.*, 106, 474, 2004.
8. Schmitz, A., Seale, J.L., and Schmitz, T.G., Determinants of Brazil's ethanol sugar blend ratios, *Int. Sugar J.*, 106, 586, 2004.
9. Mitchell, P., Grow your own oil, *Chem. Ind.*, 2, 15, 2005.
10. Iogen Energy Canada, demo-plant near Ottawa since 2004, commercial plant dealing with 700,000 t straw/y expected in 2007. Shell.
11. De Boks, P.A. et al., Process of the continuous conversion of glucose to hydrocarbons, *Biotechnol. Lett.*, 4, 447, 1982.
12. Haw, J.F. and Marcus, D.M., Methanol to hydrocarbon catalysis, in *Handbook of Zeolite Science and Technology*, Auerbach, S.M., Carrado, K.A., and Dutta, P.K., Eds., Marcel Dekker, New York, 2003, p. 833.
13. Oudejans, J.C., van den Oosterkamp, P.F., and van Bekkum, H., Conversion of ethanol over zeolite H-ZSM-5 in the presence of water, *Appl. Catal.*, 3, 109, 1982.
14. Dumesic, J.A., Aqueous-phase catalytic processes for the production of hydrogen and alkanes from biomass-derived compounds over metal catalysts, paper presented at the 6th Netherlands Catalysis and Chemistry Conference, Noordwijkerhout, The Netherlands, March 7–9, 2005.
15. Huber, G.W., Slabaker, J.W., and Dumesic, J.A., Raney Ni-Sn catalyst for H_2 production from biomass-derived hydrocarbons, *Science*, 300, 2075, 2003.

16. Moreau, C., Belgacem, M.N., and Gandini, A., Recent catalytic advances in the chemistry of substituted furans from carbohydrates, *Top. Catal.*, 27, 11, 2004.
17. Luijkx, G.C.A., Hydrothermal conversion of carbohydrates and related compounds, Ph.D. thesis, Delft University of Technology, The Netherlands, 1994.
18. Papachristos, M.J. et al., The effect of the molecular structure of antiknock additives on engine performance, *J. Inst. Energy*, 64, 113, 1991.
19. Fagan, P.J. et al., Preparation of levulinic acid esters and formic acid esters from biomass and olefins, PCT international patent application WO 03 85071, 2003.
20. Kadakia, A.M., Alcohol-based industry marching into the 21st century, *Chem. Weekly*, Aug. 5, 1997, p. 57.
21. le Febre, R.A., Hoefnagel, A.J., and van Bekkum, H., The reaction of ammonia and ethanol or related compounds towards pyridines over high-silica medium pore zeolites, *Recl. Trav. Chim. Pays-Bays*, 115, 511, 1996.
22. Swift, K.A.D., Catalytic transformations of the major terpene feedstocks, *Top. Catal.*, 27, 143, 2004; Ravasio, N. et al., Mono- and bifunctional heterogeneous catalytic transformation of terpenes and terpenoids, *Top. Catal.*, 27, 157, 2004; Monteiro, J.L.F. and Velosos, C.O., Catalytic conversions of terpenes into fine chemicals, *Top. Catal.*, 27, 169, 2004.
23. Hölderich, W.F. and Laufer, M.C., Zeolites and nonzeolitic molecular sieves in the synthesis of fragrances and flavours, in *Zeolites for Cleaner Technologies*, Guisnet, M. and Gilson, J.-P., Eds., Imperial College Press, London, 2002, p. 301.
24. Kunkeler, P.J. et al, Application of zeolite titanium beta in the rearrangement of α-pinene oxide to campholenic aldehyde, *Catal. Lett.*, 53, 135, 1998.
25. Altmann, B.-R., Biokraftstoffe, Anforderungen, Eigenschaften, Auswirkungen von Beimischungen zu Konventionellen Fraftstoffen, *Erdöl, Erdgas, Kohle*, 121, 156, 2005.
26. Hutchings, G.J. et al., New directions in gold catalysis, *Gold Bull.*, 37, 1, 2004; Oxidation of glycerol using supported gold catalysts, *Top. Catal.*, 27, 131, 2004.
27. Yoo, J.-W. et al., Process for the preparation of glycidol, U.S. patent 6,316,641, 2001.
28. Yokoyama, T. and Setoyama, T., Carboxylic acids and derivatives, in *Fine Chemicals through Heterogeneous Catalysis*, Sheldon, R.A. and van Bekkum, H., Eds., Wiley-VCH, 2001, Weinheim, Germany, p. 370.
29. Mol, J.C., Catalytic metathesis of unsaturated fatty acid esters and oils, *Top. Catal.*, 27, 97, 2004.
30. Zhang, S., Zonchao, S., and Steichen, D., Skeletal isomerization of fatty acids with a zeolite catalyst, patent application WO 03 082464, 2003.
31. Mirasol, F., Focus on nutrition gives boost to hydrocolloid sector as supply tightens, *Chem. Mark. Rep.*, March 28, 2005, p. 32.
32. Nishinari, K., Ed., *Hydrocolloids*, Elsevier, Amsterdam, 2000.
33. *State of the World*, Earthscan Publ., London, 1999, p. 62.
34. Swatloski, R.P. et al., Dissolution of cellulose with ionic liquids, *J. Am. Chem. Soc.*, 124, 4974, 2002.
35. Janssen, M.H.A. et al., Room temperature ionic liquids that dissolve carbohydrates in high concentrations, *Green Chem.*, 7, 39, 2005.
36. Kamide, K., *Cellulose and Cellulose Derivatives*, Elsevier, Amsterdam, 2005.
37. Isogai, A. and Kato, Y., Preparation of polyuronic acid form cellulose by TEMPO-mediated oxidation, *Cellulose*, 5, 153, 1998.
38. De Guzman, D., Bio-based mannitol closer to market, *Chem. Mark. Rep*, Apr. 18, 2005, p. 40.

39. Jacobs, P.A. and Hinnekens, H., Single-step catalytic process for the direct conversion of polysaccharides to polyhydric alcohols by simultaneous hydrolysis and hydrogenation, European patent application 329,923, 1989.

40. Bragd, P.L., van Bekkum, H., and Besemer, A.C., TEMPO-mediated oxidation of polysaccharides: survey of methods and applications, *Top. Catal.*, 27, 49, 2004.

41. Kachkarova-Sorokina, S.L., Gallezot, P., and Sorokin, A.B., A novel clean catalytic method for waste-free modification of polysaccharides by oxidation, *Chem. Commun.*, 2844, 2004.

42. Gotlieb, K.F. and Capelle, A., *Starch Derivatization: Fascinating and Unique Industrial Opportunities*, Wageningen Academic Publishers, The Netherlands, 2005.

43. van der Poel, P.W., Schiweck, H., and Schwartz, T., Eds., *Sugar Technology*, Verlag Bartens KG, Berlin, 1998.

44. Hough, L., Applications of the chemistry of sucrose, in *Carbohydrates as Organic Raw Materials*, Vol. I, Lichtenthaler, F.W., Ed., VCH Weinheim, Germany, 1991, p. 33.

45. de Leenheer, L., Production and use of inulin: industrial reality with a promising future, in *Carbohydrates as Organic Raw Materials*, Vol. III, VCH/CRF, The Hague, 1996, p. 67.

46. Stolzenburg, K., Topinambur: Rohstoff für die Inulin- und Fructosegewinnung, *Zuckerind.*, 130, 193, 2005.

47. Stevens, C.V., Meriggi, A., and Booten, K., Chemical modification of a valuable renewable resource: inulin and its industrial applications, *Biomacromolecules*, 2, 1, 2001.

48. Verraest, D.L. et al., Distribution of substituents in O-carboxymethyl and O-cyanoethyl ethers of inulin, *Carbohydr. Res.*, 302, 203, 1997.

49. Verraest, D.L. et al., Carboxymethyl inulin: a new inhibitor for calcium carbonate precipitation, *J. Am. Oil Chem. Soc.*, 73, 55, 1996.

50. Booten, K. and Levecke, B., Polymer carbohydrate-based surfactants and their use in personal care applications, *SOFW-J.*, 130 (8), 10, 2004.

51. Besemer, A.C. and van Bekkum, H., The relation between calcium sequestering capacity and oxidation degree of dicarboxy-starch and dicarboxy-inulin, *Starch*, 46, 419, 1994.

52. Rogge, T.M. et al., Improved synthesis and physicochemical properties of alkoxylated inulin, *Top. Catal.*, 27, 39, 2004.

53. Timmermans, E., Lactose: its manufacture and physico-chemical properties, in *Carbohydrates as Organic Raw Materials*, Vol. III, VCH/CRF, The Hague, 1996, p. 3.

54. Abbadi, A. et al., New food antioxidant additive based on hydrolysis products of lactose, *Green Chem.*, 5, 74, 2003.

3 Industrial Chemistry with Nature-Based Bioprocesses

Bernard Witholt

CONTENTS

3.1 INTRODUCTION

The life sciences are central to a growing number of industrial sectors. Figure 3.1 shows areas where biotechnology has already had an impact, or is likely to become important in the next few decades. "Red biotech" encompasses the pharmaceutical sector, an extremely well-developed biotechnology area because

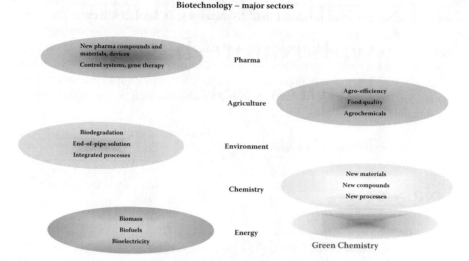

Biotechnology – major sectors

New pharma compounds and
materials, devices

Control systems, gene therapy   Pharma

                                                        Agro-efficiency
                                                        Food quality
                            Agriculture   Agrochemicals

Biodegradation
End-of-pipe solution
Integrated processes   Environment

                                                        New materials
                            Chemistry   New compounds
                                                        New processes

Biomass
Biofuels
Bioelectricity   Energy

                                         Green Chemistry

FIGURE 3.1 The many shades of biotechnology. White or industrial biotechnology is often seen as close to green chemistry, with similar goals such as long-term sustainability, minimization of toxic by-products, and a reduction of solvent use.

of its importance in generating successful new drugs, antibiotics, and vaccines. "Green biotech" has also had a major impact due to the development of new disease- and insect-resistant crops. This area has raised significant concern, especially in Western Europe, where genetically modified crops are generally not accepted, in contrast to the United States, Canada, Brazil, Argentina, and, lately, China as well, where such crops contribute a sizable portion of total corn, soya, and cotton production.

There are two areas that have become quite popular in the last few years. "White biotechnology" deals with the production of chemicals via biocatalytic processes, an area that is now seen as a significant contributor to new processes in the chemical industry.* "Black biotechnology" focuses on biofuels, which are produced from biomass. Major examples include bioethanol from sugar cane (Brazil, Australia) and corn starch (United States) and biodiesel from plant oils (Brazil and Germany; developing in the United States).

3.1.1 White Biotechnology

The development of industrially interesting bioconversions will in many cases involve reactions of organic compounds, most of which are not soluble in aqueous media. At the same time, the enzyme systems or whole cells that will be developed for such biocatalytic reactions will typically function optimally in aqueous media. Thus, it is necessary to develop (a) solutions that allow the

---

* (McKinsey 2004), see http://www.mckinsey.com/clientservice/chemicals/pdf/BioVision_Booklet_final.pdf and http://www.mckinsey.com/clientservice/chemicals/potentialprofit.asp.

combination of biocatalysts that function best in aqueous environments and (b) reaction substrates and products that dissolve best in apolar solvents. The simplest approach is to combine such media in a single reactor. Apolar solvents and aqueous media are usually poorly or not at all miscible. However, when they are mixed vigorously, emulsions are formed, much like those produced when olive oil and vinegar are mixed to produce a salad dressing. An interesting question is whether enzymes and microorganisms can function in apolar media and emulsions.

3.1.2 BIOTRANSFORMATIONS IN THE PRESENCE OF SOLVENTS

Many solvents are chemically inert with respect to biological systems. Their mode of action is due not to chemical effects of the individual molecules, but to their bulk-solvent properties, which may or may not affect a particular macromolecular assembly. Thus, biocatalysis need not be restricted to aqueous systems (Klibanov 2000), and it is not surprising that there are enzymes that catalyze reactions of nonwater-soluble compounds. These compounds occur in nature, and many of these are therefore likely to be synthesized and degraded by some living systems. In addition, the past decades of membrane research have demonstrated that many enzymes operate within or near biomembranes and are therefore exposed (to a large extent) to an apolar rather than an aqueous environment.

Klibanov and his coworkers at MIT have capitalized on these ideas and shown that there are enzymes that are perfectly capable of functioning in bulk organic solvents that contain only a few percent water (Zaks 1985). Such enzymes may or may not be soluble in organic solvents, depending on the nature of the enzyme surface. Arnold and her coworkers have shown that it is possible to modify an enzyme surface by substituting apolar amino acids for polar surface residues, such that the enzyme becomes soluble and more active in a bulk organic solvent (Chen and Arnold 1993). Therefore, there is no intrinsic reason why enzymatic reactions should be restricted to water-soluble compounds or should occur in aqueous environments only (Dordick 1988).

Clearly then, some enzymes operate on apolar compounds, and some enzymes function in apolar environments. Indeed, the development of enzymatic reactions that occur in purely nonaqueous media is conceivable. Ultimately, bioconversions of apolar organic compounds may be based on biocatalysis with inexpensive enzymes, tailored to specific reactions, with high selectivity and productivities, long useful lifetimes, and easy introduction into existing bioreactor and downstream processing systems. In time, such biocatalysis will blend with chemical catalysis, the only distinguishable characteristic perhaps being the nature of the catalyst.

For complex enzyme systems and for enzymes that require energy input via cosubstrates and cofactors, this situation is not likely to be attained soon. Here, whole cells will continue to be the mainstay of bioprocesses. Such cells will have to meet several key requirements: solvent tolerance, high selectivity and productivity of relevant intracellular enzymes, cofactor generation where needed, and prolonged cell survival and enzyme function in large-scale bioprocesses.

3.2 WHOLE-CELL BIOCONVERSIONS OF ORGANIC COMPOUNDS IN TWO-LIQUID-PHASE MEDIA WITH SOLVENT-RESISTANT ORGANISMS

3.2.1 MICROBIAL GROWTH IN SOLVENTS AND EMULSIONS

It has been known for at least a century that some microorganisms are capable of growing in the presence of organic solvents, and in some cases they can metabolize solvents as sources of energy and carbon (Fairhall 1920; Hopkins and Chibnall 1932). In nature, such organisms play a role in the biodegradation of aliphatic and aromatic components in oil. They have been isolated from natural oil deposits or soil, from human-made oil wells, from oil spills, and in general from oil-rich locations, such as machine shops or factories (Johnson 1967).

There is considerable variation in solvent resistance among microorganisms. Earlier work in our laboratory (Rajagopal 1996) has shown that it is possible to determine a unique and reproducible $logP_{50}$ for a given microbial strain. This $logP_{50}$ defines the solvent polarity (logP) that allows 50% survival of the microbes.[†] Cells survive fully in the presence of solvents with logP only slightly higher than $logP_{50}$, while they die in the presence of solvents with a logP that is slightly lower than $logP_{50}$. Thus, wild-type *Escherichia coli* W3110 — a K12 strain (Bachmann 1972) — can grow perfectly well on glucose in the presence of large volumes of long- and medium-chain-length alkanes down to n-octane (logP = 4.5), cyclooctane (logP = 4.1), and even n-heptane (logP = 4.0), but it cannot grow in the presence of 1-octene (logP = 3.7), n-hexane (logP = 3.5), cyclohexane (logP = 3.2), and other solvents with logP < 3.7 (Favre-Bulle et al. 1991).

An interesting example of the variability of biological responses to solvents is that provided by toluene. Toluene has been used extensively since the 1950s as a tool in rendering cells permeable to small molecules, thereby permitting enzyme assays of whole cells (Herzenberg 1959), and toluene has been quite useful in the study of multienzyme complexes that could not be isolated and studied *in vitro*. Early studies in our laboratory have shown that toluene extracts lipids from the cellular membranes (de Smet et al. 1978). Toluene-treated cells lose their intracellular metabolite pool in seconds, so that these cells are no longer able to metabolize external carbon and energy sources. Thus, despite the fact that most, if not all, intracellular enzymes remain active after toluene treatment, cells are killed by toluene because the membrane bilayer is impaired.

If cells could withstand toluene-engendered damage at the ultrastructural level, they might well survive. In fact, today many examples of such cell types are known. In 1989, Horikoshi and his coworkers first described a *Pseudomonas* isolate that is capable of growing in the presence of 50% (v/v) toluene, despite the fact that only 1% (v/v) toluene is lethal to other *Pseudomonads* (Inoue and Horikoshi 1983). Work is now underway to understand solvent resistance, and although it is not yet clear

---

[†] The partition coefficient (P_{oct}) for a given compound is equal to the ratio of the solubility of the compound in octanol (or some other standard solvent, to be denoted) to its solubility in water, under a specific set of standard conditions. The resulting ratio is usually expressed as the logarithm of the partition coefficient, denoted as $logP_{oct}$ or simply logP.

which factors account for the differences in toluene resistance between Horikoshi's solvent-resistant strain and most other solvent-sensitive microorganisms, likely factors include alterations in membrane and cell-wall composition and architecture, alterations in lipid synthesis and degradation, as well as active export of toluene by solvent pumps (Godoy et al. 2004; Isken and de Bont 1996, 1998; Kieboom et al. 1998a, 1998b; Ramos et al. 2002; Segura et al. 2003; Weber and de Bont 1996).

3.2.2 WHOLE-CELL BIOCATALYSIS IN TWO-LIQUID-PHASE MEDIA

The practical application of biooxidation reactions requires that mono- and dioxygenases operate in functional and metabolically active cells (de Smet et al. 1981, 1983b; Held et al. 1998, 1999; Panke et al. 1999; Schmid et al. 1998b; Staijen and Witholt 1998). This is so because of the requirement of these enzymes for cofactors and cofactor regeneration. The nature of the two-liquid-phase system in which these cells must function will be dictated by the substrates used and the products formed; the conditions necessary for effective downstream processing; and the overall economics of the process as it emerges from the research and pilot development stages (Mathys et al. 1999; Panke et al. 1999). These requirements may determine the nature and quality of the solvent to be used as the apolar phase (the substrate may or may not be the solvent as well); the culture conditions; and in some cases, the state of recirculated separate phases, one or both of which may return from the downstream processing system. These various parameters in turn impose limitations on the host strain and the specific oxidation enzyme system to be used.

One particularly interesting strain was first isolated in the late 1930s by Lee and Chandler, then at the Rice Institute in Houston (Lee 1941). They examined water-oil emulsions at the Hughes Tool Co., also in Houston. Large amounts of emulsion were pumped over tools and metal objects during machining (milling and lathing) operations in company machine shops. The emulsion flowed over pieces as they were machined and then into collecting basins under the metalworking equipment. Workers stood in these basins and sometimes developed pustules and skin infections on their hands, which led to the notion that the emulsion might contain pathogenic microorganisms. Lee and Chandler set out to find and identify these organisms and found one single predominating organism that had all of the characteristics of a pseudomonad. They named it *Pseudomonas oleovorans*, an organism that has been worked with intensively since.

Significant interest in the biocatalytic potential of this and similar organisms began to develop in the early 1960s, when laboratory studies by van Linden and coworkers at the Shell laboratories in Amsterdam showed that certain pseudomonads are capable of oxidizing and utilizing alkanes for growth (van der Linden 1965). Subsequently, Coon and his coworkers at Michigan State University isolated an enzyme system responsible for the terminal oxidation of *n*-alkanes and fatty acids from *Pseudomonas oleovorans* in the 1960s and 1970s (Peterson et al. 1966; Ruettinger et al. 1977). In this same period, a growing group of researchers began to study the P450 monooxygenases from eukaryotic and bacterial cells. These monooxygenases oxidize a wide range of organic compounds, among them alkanes and complex cyclic compounds, including aromatic ring systems and steroids. Several hundred P450 monooxygenases have been described to date (Lewis 2003).

3.2.3 TWO-LIQUID-PHASE WHOLE-CELL BIOCATALYSIS REACTOR

In the biocatalysis reactor, shown schematically in Figure 3.2, the emulsion is stirred sufficiently well that the apolar phase forms microdroplets with diameters on the order of 5 to 15 μm, as shown in Figure 3.3. As a result there is a very large surface layer between the aqueous phase and the apolar droplet phase. This permits substrates dissolved in the apolar droplets to exchange rapidly with the continuous aqueous phase, which contains the microorganisms that will take up the substrate and convert it to product, which will again migrate to the apolar droplets. Although the exchange of substrates and products is rapid, their concentration in the apolar and aqueous phases is determined by the partitioning of these compounds between the two phases. Many of the substrates of interest in bioreactions have logPs in the range of 3 to 6, which means that the solubility of these compounds in water is 10^{-3} to 10^{-6} lower than that in octanol or many other apolar organic solvents. .

Consequently, although substrates and products exchange rapidly between the two phases in a two-liquid-phase bioreaction system, the concentration of these

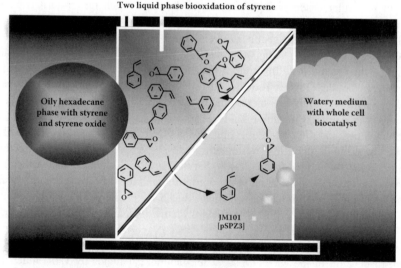

Two liquid phase biooxidation of styrene

Oily hexadecane phase with styrene and styrene oxide

Watery medium with whole cell biocatalyst

JM101
[pSPZ3]

FIGURE 3.2 Bioconversion in a two-liquid-phase culture. A recombinant *E. coli JM101* host, equipped with a styrene-oxidation system on plasmid *pSPZ3*, converts styrene to styrene oxide. Both the substrate (styrene) and the product (styrene oxide) dissolve in the organic phase (golden brown), which in this experiment is hexadecane. A very small amount of styrene partitions into the aqueous phase (light blue) and enters the recombinant cells, where it is oxidized to styrene oxide. This product then partitions back into the apolar phase because it is much more soluble in hexadecane than in the aqueous phase. As a result, the cells are exposed to a very low concentration of the toxic substrate and product, and the cells remain biocatalytically active for 10 to 20 h. (Experiments reported by Sven Panke and Marcel Wubbolts, Institute of Biotechnology, ETH, Zurich, 1996.)

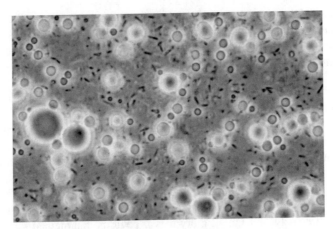

FIGURE 3.3 Light microscopy of *Pseudomonas oleovorans* growing in a two-liquid-phase culture, such as shown in Figure 3.2. The aqueous phase is the continuous phase and contains the rod-shaped bacteria. The organic phase consists of droplets that have a diameter of 5 to 15 µm. (This and similar photos were made by Marie-Jose de Smet [University of Groningen, 1980] and Andrew Schmid [Institute of Biotechnology, ETH, Zurich, 1995]).

compounds in the aqueous phase is quite low. This means that even when such compounds are toxic for microorganisms, this toxicity is generally not a problem, because the effective concentrations of either substrates or products in the aqueous phase remain quite low. Thus, two-liquid-phase bioreaction systems are likely to play a significant role in future developments of white biotechnology.

With the basic problem of adapting biocatalytic processes to classical organic chemistry largely solved, there are several other requirements that must be met for all biocatalytic and chemocatalytic processes.

First, it is necessary to find and develop catalysts that carry out the desired reaction with the needed chemo-, regio-, and stereospecificities. This is dealt with via a variety of screening and selection methods, and once a promising biocatalyst is identified, attempts are made to improve it further via site-directed mutagenesis, DNA shuffling techniques, directed evolution, or *in silico* protein design techniques.

Second, high catalytic activities per unit bioreactor volume must be attained and maintained for long periods of time. One approach to meeting these requirements for two-liquid-phase biocatalytic processes is to maximize cellular enzyme activities by high-level expression of active enzymes. In addition, we attempt to grow cells to high densities in the aqueous phase of two-liquid-phase media.

3.2.4 ORGANISMS SUITABLE AS HOSTS FOR BIOCONVERSIONS IN TWO-LIQUID-PHASE MEDIA

Although organic solvent with low $logP_{oct}$ (<3.5) are generally lethal to bacteria (Rajagopal 1996), there are many organisms, including *E. coli*, that function reasonably well in the presence of organic solvents with intermediate $logP_{oct}$ (3.5 to 5.0) (Favre-Bulle et al. 1991). As work in this area develops in the next decade, it is likely that a few especially suitable species will emerge, which will gradually be improved to become standard two-liquid-phase bioconversion hosts. Such hosts might be endowed not only with good survival characteristics in a range of two-liquid-phase environments, but they are likely also to contain adequate cofactor production and regeneration systems (Wubbolts et al. 1990) and properties that favor high oxygen-transfer rates (Khosla et al. 1990; Magnolo et al. 1991) and facilitate downstream processing.

We have also developed continuous cultures in two-liquid-phase bioreactors, again for both *P. oleovorans* (Durner et al. 2000; Preusting 1993a) and *E. coli* recombinants. These cultures exhibit stable alkane oxidation activity for periods of a week to a month. Cell densities are lower (1 to 10 g dry mass per l), and cellular activities are generally higher than those found for the fed-batch system. For instance, for the production of polyhydroxyalkanoates (PHAs) with continuous growth on n-octane, a cell density of 10 to 12 g dry mass/l could be maintained for at least one month, with cellular PHA contents of 20 to 25% (m/m) (Favre-Bulle et al. 1993; Preusting 1993a).

3.2.5 SPACE-TIME YIELDS AND LIMITATIONS

The maximum productivity attainable is limited by the rate of oxygen transfer to cells under practical conditions. We usually try to improve the performance of the biocatalytic system through modifications of the biocatalysis enzyme-host combination and through technical efforts to maximize cell densities and performance in continuous and fed-batch cultures, while at the same time minimizing growth rates, since growth requires continuous input of nutrients. See Mathys et al. (1999) for a detailed analysis of large-scale bioconversion and downstream processing systems. In addition, it is necessary to channel the largest possible fraction of the substrate into product rather than cell-mass production, which we do by providing an aqueous-medium-based carbon and energy source, such as glucose or glycerol.

With fed-batch cultures, we have been able to reach cell densities of 40 g dry mass per l, both for *P. oleovorans* (Preusting 1993b) and for *E. coli* recombinants (Wubbolts 1996). Unfortunately, it is not trivial to retain high enzyme activities when cell densities are increased. Enzyme activities (U/g cell dry mass) typically decrease when cells are grown to high densities. In practice, this means that an optimum combination of cell density and enzyme activity must be sought, resulting in an overall volumetric activity (units of enzyme activity per liter of bioreactor medium). Typical maximum activities seen in the biooxidation systems we have worked with a range from 200 to 500 units per liter of aqueous medium component for 5 to 15 h (Buhler et al. 2003b; Panke et al. 2002). If such systems can be scaled up to industrial production with equivalent yearly activities for 40% of the time, this will result in space-time yields between 5 and 15 tons/m^3 (aqueous volume)/yr.

3.2.6 DOWNSTREAM PROCESSING

The above activities result in two-liquid-phase bioconversion systems of varying efficacy. The next step has been to develop downstream processes to separate products from substrates and possible other (contaminating) compounds present in the fermentation or enzyme medium. Such processes have been developed to function either in batch mode or in series with a continuous culture (Mathys et al. 1999). They consist of several steps, beginning with a phase separation of the complete culture medium.

Figure 3.4 shows an example of a continuous two-liquid-phase culture of *P. oleovorans* (left panel), which is gradually collected in a large flask (right panel), where there is a slow phase separation into an aqueous (lower) phase and an organic (upper) phase that still contains aqueous medium, forming a mousse that must be further separated into an organic and an aqueous phase. Following this second separation, the product is removed from the cell-free organic phase by fractional distillation (Mathys et al. 1998a, 1998b) or large-scale high-performance liquid chromatography (HPLC). The remaining organic phase might consist of the substrate (Favre-Bulle and Witholt 1992), or it might consist of a carrier solvent such as dodecane (Wubbolts et al. 1996), hexadecene (Schmid et al. 1998a), or bis(2-ethylhexyl)phthalate (Panke et al. 2002).

FIGURE 3.4 Two-liquid-phase continuous culture of *Pseudomonas oleovorans*, consisting of 15% (v/v) octane and 85% (v/v) aqueous medium containing ca. 10 g (cdm)/l cells (left panel). The culture is stirred at 1500 rpm, which results in an emulsion of small octane droplets (see Figure 3.3). The culture is harvested continuously (right panel). The aqueous phase with bacteria sinks to the bottom. The octane phase with some trapped water floats to the top and forms a mousse. Most of the water can be separated from the apolar octane phase by centrifugation, and the apolar phase can then be subjected to distillation to separate biocatalysis products from octane. (Photos supplied by Wil Hazenberg and Andrew Schmid, Institute of Biotechnology, ETH, Zurich, 1995.)

For industrial cost optimization, the remaining organic phase (either substrate or carrier or both, and generally the major fraction of the initially used organic phase) will be returned to the continuous reactor (Mathys et al. 1999). It may be necessary to remove inhibitory compounds that affect the bioconversion efficiency or other undesirable contaminants that may cause problems in repeated cycles through the complete bioconversion-downstream processing system.

In some cases, the bioconversion of organic compounds results in water-soluble compounds such as acids (Favre-Bulle and Witholt 1992), which need to be processed via precipitation, crystallization, filtration, extraction, chromatographic techniques, or other classical methods.

3.2.7 Integrated Bioconversion-Bioprocessing System

Each of the above topics provides ingredients for an integrated bioconversion system. Optimization of the entire biosynthesis and processing chain requires that all of the components in this complex process be sufficiently understood so that the activity of each component can be modulated with respect to that of all other steps such that the overall process is optimized. As an example, maximizing product accumulation in the two-liquid-phase bioreactor might appear desirable, since it facilitates downstream processing. However, some products, e.g., alkanoic acids, which accumulate in the water phase (Favre-Bulle et al. 1993), inhibit cell growth and enzyme activity. It might be preferable therefore to aim for a high substrate conversion rate, with a rather low steady-state product accumulation level. This has advantages for the volumetric productivity in the reactor, but now it becomes more difficult to remove the product from the reactor efflux. Here, it is useful to consider the product removal process. If the efflux stream is recycled after product removal, the extent of the removal need not be complete. It should be very specific, in the sense that the product removed meets the required purity criteria, but whatever is not removed from the efflux stream is returned to the reactor, where a steady state is maintained via (a) the continuous removal of an efflux stream containing the product and (b) the continuous influx of the remaining product via the return stream from the downstream processing unit plus the newly synthesized product in the bioreactor (Mathys et al. 1999). In the steady state, there is material (liquid phases, product in one of these phases, cell mass) cycling in the bioreactor-downstream processing unit, resulting in a net steady-state concentration of product in the bioreactor, and in a differential across the downstream processing unit. Optimizing the complete system requires finding the steady-state conditions that yield maximum overall productivities at required product purity criteria.

3.3 SPECIFIC COMPOUNDS OF INTEREST

3.3.1 Synthesis of Poly-3-Hydroxyalkanoates (PHAs) by Pseudomonas Spp.

PHAs are biocompatible and biodegradable thermoplastic polyesters that are made by many bacteria when these are grown on ample supplies of carbon sources and

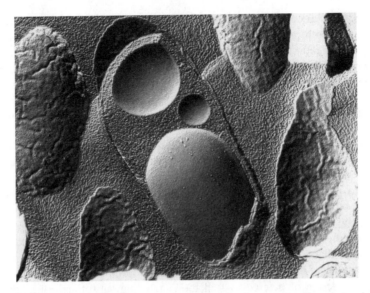

FIGURE 3.5 Freeze fracture electron micrograph of *P. oleovorans* cells growing in a two-liquid-phase medium containing 85% aqueous phase and 15% octane. (EM photo by Jaap Kingma and Hans Preusting, University of Groningen, The Netherlands, 1990.)

limiting amounts of nitrogen, sulfur, phosphorus, or magnesium (de Smet et al. 1983b; Hartmann et al. 2006; Kessler and Witholt 1998; Lageveen et al. 1988; Zinn et al. 2001). The composition of the PHAs formed under such conditions depends on the growth substrates, culture conditions, and the specificity of the PHA-synthesizing enzymes (Huisman et al. 1989; Lageveen et al. 1988). *Pseudomonas oleovorans* produces two PHA polymerases that show a preference for C_7 to C_9 3-hydroxyacyl (CoA) monomers, probably formed by beta-oxidation of fatty acids and oxidized alkanes or by *de novo* fatty acid biosynthesis during growth on carbohydrates (Ren et al. 2005a, 2005b). Many other (CoA activated) fatty acyl derivatives can also be utilized as cosubstrates by *P. oleovorans*, permitting the biosynthesis of a wide range of PHA copolymers (Kessler et al. 2001; van der Walle et al. 2001). PHA chains form intracellular granules of 200- to 500-nm diameter. Figure 3.5 shows some of these granules as they develop in the cytoplasm of *P. oleovorans*. PHAs are of interest to synthetic and polymer chemists because it is still difficult, if not impossible, to produce synthetic biodegradable polymers, and because the monomers may be useful chiral synthons for further chemical synthesis (de Roo et al. 2002). The bacterial polymer granules are also of interest to polymer chemists and materials scientists as doping materials in polymer blends because it is difficult to create small synthetic polymer granules.

The production of useful amounts of a variety of PHAs with specific chemical, physical, and mechanical properties requires adequate intracellular PHA accumulation as well as simultaneous high cell-mass production on the different substrates used (Kellerhals 1999; Schmid et al. 1998b).

3.3.2 SMALL MOLECULES: ORGANIC CHEMISTRY WITH WHOLE-CELL BIOCATALYSTS

We have examined the oxidation of aliphatic (Chang et al. 2000; Duetz et al. 2003; Smits et al. 1999, 2003; van Beilen et al. 2001, 2002, 2003a, 2003b; Whyte et al. 2002) and aromatic (Buhler et al. 2000; Li et al. 2002; Panke et al. 2000; Wubbolts et al. 1994) compounds by bacterial enzymes such as monooxygenases, oxidases, and dehydrogenases.

To extend the number of potentially interesting biocatalysts, we have developed a strain library of alkane- and aromatic-compound-oxidizing organisms as well as a rapid screening system to identify desired enzymatic activities (Duetz et al. 2000; Duetz and Witholt 2001, 2004).[‡] Using this library, we have explored the genetics of alkane oxidation by a large number of environmental microorganisms in more detail (Smits et al. 2002; van Beilen et al. 2002; van Beilen 2003), which provides a basis for the development of additional biocatalytic systems for the oxidation of alkanes and alkane derivatives.

We have also used this strain library to develop biocatalysts for the oxidation and hydroxylation of heterocyclic alkanes, including substituted pyrrolidines and pyrrolidinones (Li 1999; Li et al. 2001, 2002). Similarly, we have studied the conversion of D-limonene, widely available as a by-product of the orange juice industry, to (+) trans-carveol (Duetz et al. 2001b, 2003) and perillyl alcohol (van Beilen et al. 2005). Examples of biooxidations of aromatic compounds include the epoxidation of styrene to styrene epoxide (Panke et al. 1999; Wubbolts 1994) and the production of benzaldehyde (Buhler et al. 2003a, 2003b).

The enzymes involved in the above bioconversions generally require cofactors (microtimamide aolenine olinucleotide (phosphate) [NAD(P)], flavin aolenine dinucleotide [FAD], pyrroloquinoline quinone [PQQ]) or electron-transfer proteins, which is the reason why such reactions must be carried out in whole cells (Duetz et al. 2001a). The performance of these enzymes depends on the state of the host cells (Chen et al. 1996; de Smet et al. 1983; Staijen et al. 2000). Thus, we have studied the characteristics of intact cells in two-liquid-phase media, and we have determined how these affect the performance of relevant enzymes (Favre-Bulle and Witholt 1992). Many of these studies have utilized *P. oleovorans*, the organism first isolated by Lee and Chandler in Houston, which, as might be expected given the source of this organism (Lee 1941), performs very nicely in two-liquid-phase media (Witholt et al. 1990).

3.4 INDUSTRIAL POTENTIAL OF TWO-LIQUID-PHASE BIOCATALYSIS

Biocatalysis has long been taken to be a promising tool in synthesis because of the expected regio- and enantioselectivity of enzymes. Synthetic catalysts with steadily improving selectivities are meanwhile emerging, based in part on lessons

---

[‡] See www.enzyscreen.com for more information on the strain library and screening equipment and approaches.

learned from protein structure and enzyme catalytic mechanisms (Ligtenbarg et al. 2003).

What then is the industrial and economic potential of biocatalytic processes? Are they limited to expensive fine chemicals only, or can they be expected to spread into other areas? What is the lower limit of production and processing costs per unit product manufactured in a two-liquid-phase bioconversion system?

Answers to these questions require more complete information than available at present. A partial answer is that several major chemical companies, including BASF, DSM, and Lonza, have begun to explore two-liquid-phase bioconversions of apolar compounds in the past one or two decades (Schmid et al. 2001). Together with DuPont, Dow, Cargill, and several other companies, these companies have defined white biotechnology as the next frontier of the developing biotechnology sector (see footnote 1).

Another partial answer is that there is considerable experience with large-scale installations for the production of single-cell protein (SCP) by microorganisms from various bulk carbon sources, including agricultural wastes, starches, and oils, both vegetable and mineral (Scrimshaw 1966). Until 20 years ago, SCP was seen as a potentially interesting food supplement (Kharatyan 1978; Kihlberg 1972). Work on SCP led to the development of large-scale two-liquid-phase systems for the utilization of n-paraffins and mineral oil by yeast, both by Western oil companies and by companies in the former Soviet Union and former East Germany (Calvert 1976). SCP ultimately failed to develop because the economics of the process were disappointing and the need for food from oil was less pressing than believed in the 1960s and 1970s. The experience with these installations will, however, be useful in the development of industrial-scale two-liquid-phase biocatalytic processes with organic compounds.

Thus, as biocatalyst- and synthetic-catalyst-based processes improve, differences in process performance or product quality may, in many cases, fade away. Increasingly, multistep synthesis routes combine both of these approaches (Schoemaker et al. 2003), with the choice of specific catalytic steps being guided by product life-cycle analysis (Saling et al. 2005; Sugiyama et al. 2006) and the economics of overall reaction sequences.

ACKNOWLEDGMENTS

I have enjoyed working with a large number of collaborators, each of whom has contributed to the concepts and data discussed in this chapter. I would like to particularly mention Marie-Jose de Smet, Roland Lageveen, Jaap Kingma, Eddy Kool van Langenberghe, Olivier Favre-Bulle, Hans Preusting, Marcel Wubbolts, Andrew Schmid, Renata Mathys, Sven Panke, Bruno Buhler, Jin Byung Park, and Andreas Schmid for their work and very enjoyable discussions regarding two-liquid-phase bioreaction systems. I am also grateful to Wouter Duetz, Jan van Beilen, Theo Smits, Zhi Li, and Dongliang Chang for all of their efforts in the search for and analysis of new substrate-strain-enzyme combinations.

REFERENCES

Bachmann, B.J., Pedigrees of some mutant strains of *Escherichia coli* K-12, *Bacteriol. Rev.*, 36, 525, 1972.

Buhler, B., Schmid, A., Hauer, B., and Witholt, B., Xylene monooxygenase catalyzes the multistep oxygenation of toluene and pseudocumene to corresponding alcohols, aldehydes, and acids in *Escherichia coli* JM101, *J. Biol. Chem.*, 275, 10085, 2000.

Buhler, B., Bollhalder, I., Hauer, B., Witholt, B., and Schmid, A., Use of the two-liquid phase concept to exploit kinetically controlled multistep biocatalysis, *Biotechnol. Bioeng.*, 81, 683, 2003a.

Buhler, B., Bollhalder, I., Hauer, B., Witholt, B., and Schmid, A., Chemical biotechnology for the specific oxyfunctionalization of hydrocarbons on a technical scale, *Biotechnol. Bioeng.*, 82, 833, 2003b.

Calvert, C.C., Systems for the indirect recycling by using animal and municipal wastes as a substrate for protein production, paper presented at FAO Animal Production and Health, Proceedings of a Technical Consultation, Rome, 1976.

Chang, D., Witholt, B., and Li, Z., Preparation of (S)-N-substituted 4-hydroxy-pyrrolidin-2-ones by regio- and stereoselective hydroxylation with *Sphingomonas* spp. HXN-200, *Org. Lett.*, 2, 3949, 2000.

Chen, K. and Arnold, F.H., Tuning the activity of an enzyme for unusual environments: sequential random mutagenesis of subtilisin E for catalysis in dimethylformamide, *Proc. Natl. Acad. Sci. USA*, 90, 5618, 1993.

Chen, Q., Janssen, D.B., and Witholt, B., Physiological changes and alk gene instability in *Pseudomonas oleovorans* during induction and expression of alk genes, *J. Bacteriol.*, 178, 5508, 1996.

de Roo, G., Kellerhals, M.B., Ren, Q., Witholt, B., and Kessler, B., Production of chiral R-3-hydroxyalkanoic acids and R-3-hydroxyalkanoic acid methylesters via hydrolytic degradation of polyhydroxyalkanoate synthesized by pseudomonads, *Biotechnol. Bioeng.*, 77, 717, 2002.

de Smet, M.J., Kingma, J., and Witholt, B., The effect of toluene on the structure and permeability of the outer and cytoplasmic membranes of *Escherichia coli*, *Biochim. Biophys. Acta*, 506, 64, 1978.

de Smet, M.J., Wynberg, H., and Witholt, B., Synthesis of 1,2-epoxyoctane by *Pseudomonas oleovorans* during growth in a two-phase system containing high concentrations of 1-octene, *Appl. Environ. Microbiol.*, 42, 811, 1981.

de Smet, M.J., Eggink, G., Witholt, B., Kingma, J., and Wynberg, H., Characterization of intracellular inclusions formed by *Pseudomonas oleovorans* during growth on octane, *J. Bacteriol.*, 154, 870, 1983a.

de Smet, M.J., Kingma, J., Wynberg, H., and Witholt, B., *Pseudomonas oleovorans* as a tool in bioconversions of hydrocarbons: growth, morphology and conversion characteristics in different two-phase systems, *Enzyme Microbial Technol.*, 5, 352, 1983b.

Dordick, J.S., Biocatalysis in nonaqueous media: patents and literature, *Appl. Biochem. Biotechnol.*, 19, 103, 1988.

Duetz, W.A., Ruedi, L., Hermann, R., O'Connor, K., Buchs, J., and Witholt, B., Methods for intense aeration, growth, storage, and replication of bacterial strains in microtiter plates, *Appl. Environ. Microbiol.*, 66, 2641, 2000.

Duetz, W.A. and Witholt, B., Effectiveness of orbital shaking for the aeration of suspended bacterial cultures in square-deepwell microtiter plates, *Biochem. Eng. J.*, 7, 113, 2001.

Duetz, W.A., van Beilen, J.B., and Witholt, B., Using proteins in their natural environment: potential and limitations of microbial whole-cell hydroxylations in applied biocatalysis, *Curr. Opin. Biotechnol.*, 12, 419, 2001a.

Duetz, W.A., Fjallman, A.H., Ren, S., Jourdat, C., and Witholt, B., Biotransformation of D-limonene to (+) trans-carveol by toluene-grown *Rhodococcus opacus* PWD4 cells, *Appl. Environ. Microbiol.*, 67, 2829, 2001b.

Duetz, W.A., Bouwmeester, H., van Beilen, J.B., and Witholt, B., Biotransformation of limonene by bacteria, fungi, yeasts, and plants, *Appl. Microbiol. Biotechnol.*, 61, 269, 2003.

Duetz, W.A. and Witholt, B., Oxygen transfer by orbital shaking of square vessels and deepwell microtiter plates of various dimensions, *Biochem. Eng. J.*, 17, 181, 2004.

Durner, R., Witholt, B., and Egli, T., Accumulation of Poly[(R)-3-hydroxyalkanoates] in *Pseudomonas oleovorans* during growth with octanoate in continuous culture at different dilution rates, *Appl. Environ. Microbiol.*, 66, 3408, 2000.

Fairhall, L.T. and Bates, P.M., The sterilization of oils by means of ultraviolet rays, *J. Bacteriol.*, 5, 49, 1920.

Favre-Bulle, O., Schouten, T., Kingma, J., and Witholt, B., Bioconversion of n-octane to octanoic acid by a recombinant *Escherichia coli* cultured in a two-liquid phase bioreactor, *Biotechnology (NY)*, 9, 367, 1991.

Favre-Bulle, O., Weenink, E., Vos, T., Preusting, H., and Witholt, B., Continuous bioconversion of n-octane to octanoic acid by recombinant *Escherichia coli* (alk') growing in a two-liquid-phase chemostat, *Biotechnol. Bioeng.*, 41, 263, 1993.

Favre-Bulle, O. and Witholt, B., Biooxidation of n-octane by a recombinant *Escherichia coli* in a two-liquid-phase system: effect of medium components on cell growth and alkane oxidation activity, *Enzyme Microbial Technol.*, 14, 931, 1992.

Godoy, P., Ramos-Gonzalez, M.I., and Ramos, J.L., *Pseudomonas putida* mutants in the exbBexbDtonB gene cluster are hypersensitive to environmental and chemical stressors, *Environ. Microbiol.*, 6, 605, 2004.

Hartmann, R., Hany, R., Pletscher, E., Ritter, A., Witholt, B., and Zinn, M., Tailor-made olefinic medium-chain-length poly([R]-3-hydroxyalkanoates) by *Pseudomonas putida* GPo1: batch versus chemostat production, *Biotechnol. Bioeng.*, 93, 737, 2006.

Held, M., Schmid, A., Kohler, H.P., Suske, W., Witholt, B., and Wubbolts, M.G., An integrated process for the production of toxic catechols from toxic phenols based on a designer biocatalyst, *Biotechnol. Bioeng.*, 62, 641, 1999.

Held, M., Suske, W., Schmid, A., Engesser, K.H., Kohler, H.P.E., Witholt, B., and Wubbolts, M.G., Preparative scale production of 3-substituted catechols using a novel monooxygenase from *Pseudomonas azelaica* HBP 1, *J. Molecular Catalysis B-Enzymatic*, 5, 87, 1998.

Herzenberg, L.A., Studies on the induction of beta-galactosidase in a cryptic strain of *Escherichia coli*, *Biochimica Biophysica Acta*, 31, 525, 1959.

Hopkins, S.J. and Chibnall, A.C., XVI: Growth of *Aspergillus versicolor* on higher paraffins, *Biochem. J.*, 26, 133, 1932.

Huisman, G.W., de Leeuw, O., Eggink, G., and Witholt, B., Synthesis of poly-3-hydroxy-alkanoates is a common feature of fluorescent pseudomonads, *Appl. Environ. Microbiol.*, 55, 1949, 1989.

Inoue A. and Horikoshi, K., A *Pseudomonas* thrives in high concentrations of toluene, *Nature*, 338, 264, 1989.

Isken, S. and de Bont, J.A., Active efflux of toluene in a solvent-resistant bacterium, *J. Bacteriol.*, 178, 6056, 1996.

Isken, S. and de Bont, J.A., Bacteria tolerant to organic solvents, *Extremophiles*, 2, 229, 1998.

Johnson, M.J., Growth of microbial cells on hydrocarbons, *Science*, 155, 1515, 1967.

Kellerhals, M., Hazenberg, W., and Witholt, B., High cell density fermentations of *Pseudomonas oleovorans* for the production of mcl-PHAs in two-liquid phase media, *Enzyme Microbial Technol.*, 24, 111, 1999.

Kessler, B. and Witholt, B., Synthesis, recovery and possible application of medium-chain-length polyhydroxyalkanoates: a short overview, *Macromolecular Symposia*, 130, 245, 1998.

Kessler, B., Weusthuis, R., Witholt, B., and Eggink, G., Production of microbial polyesters: fermentation and downstream processes, *Adv. Biochem. Eng. Biotechnol.*, 71, 159, 2001.

Kharatyan, S.G., Microbes as food for humans, *Annu. Rev. Microbiol.*, 32, 301, 1978.

Khosla, C., Curtis, J.E., DeModena, J., Rinas, U., and Bailey, J.E., Expression of intracellular hemoglobin improves protein synthesis in oxygen-limited *Escherichia coli*, *Biotechnology (NY)*, 8, 849, 1990.

Kieboom, J., Dennis, J.J., de Bont, J.A., and Zylstra, G.J., Identification and molecular characterization of an efflux pump involved in *Pseudomonas putida* S12 solvent tolerance, *J. Biol. Chem.*, 273, 85, 1998a.

Kieboom, J., Dennis, J.J., Zylstra, G.J., and de Bont, J.A., Active efflux of organic solvents by *Pseudomonas putida* S12 is induced by solvents, *J. Bacteriol.*, 180, 6769, 1998b.

Kihlberg, R., Microbe as a source of food, *Annu. Rev. Microbiol.*, 26, 427, 1972.

Klibanov, A.M., Answering the question: "Why did biocatalysis in organic media not take off in the 1930s?" *Trends Biotechnol.*, 18, 85, 2000.

Lageveen, R.G., Huisman, G.W., Preusting, H., Ketelaar, P., Eggink, G., and Witholt, B., Formation of polyesters by *Pseudomonas oleovorans*: effect of substrates on formation and composition of poly-(R)-3-hydroxyalkanoates and poly-(R)-3-hydroxyalkenoates, *Appl. Environ. Microbiol.*, 54, 2924, 1988.

Lee, M.C., A study of the nature, growth and control of bacteria in cutting compounds, *J. Bacteriol.*, 41, 373, 1941.

Lewis, D.F., P450 structures and oxidative metabolism of xenobiotics, *Pharmacogenomics*, 4, 387, 2003.

Li, Z., Feiten, H.-J., van Beilen, J.B., Duetz, W., and Witholt, B., Preparation of optically active N-benzyl-3-hydroxypyrrolidine by enzymatic hydroxylation, *Tetrahedron: Asymmetry*, 10, 1323, 1999.

Li, Z., Feiten, H.J., Chang, D., Duetz, W.A., van Beilen, J.B., and Witholt, B., Preparation of (R)- and (S)-N-protected 3-hydroxypyrrolidines by hydroxylation with *Sphingomonas* spp. HXN-200, a highly active, regio- and stereoselective, and easy to handle biocatalyst, *J. Org. Chem.*, 66, 8424, 2001.

Li, Z., van Beilen, J.B., Duetz, W.A., Schmid, A., de Raadt, A., Griengl, H., and Witholt, B., Oxidative biotransformations using oxygenases, *Curr. Opin. Chem. Biol.*, 6, 136, 2002.

Ligtenbarg, A.G.J., Hage, R., and Feringa, B.L., Catalytic oxidations by vanadium complexes, *Coordination Chem. Rev.*, 237, 89, 2003.

Magnolo, S.K., Leenutaphong, D.L., DeModena, J.A., Curtis, J.E., Bailey, J.E., Galazzo, J.L., and Hughes, D.E., Actinorhodin production by *Streptomyces coelicolor* and growth of *Streptomyces lividans* are improved by the expression of a bacterial hemoglobin, *Biotechnology (NY)*, 9, 473, 1991.

Mathys, R.G., Kut, O.M., and Witholt, B., Alkanol removal from the apolar phase of a two-liquid phase bioconversion system; part 1: comparison of a less volatile and a more volatile *in situ* extraction solvent for the separation of 1-octanol by distillation, *J. Chem. Technol. Biotechnol.*, 71, 315, 1998a.

Mathys, R.G., Schmid, A., Kut, O.M., and Witholt, B., Alkanol removal from the apolar phase of a two-liquid phase bioconversion system; part 2: effect of fermentation medium on batch distillation, *J. Chem. Technol. Biotechnol.*, 71, 326, 1998b.

Mathys, R.G., Schmid, A., and Witholt, B., Integrated two-liquid phase bioconversion and product-recovery processes for the oxidation of alkanes: process design and economic evaluation, *Biotechnol. Bioeng.*, 64, 459, 1999.

Panke, S., de Lorenzo, V., Kaiser, A., Witholt, B., and Wubbolts, M.G., Engineering of a stable whole-cell biocatalyst capable of (S)-styrene oxide formation for continuous two-liquid-phase applications, *Appl. Environ. Microbiol.*, 65, 5619, 1999.

Panke, S., Wubbolts, M.G., Schmid, A., and Witholt, B., Production of enantiopure styrene oxide by recombinant *Escherichia coli* synthesizing a two-component styrene monooxygenase, *Biotechnol. Bioeng.*, 69, 91, 2000.

Panke, S., Held, M., Wubbolts, M.G., Witholt, B., and Schmid, A., Pilot-scale production of (S)-styrene oxide from styrene by recombinant *Escherichia coli* synthesizing styrene monooxygenase, *Biotechnol. Bioeng.*, 80, 33, 2002.

Peterson, J.A., Basu, D., and Coon, M.J., Enzymatic omega-oxidation; I: electron carriers in fatty acid and hydrocarbon hydroxylation, *J. Biol. Chem.*, 241, 5162, 1966.

Preusting, H., Hazenberg, W., and Witholt, B., Continuous production of poly(3-hydroxy-alkanoates) by *Pseudomonas oleovorans* in a high-cell-density, two-liquid-phase chemostat, *Enzyme Microbial Technol.*, 15, 311, 1993a.

Preusting, H., van Houten, R., Hoefs, A., Kool van Langenberghe, E., Favre-Bulle, O., and Witholt, B., High cell density cultivation of *Pseudomonas oleovorans*: growth and production of poly(3-hydroxyalkanoates) in two-liquid phase batch and fed-batch systems, *Biotechnol. Bioeng.*, 41, 550, 1993b.

Rajagopal, A., Growth of gram-negative bacteria in the presence of organic solvents, *Enzyme Microbial Technol.*, 19, 606, 1996.

Ramos, J.L., Duque, E., Gallegos, M.T., Godoy, P., Ramos-Gonzalez, M.I., Rojas, A., Teran, W., and Segura, A., Mechanisms of solvent tolerance in gram-negative bacteria, *Annu. Rev. Microbiol.*, 56, 743, 2002.

Ren, Q., de Roo, G., van Beilen, J.B., Zinn, M., Kessler, B., and Witholt, B., Poly(3-hydroxyalkanoate) polymerase synthesis and *in vitro* activity in recombinant *Escherichia coli* and *Pseudomonas putida*, *Appl. Microbiol. Biotechnol.*, 69, 286, 2005a.

Ren, Q., van Beilen, J.B., Sierro, N., Zinn, M., Kessler, B., and Witholt, B., Expression of PHA polymerase genes of *Pseudomonas putida* in *Escherichia coli* and its effect on PHA formation, *Antonie Van Leeuwenhoek*, 87, 91, 2005b.

Riese, J. and Bachmann, R., Industrial biotechnology: Turning the potential into profits, *Chemical Market Reporter*, McKinsey&Company, December1 , 2004.

Ruettinger, R.T., Griffith, G.R., and Coon, M.J., Characterization of the omega-hydroxylase of *Pseudomonas oleovorans* as a nonheme iron protein, *Arch. Biochem. Biophys.*, 183, 528, 1977.

Saling, P., Maisch, R., Silvani, M., and Konig, N., Assessing the environmental-hazard potential for life cycle assessment, eco-efficiency and SEEbalance (R), *Int. J. Life Cycle Assessment*, 10, 364, 2005.

Schmid, A., Sonnleitner, B., and Witholt, B., Medium chain length alkane solvent-cell
    transfer rates in two-liquid phase, *Pseudomonas oleovorans* cultures, *Biotechnol.
    Bioeng.*, 60, 10, 1998a.
Schmid, A., Kollmer, A., Mathys, R.G., and Witholt, B., Developments toward large-
    scale bacterial bioprocesses in the presence of bulk amounts of organic solvents,
    Extremophiles, 2, 249, 1998b.
Schmid, A., Dordick, J.S., Hauer, B., Kiener, A., Wubbolts, M.G., and Witholt, B.,
    Industrial biocatalysis today and tomorrow, *Nature*, 409, 258, 2001.
Schoemaker, H.E., Mink, D., and Wubbolts, M.G., Dispelling the myths: biocatalysis in
    industrial synthesis, *Science*, 299, 1694, 2003.
Scrimshaw, N.S., Applications of nutritional and food science to meeting world food
    needs, *Proc. Natl. Acad. Sci. USA*, 56, 352, 1966.
Segura, A., Rojas, A., Hurtado, A., Huertas, M.J., and Ramos, J.L., Comparative genomic
    analysis of solvent extrusion pumps in *Pseudomonas* strains exhibiting different
    degrees of solvent tolerance, *Extremophiles*, 7, 371, 2003.
Smits, T.H., Rothlisberger, M., Witholt, B., and van Beilen, J.B., Molecular screening
    for alkane hydroxylase genes in gram-negative and gram-positive strains, *Environ.
    Microbiol.*, 1, 307, 1999.
Smits, T.H., Balada, S.B., Witholt, B., and van Beilen, J.B., Functional analysis of alkane
    hydroxylases from gram-negative and gram-positive bacteria, *J. Bacteriol.*, 184,
    1733, 2002.
Smits, T.H., Witholt, B., and van Beilen, J.B., Functional characterization of genes
    involved in alkane oxidation by *Pseudomonas aeruginosa*, *Antonie Van Leeuwen-
    hoek*, 84, 193, 2003.
Staijen, I.E. and Witholt, B., Synthesis of alkane hydroxylase of *Pseudomonas oleovorans*
    increases the iron requirement of alk + bacterial strains, *Biotechnol. Bioeng.*, 57,
    228, 1998.
Staijen, I.E., Van Beilen, J.B., and Witholt, B., Expression, stability and performance of
    the three-component alkane mono-oxygenase of *Pseudomonas oleovorans* in
    Escherichia coli, *Eur. J. Biochem.*, 267, 1957, 2000.
Sugiyama, H., Hirao, M., Mendivil, R., Fischer, U., and Hungerbuhler, K., A hierarchical
    activity model of chemical process design based on life cycle assessment, *Process
    Saf. Environ. Prot.*, 84, 63, 2006.
van Beilen, J.B., Panke, S., Lucchini, S., Franchini, A.G., Rothlisberger, M., and Witholt,
    B., Analysis of *Pseudomonas putida* alkane-degradation gene clusters and flank-
    ing insertion sequences: evolution and regulation of the alk genes, *Microbiology*,
    147, 1621, 2001.
van Beilen, J.B., Smits, T.H., Whyte, L.G., Schorcht, S., Rothlisberger, M., Plaggemeier,
    T., Engesser, K.H., and Witholt, B., Alkane hydroxylase homologues in gram-
    positive strains, *Environ. Microbiol.*, 4, 676, 2002.
van Beilen, J.B., Li, Z., Duetz, W.A., Smits, T.H.M., and Witholt, B., Diversity of alkane
    hydroxylase systems in the environment, *Oil Gas Sci. Technol., Rev. IFP*, 58, 427,
    2003.
van Beilen, J.B., Duetz, W.A., Schmid, A., and Witholt, B., Practical issues in the
    application of oxygenases, *Trends Biotechnol.*, 21, 170, 2003a.
van Beilen, J.B., Mourlane, F., Seeger, M.A., Kovac, J., Li, Z., Smits, T.H., Fritsche, U.,
    and Witholt, B., Cloning of Baeyer-Villiger monooxygenases from *Comamonas*,
    Xanthobacter and *Rhodococcus* using polymerase chain reaction with highly
    degenerate primers, *Environ. Microbiol.*, 5, 174, 2003b.

van Beilen, J.B., Holtackers, R., Luscher, D., Bauer, U., Witholt, B., and Duetz, W.A., Biocatalytic production of perillyl alcohol from limonene by using a novel *Mycobacterium* Spp. cytochrome P450 alkane hydroxylase expressed in *Pseudomonas putida*, *Appl. Environ. Microbiol.*, 71, 1737, 2005.

van der Linden, A.C. and Thijsse, G.J., The mechanisms of microbial oxidations of petroleum hydrocarbons, *Adv. Enzymol. Relat. Areas Mol. Biol.*, 27, 469, 1965.

van der Walle, G.A., de Koning, G.J., Weusthuis, R.A., and Eggink, G., Properties, modifications and applications of biopolyesters, *Adv. Biochem. Eng. Biotechnol.*, 71, 263, 2001.

Weber, F.J. and de Bont, J.A., Adaptation mechanisms of microorganisms to the toxic effects of organic solvents on membranes, *Biochim. Biophys. Acta*, 1286, 225, 1996.

Whyte, L.G., Smits, T.H., Labbe, D., Witholt, B., Greer, C.W., and van Beilen, J.B., Gene cloning and characterization of multiple alkane hydroxylase systems in *Rhodococcus* strains Q15 and NRRL B-16531, *Appl. Environ. Microbiol.*, 68, 5933, 2002.

Witholt, B., de Smet, M.J., Kingma, J., van Beilen, J.B., Kok, M., Lageveen, R.G., and Eggink, G., Bioconversions of aliphatic compounds by *Pseudomonas oleovorans* in multiphase bioreactors: background and economic potential, *Trends Biotechnol.*, 8, 46, 1990.

Wubbolts, M.G., Terpstra, P., van Beilen, J.B., Kingma, J., Meesters, H.A., and Witholt, B., Variation of cofactor levels in *Escherichia coli*: sequence analysis and expression of the pncB gene encoding nicotinic acid phosphoribosyltransferase, *J. Biol. Chem.*, 265, 17665, 1990.

Wubbolts, M.G., Hoven. J., Melgert, B., and Witholt, B., Efficient production of optically active styrene epoxides in two-liquid phase cultures, *Enzyme Microbial Technol.*, 16, 887, 1994.

Wubbolts, M.G., Reuvekamp, P., and Witholt, B., TOL plasmid-specified xylene oxygenase is a wide substrate range monooxygenase capable of olefin epoxidation, *Enzyme Microbial Technol.*, 16, 608, 1994.

Wubbolts, M.G., Favre-Bulle, O., and Witholt, B., Biosynthesis of synthons in two-liquid-phase media, *Biotechnol. Bioeng.*, 52, 301, 1996.

Zaks, A., Enzyme-catalyzed processes in organic solvents, *Proc. Natl. Acad. Sci. USA*, 82, 3192, 1985.

Zinn, M., Witholt, B., and Egli, T., Occurrence, synthesis and medical application of bacterial polyhydroxyalkanoate, *Adv. Drug Deliv. Rev.*, 53, 5, 2001.

4 Catalytic Conversion of Carbohydrates to Oxygenates

Hasan Mehdi, Róbert Tuba, László T. Mika,
Andrea Bodor, Kornél Torkos,
and István T. Horváth

The sustainability of human civilization primarily depends on whether the energy requirements of its increasing population can be satisfied in the future. While the establishment of the exact date of the depletion of fossil fuels seems difficult, skyrocketing oil and gas prices could come much earlier than 2050, as predicted by several studies.[1] Efficient conversion of solar energy to electricity could open the way to the production of increasing amounts of hydrogen and finally to the development of a hydrogen economy.[1] Of course, the increasing population will demand increasingly larger volumes of carbon-based consumer products. Carbon dioxide is the simplest renewable carbon source, and there are several studies on the hydrogenation of CO_2.[2] Since there is no effective direct CO_2-based process known to produce large amounts of organic chemicals, nature can help to convert CO_2 to biomass, which could serve as the renewable carbon resource. The hydrogenation of carbohydrates could be one of the approaches to produce key carbon-based intermediates.[3] If the currently forecasted transition to a hydrogen economy materializes in the coming decades,[1] hydrogen would be available in large quantities at reasonable prices for the conversion of carbohydrates to a variety of industrial chemical products, including oxygenates or even hydrocarbons such as hexanes and pentanes. The latter could truly be considered as a "renewable oil," ready to be used by existing technologies.

The acid-catalyzed dehydration of carbohydrates giving 5-hydroxymethylfurfural[4] (Figure 4.1) and levulinic acid[5] (Figure 4.2) is one of the most studied and used conversions of carbohydrates. Since it is known that these acid-catalyzed dehydrations proceed through C,C- and C,O-double bonds containing intermediates, we have been investigating the possibility of combining the dehydration of carbohydrates with *in situ* hydrogenation of the reaction intermediates.[6]

Since carbohydrates are water soluble and the expected products have lower solubility in aqueous phase, the use of water-soluble catalysts for hydrogenation offers the possibility of facile product separation and catalyst recycling.[7] When sucrose (0.6 mol/l) was dissolved in an aqueous solution of sulfuric acid (1.8 mol/l) and treated with hydrogen (85 bar) at 140°C for 4 h in the presence of an *in situ*-generated Ru

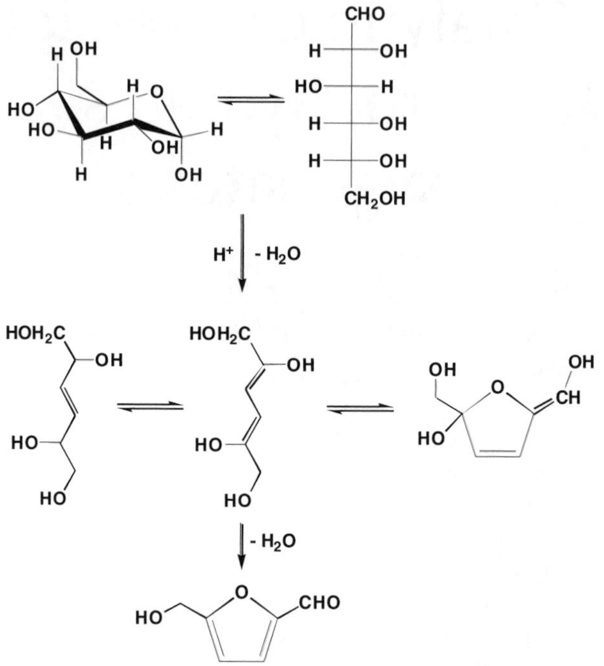

FIGURE 4.1 Formation of 5-hydroxymethylfurfural from glucose.

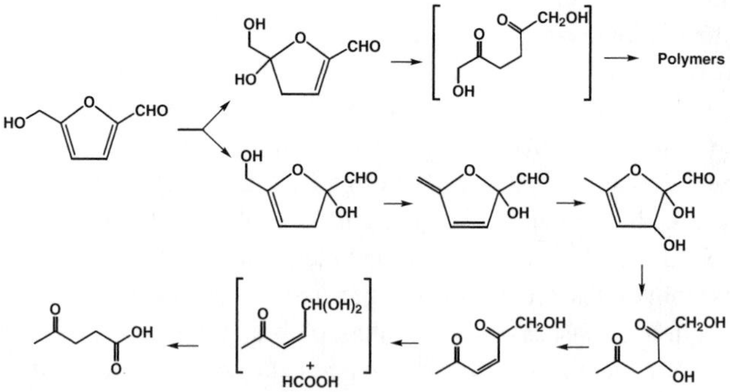

FIGURE 4.2 Formation of levulinic acid from 5-hydroxymethylfurfural.

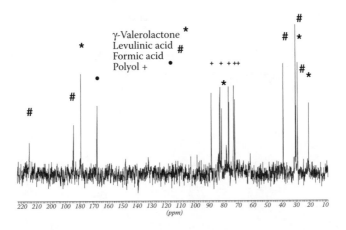

FIGURE 4.3 Catalytic conversion of sucrose.

catalyst (RuCl$_3$, 0.006 mol/l; tris(3-sulfonatophenyl)phosphane trisodium salt (TPPTS), 0.02 mol/l; NaI, 0.008 mol/l), the formation of levulinic acid, formic acid, γ-valerolactone, and a polyol was observed by high-pressure nuclear magnetic resonance (NMR) (Figure 4.3). We have confirmed that the formation of levulinic acid is due to sulfuric acid-catalyzed dehydration of sucrose. It should be noted that numerous homogeneous and heterogeneous catalytic systems have been reported for the hydrogenation of levulinic acid to γ-valerolactone.[8] We have also observed that levulinic acid could be converted to γ-valerolactone completely under similar conditions using the same catalyst. The final product γ-valerolactone can be readily extracted from the aqueous phase with ethyl acetate.

Only a few catalytic systems are known for the reduction of γ-valerolactone.[9] Its hydrogenation to 2-methyltetrahydrofuran (2-Me-THF) was achieved using a ruthenium catalyst that has previously been used to hydrogenate several other lactones to their corresponding diols.[10] Complete reduction was observed when γ-valerolactone (12.6 mmol) was treated with 70 bar H$_2$ at 200°C in the presence of Ruacetylacetonato(acac)$_3$ (0.03 mmol), PBu$_3$ (1.0 mmol) as catalyst, and NH$_4$PF$_6$ (0.53 mmol) as cocatalyst for 46 h. It is probable that 1,4-pentanediol is an intermediate, as it has been shown to give 2-Me-THF via its dehydration in acidic media at elevated temperatures. It should be noted that the conversion of 2-Me-THF to hydrocarbons was achieved when it (2.32 mmol) was dissolved in CF$_3$SO$_3$H (9.30 mmol) and reacted with hydrogen (70 bar) in the presence of Cl$_2$Pt(2,2′-bipyrimidine)[11] (0.023 mmol) at 150°C. *In situ* high-pressure nuclear magnetic resonance (NMR) experiments have shown that the completely protonated 2-methyltetrahydrofuran (Figure 4.4) was converted to oxygen-containing carbocations and alkanes in 15 h. Additional heating of the reaction mixture for 5 h resulted in the formation of a biphasic system. NMR, gas chromatography (GC), and GC-mass spectrometry (GC-MS) measurements have shown that the colorless upper phase consists of a mixture of hydrocarbons, mainly butane, isobutane, pentane, and isopentane.

We have shown that carbohydrates can be converted to different C$_5$-oxygenates (including γ-valerolactone and 2-methyltetrahydrofuran) and a mixture of hydrocar-

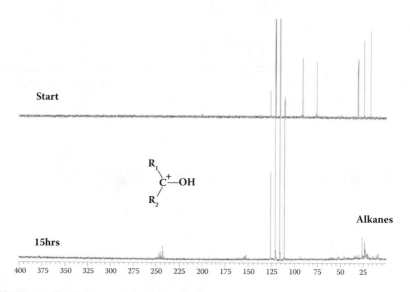

FIGURE 4.4 Catalytic conversion of 2-Me-THF.

bons by using homogeneous transition-metal catalysts. While the conversion of C_5-oxygenates to other organic intermediates has yet to be developed, the hydrocarbons produced from carbohydrates could be considered as a "renewable oil," ready to be converted by existing technologies to all products used today.

SC004x002.eps

SC004x001.eps

ACKNOWLEDGMENTS

This work was funded by the Hungarian National Scientific Research Fund (T047207). High-pressure NMR tubes were donated by Exxon-Mobil Research and Engineering Company. We are grateful to the Department of Chemistry at Princeton University for donating a Bruker AC 250 NMR spectrometer and to the American Chemical Society for covering the shipping cost from Princeton to Budapest.

REFERENCES

1. Aleklett, K., Campell, C.J., "The Peak and Decline of World Oil and Gas Production," *Minerals & Energy* 18, 5, 2003.
2. National Research Council, *The Hydrogen Economy: Opportunities, Costs, Barriers and R&D Needs*, National Academy Press, Washington, DC, 2004; European Hydrogen and Fuel Cell Technology Platform, Strategic Overview, "Hydrogen and Fuel Cell Technology Platform & Joint Technology Initiative", available on-line at https://www.hfpeurope.org/hfp/news#859, 2005.
3. Arakawa H. et al., "Catalysis research of relevance to carbon management: Progress, challenges, and opportunities", *Chem.Rev.*, 101, 953, 2001.
4. Lichtenthaler, F.W., *Carbohydrates as Organic Raw Materials*, VCH, Weinheim, Germany, 1991.
5. Rapp, K.M., "Process for preparing pure 5-hydroxymethylfurfuraldehyde", U.S. patent 4,740,605 (1998); Armaroli, T., Busca, G., Carlini, C., Giuttari, M., Raspolli Galletti, A.M., and Sbrana, G., "Acid sites characterization of niobium phosphate catalysts and their activity in fructose dehydration to 5-hydroxymethyl-2-furaldehyde", *J.Mol.Cat.A: Chemical*, 151, 233, 2000.
6. Mulder G.J., "Untersuchungen über die Humussubstanzen", *J.Prakt.Chem.*, 21, 203, 1840; Thomas, R.W. and Schuette, H.A., "Levulinic acid. I. Its preparation from carbohydrates by digestion with hydrochloric acid under pressure", *J.Am.Chem.Soc.*, 53, 2324, 1931; Leonard, R.H., "Levulinic acid as a Basic Raw Material", *Ind.Eng.Chem.*, 48, 1331, 1956; Bozell, J.J., Moens, L., Elliott, D.C., Wang, Y., Neuenschwander, G.G., Fitzpatrick, S.W., Bilski, R.J., and Jarnefeld, J.L., "Production of levulinic acid and use as a platform chemical for derived products", *Res.Con-serv.Recycl.*, 28, 227, 2000; Srokol, Z., Bouche, A.-G., Von Estrik, A., Strik, R.C.J., Maschmeyer, T., and Peters, J.A., "Hydrothermal upgrading of biomass to biofuel; studies on some monosaccharide model compounds", *Carbohydr.Res.*, 339, 1717, 2004.
7. Mehdi, H., Bodor, A., Tuba, R., and Horváth, I.T., "Conversion of carbohydrates to oxygenates and/or hydrocarbons", Abstracts of Papers of the American Chemical Society, Vol. 226, U721 310-INOR, Part 1, Sep. 2003; Mehdi, H., Bodor, A., and Horváth, I.T., "Dehydration and hydrogenation of carbohydrates with aqueous biphase catalysts", *Abstracts of Papers of the American Chemical Society*, Vol. 227, 095-CELL, Part 1, Mar. 2004.
8. Joó, F., *Encyclopedia of Catalysis*, Vol. 1, Wiley, New York, 2002, p. 737.

9. Schuette, H.A. and Thomas, R.W., "Valerolactone. III. Preparation by the catalytic reduction of levulinic acid with hydrogen in the presence of platinum oxide", *J.Am.Chem.Soc.*, 52, 3010, 1930; Thomas, R.W., Schuette, H.A., and Cowley, M.A., "Levulinic acid. III. Hydrogenation of certain of its alkyl esters in the presence of platinum catalysts", *J.Am.Chem. Soc.*, 53, 3861, 1931; Christian, R.V., Brown, H.D., and Hixon, R.M., "Derivatives of γ-valerolactone, 1,4-pentanediol, and 1,4-bis(2-cyanoethoxy)pentane", *J.Am.Chem.Soc.*, 69, 1961, 1947; Broadbent, H.S. and Selin, T.G., "Rhenium catalysts. VI. Rhenium(IV) oxide hydrate", *J.Org.Chem.*, 28, 2343, 1963; Menzer, L.E., "Production of 5-methylbutyrolactone from levulinic acid", U.S. patent 6,617,464, (2003).

10. Christian, R.V., Brown, H.D., and Hixon, R.M., "Derivatives of γ-valerolactone, 1,4-pentanediol, and 1,4-bis(2-cyanoethoxy)pentane", *J.Am.Chem.Soc.*, 69, 1961, 1947; Elliott, D.C. and Frye, J.G., "Hydrogenated 5-carbon compound and method of making", U.S. patent 5,883,266, (1999); Behr, A. and Brehme, V.A., "Bimetallic-catalyzed reduction of carboxylic acids and lactones to alcohols and diols", *Adv.Synth.Catal.*, 344, 525, 2002.

11. Hara, Y., Inagaki, H., Nishimura, S., and Wada, K., "Selective hydrogenation of cyclic ester to alpha, omega-diol catalyzed by cationic ruthenium complexes with trialkylphosphine ligands", *Chem.Lett.*, 1983, 1992.

12. Periana, R.A., Taube, D.J., Gamble, S., Taube, H., Satoh, T., and Fujii, H., "Platinum catalysts for the high-yield oxidation of methane to a methanol derivative", *Science*, 280, 560, 1998.

Part II

Plastics and Materials from Renewable Resources

5 Developments and Future Trends for Environmentally Degradable Plastics

Emo Chiellini, Patrizia Cinelli, and Andrea Corti

CONTENTS

5.1 INTRODUCTION

5.1.1 PETRO POLYMERS AND BIOPOLYMERS: REMARKS ON ENVIRONMENTAL IMPACT

The worldwide consumption of polymeric materials and plastics rises annually by around 7 to 10%. Total consumption in 2000 was approximately 200 million tonnes (t), which corresponds to nearly 30 kg per capita, with an average of 80 to 100 kg in industrialized countries and 2 to 20 kg in emerging countries and countries in transition.[1] More than 98% of plastics are based on fossil feedstocks (crude oil), the reserves of which are predicted to last for only approximately 80 more years.[2]

Public concern about the environmental consequences bound to the production and consumption of various materials and products is increasing. These effects occur at every stage in a product's life cycle — from the extraction of the raw materials through the processing, manufacturing, and transportation phases, ending with use and disposal or recycling.[3]

At the occasion of the World Conference on Ecology in 1992 in Rio de Janeiro, the United Nations Framework Convention on Climatic Change was signed. With the Kyoto Protocol in 1997, industrialized nations undertook the initiative of reducing greenhouse gas (GHG) emission at least 5% below the amount of 1990. The European Union (EU) committed itself to cutting GHG emissions to 8% below the level quoted by the year 2008.[4]

In 2003, the consumption of polymer for plastic applications in Western Europe was approximately 40 million tonnes. Many plastics applications involve a consumption time of less than 1 year; after that, the vast majority of these plastics are discarded as waste. If fossil-based plastics were to be replaced by starch-based polymers, it is estimated that CO_2 emissions could be reduced by 0.8 to 3.2 t per t of material produced.[5]

Recent EU legislation increasingly requires the recovery of plastic waste through recycling, composting, or energy recovery. The plastics industry mostly recycles its own in-plant scraps, and although commercial-scale plastics processing has been available for many years, postconsumer plastics recycling is still very limited. Moreover, if recycling is to offer true environmental benefits, whatever the material involved, several factors must be taken into account. Most broadly, the manpower and economic resources used to collect, sort, and recycle must be less than, or at least comparable with, those used to produce virgin materials. Sufficient demand from end-market users for the recycled material is also a vital prerequisite, and a key factor for marketing these products is that the use of recycled materials does not compromise product safety or performance.[1]

Organic recycling is a specific recycling option for biodegradable waste, i.e., for biodegradable or compostable materials. It diverts biodegradable waste from landfills, preventing emissions of methane that represent a very powerful GHG generated in anaerobic conditions by landfills. Composting technologies, used for the disposal of food and yard waste that account for 25 to 30% of total municipal solid waste, are particularly suitable for the disposal of biodegradable bioplastics and plastics from fossil feedstock, together with soiled or food-contaminated paper.

The term "biopolymer" and hence the converted plastic items (bioplastics) refers to natural products that are polymeric in character as grown or can be converted to polymeric materials by conventional or enzymatic synthetic procedures.[6] Thus, under that heading, one can include *natural polymers* used as direct feedstock for plastic production as well as *artificial polymers*, such as those obtained by chemical modification of preformed natural polymers or by polymerization of monomers deriving from renewables.[7] Total demand for biodegradable polymers in the U.S., Europe, and Japan was 20 kt in 1998, valued at $95 million.[8] The European market for bioplastics, resulting from the fruits of 15 years of technological development, is

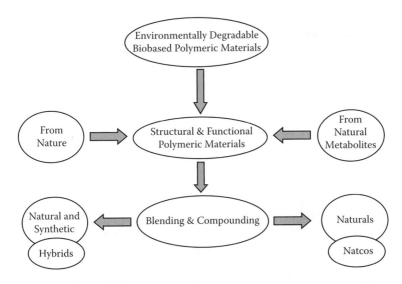

FIGURE 5.1 Main options for the production of environmentally degradable biobased polymeric materials and plastics.

growing slowly but steadily. According to the industry association (International Biodegradable Polymers Association and Working Group — IBAW), the bioplastics usage in 2004 amounted to 50 kt. Compostable rubbish bags and starch-based loose packing material constitute the major part. The IBAW estimates that one-tenth of all plastics applications in Europe could be satisfied by modern bioplastics, a figure that corresponds to approximately 5 million t of polymers. However current worldwide production capacity only amounts to 300 kt,[9] which is 0.15% of the overall present production of plastics.

Traditional natural polymeric materials are represented by polysaccharides (cellulose, starch, chitin, alginic acid, ulvans, xanthanes, guar gum), proteins (fibroin, keratin, collagen, and the polynucleotides RNA and DNA), natural rubber, lignins, and vegetable oil binders. Fibrous material derived from renewable crops, their by-products, or their industrially processed wastes can be considered a good source for the formulation of polymeric blends and composites based either on synthetic and natural components (*hybrids*) or on only natural components, which we name "natcos" for quick identification (Figure 5.1).[10,11]

Biopolymers, unless heavily modified, are biodegradable by definition and can be composted, thus promoting an environmentally compatible waste-management system. Biopolymers are derived from renewable resources and therefore produce no net increase in atmospheric CO_2 balance as part of a sustainable material cycle.

5.1.2 PETRO POLYMERS AND BIOPOLYMERS: ECONOMIC CONSIDERATIONS

In industrial production, sustainability must be achieved, but with awareness that business will fail unless some minimum profit margin is guaranteed. New bioplastics should be introduced as more appropriate options in cases where the degradation constitutes a plus in specific applications by defraying the cost inherent in managing

TABLE 5.1
Energy Content in Agricultural Plants

Plant	Energy (MJ/kg)
Wheat	15.9
Barley	15.9
Potato	14.4
Maize/corn	15.3

Note: Energy content is based on dry material.

the disposal of postconsumer items.[12,13] Plastic articles that are used once and then disposed of are targeted as the primary market areas, such as food-packaging films, foams, bags, food-service items (containers of milk, water, soft-drinks), mulching films, and transplanting .[14] It is, however, taken for granted that infrastructures have to be available to allow for their biorecycling as a final stage of their recovery. Biopolymers can, in contrast to petroleum-based materials, be produced at a fairly low cost in any country, and can therefore also be economically sustainable. A normative strategy has been proposed for resource choice and recycling to meet the criterion of sustainability.[15] Also, the use of biowaste as a resource for biobased productions has been proposed because of its high content of cellulose, hemicellulose, and lignin.[16]

A wide range of biodegradable plastics are under development.[17] If they are disposed of in anaerobic digestion or incinerated in combustion plants, their natural stored energy can be utilized for producing heat and electricity. The energy content in important agricultural plants is reported in Table 5.1.[18]

The availability of raw materials from renewable resources should not interfere with food production.[19] Only 4% of the global production of oil (7.3 Gt) is used for plastics, accounting for more than 200 Mt annually.[20] Compared with the amount of fossil oil resources combusted for energy generation (95% of total fuel feedstock), the limited amount consumed for plastics production contributes only modestly to the GHG effect. So even if it were possible to replace all of the fossil-fuel-feedstock-based plastics with those derived from renewable resources, the effect on GHG reduction would be minimal.

However, the crude oil consumption devoted to plastic production represents indeed a small amount when compared with the available biomass that could be used for biopolymer materials. The estimated annual production of biomass by biosynthesis is 170 Gt, with only about 6 Gt (3.5%) used by humans worldwide.[20] The global production of crop by-products was, for instance, 2.7 Gt in the year 2000. The production of specific, pure biopolymers is significantly lower, with the exception of cellulosic fibers and starch. The total potential mass of starch from cereals is about 1.2 Gt annually. Considering that cereal carbohydrates are currently used for food and feed, the amount available for material application is limited. But even a small fraction of the total production yields considerable amounts, which could possibly compete with commodity polymers, and certainly with specialty polymers so long as the derived plastic items provide adequate performance. This represents one of the major limiting factors for a wide practical and competitive spreading of bioplastics.

TABLE 5.2
Examples of Starch-Based Biopolymers: Producers and Brand Names

Company	Brand Name
Biotec GmbH	Bioplast™
VTT Chemical Technology	COHPOL™
Groen Granulaat	Ecoplast™
Japan Corn Starch Ltd. & Grand River Technology	Evercom™
Novamont SpA	MaterBi™
Starch Tech	ReNEW™
Supol GmbH	Supol™
Novon International	Novon™

Given these factors, academic and industrial researchers are increasingly devoting their attention and efforts to the possible use of renewable feedstocks both as energy source[21,22] and as raw materials for the production of chemicals and polymeric materials and plastics.[23,24] The potential for nonfood use differs widely, with the EU having a surplus of crops and the developing countries needing all edible biopolymers primarily for food and feed. Starch costs about 0.3 €/kg and might be available in large amounts in pure form. Starch-based materials obtained from potatoes, wheat, and the like can be formed into plastics, and these materials have come a long way, with improved properties. Applications such as mulching foils and food packaging are fairly widespread. It is estimated that 15,000 tons of starch-based materials are presently used in Europe for these purposes. The market for starch-based bioplastics in 1999 has been estimated at about 20 kt/yr, mostly in the form of soluble loose fills for packaging and flexible films. Examples of starch-based biopolymer producers and product brand names are listed in Table 5.2.

Despite the relatively low raw material price for starch, there is no starch-based plastic available on the market for less than 3 to 4 €/kg. The high conversion cost for starch is remarkable, combined with the fact that the starch-based products are indeed blends with synthetic polymers such as aliphatic or aliphatic-aromatic polyesters. Other edible biopolymers such as gelatin, alginate, and xanthan all cost 3 to 10 €/kg.[25]

Several corporations are also initiating large-scale production of polylactic acid (PLA). The PLA is produced by a combination of biotech and chemical process. From starch is obtained L-lactic acid that is afterward submitted to polymerization, leading to production of PLA at an estimated price of 1 to 2 €/kg when it is produced in a large-scale plant with an annual capacity around 150 kt. This can be compared with the price for conventional plastics in bulk at 0.5 to 1 €/kg (polyethylene [PE], polyvinylchloride [PVC], polystyrene [PS], etc.), and 2 to 4 €/kg for poly(vinyl alcohol) (PVA).

Among the candidates for biodegradable plastics, poly(hydroxyalkanoates) (PHAs) have been drawing much attention because they can be produced from renewable resources, have properties similar to conventional plastics (specifically polypropylene), and are considered completely biodegradable.[26-31] However, the

TABLE 5.3
Poly(3-hydroxybutyrate) (PHB) Production Cost: Effect of Substrate and PHB Yield

Substrate	Substrate Price (€/kg)	PHB/Yield [a] (%)	Production Cost (€/kg)
Glucose	0.40	38	1.06
Sucrose	0.23	40	0.59
Methanol	0.14	43	0.34
Acetic acid	0.48	38	1.27
Ethanol	0.41	50	0.81
Cane molasses	0.18	42	0.42
Cheese whey	0.06	33	0.18
Hemicellulose hydrolysate	0.06	20	0.28

[a] (PHB wt./substrate wt.) × 100.

Source: Madison, L.L. and Huisman, G.W., *Microbiol. Mol. Biol. Rev.*, 63, 21, 1999. With permission.

fairly high production cost of PHAs makes them substantially more expensive than synthetic commodity plastics.[28] The cost of carbon substrate significantly affects the overall economics in large production scale.[32,33] Therefore, the production cost can be considerably lowered by using agricultural wastes, highlighted by Reddy et al.[34] The production cost of PHA by using the natural microbial strain *Alcaligenes eutrophus* is US$ 16 (€13) per kg, which is ten times more expensive than polypropylene. With recombinant *Escherichia coli* as the PHA producer, price can be reduced to US$ 4 per kg (€3.3), which is close to other biodegradable plastic materials such as PLA and aliphatic polyesters.

The effect of substrate cost and yield on the production cost of poly(3-hydroxybutyrate) (PHB) was reviewed by Madison and Huisman, and relevant data are presented in Table 5.3.[35] At the 227th American Chemical Society (ACS) National Meeting in 2004, Bohlmann reported an estimated polyhydroxybutyrate (PHB) production cost of US$4/lb (€7.2/kg) for the Zeneca process, as based on glucose, propionic acid mixed feed, and wastewater by enzyme permeabilization.[36]

Commercialization of plant-derived PHA will require the creation of transgenic crop plants.[37] Production of PHA on an agronomic scale could allow synthesis of biodegradable plastics in the million-ton scale compared with fermentation, which produces material in the thousand-ton scale. PHA could potentially be produced at a cost of US$ 0.20 to 0.50/kg (€0.16 to 0.41) if it could be synthesized in plants to a level of 20 to 40% dry weight and thus be competitive with the petroleum-based plastics. However, problems such as the need for transgenic plants and the competition between food production and plastic production makes this pathway unappealing for European industries.

Among the synthetic petro-based polymers, poly(vinyl alcohol) (PVA) represents one of the most interesting materials to be utilized for several eco-compatible uses in many fields of application. PVA is composed of repeating units of vinyl alcohol monomer. PVA cannot be prepared by the direct polymerization of its corresponding vinyl alcohol monomer, which is the acetaldehyde enolic form, but is usually produced indirectly by alkaline hydrolysis of polyvinyl acetate. The physical properties of commercial PVA are related to molecular weight, degree of hydrolysis, degree and type of chain branching, amount of cross-linking between polymer chains, and the type and concentration of additives. By varying these factors, useful properties such as strength, stiffness, gas-barrier characteristics, and water solubility can be controlled. Because of its water-solubility characteristics, PVA is used in several specific fields such as colloidal protection of other chemical substances in the textile and paper industries, hospital laundry bags, single-dose packaging, etc. The Italian company Idroplax has developed a technology, using a base of PVA, that allows the manufacture of products, particularly extruded films in a melt-blown process, with good mechanical properties as well as very good hygrometric characteristics. The product is marketed as Hydrolene.

Among the biodegradable polymers of synthetic origin, PVA is a particularly well-suited material for the formulation of blends with natural polymers, since it is highly polar and can also be manipulated in water solutions and, depending upon the grade, in polar nonprotic and protic functional organic solvents.[38-45] Ongoing investigation in our laboratories on the formulation and applicability of mixtures of PVA as synthetic, water-soluble, polymeric materials and biobased "fillers" from agroindustrial surplus and waste has highlighted the potential of attaining ecocompatible articles that undergo environmental degradation at the end of their service life and whose degradation can eventually be programmed.

Ongoing cooperation between a research group formerly at the U.S. Department of Agriculture (USDA) station in Peoria, IL, and presently in Albany, CA, and our research group at the University of Pisa has led to the development of several blends and composites based on PVA and lignocellulosic components derived from waste of agroindustrial dairy processing such as sugar cane, fruits, corn, wheat, and wood.[46-48]

In the last decade, researchers and industrial laboratories have devoted much attention to the production of hybrid composites based on synthetic polymeric matrices and natural fillers, with the aim of reducing production costs as well as promoting the ecological performance of final products. In these regards, the investigation of the ultimate fate of the synthetic polymeric matrices utilized for the hybrid composite materials represents an important aspect to establish the true ecocompatibility of these innovative materials.

In this chapter, we wish to synthetically review some of the recent results obtained in the formulation and utilization of blends and composites based on (a) the combination of synthetic and natural polymeric components (hybrid systems) and (b) materials based only on natural components (natcos systems). We also report the investigation of the environmental degradation of two full-carbon-backbone polymers — poly(vinyl alcohol) (PVA) and reengineered polyethylene (PE) — which, on the basis of cost-performance characteristics, constitute very promising synthetic polymeric matrices because of the feasible production technology and the likelihood of market acceptance by customers.

5.1.3 POLYVINYLICS BIODEGRADATION

Of the current worldwide production of synthetic polymers, nearly 90% is represented by full-carbon-backbone macromolecular systems (polyvinylics and polyvinylidenics),[49] and 35 to 45% of production is for one-time-use items (disposables and packaging). Therefore, it is reasonable to envisage a dramatic environmental impact attributable to the accumulation of plastic litter and waste constituted by full-carbon-backbone polymers, which are recalcitrant to physical, chemical, and biological degradation processes.

In contrast to the "hydro-biodegradation" process of natural and synthetic polymers containing hetero atoms in the main chain (polysaccharides, proteins, polyesters, polyamides, polyethers), the mechanism of biodegradation of full-carbon-backbone polymers requires an initial oxidation step, mediated or not by enzymes, followed by fragmentation, again mediated or not by enzymes, with substantial reduction in molecular weight. The functional fragments then become vulnerable to microorganisms present in different environments, with production (under aerobic conditions) of carbon dioxide, water, and cell biomass. Figure 5.2 outlines the general features of environmentally degradable polymeric materials, which are classified as hydrobiodegradables and oxobiodegradables. Typical examples of the so-called oxo-biodegradable polymers are represented by polyethylene, poly(vinyl alcohol), and lignin (a natural heteropolymer).[50] The major biodegradation mechanism of PVA in aqueous media is represented by the oxidative random cleavage of the polymer chains, the initial step being associated with the specific oxidation of methylene carbon bearing the hydroxyl group, as mediated by *oxidase-* and *dehydrogenase-*type enzymes, to give β-hydroxyketone as well as 1,3-diketone moieties. The latter groups are susceptible to carbon-carbon bond cleavage promoted by specific β-diketone hydrolase, leading to the formation of carboxyl and methyl ketone end groups.[51,52]

The ultimate biochemical fate of partially hydrolyzed PVA samples has been recently described by using *Pseudomonas vesicularis* PD strain, a specific PVA-assimilating bacterium.[53] This bacterium metabolizes PVA by a secondary alcohol oxidase throughout the oxidation of the hydroxyl groups followed by hydrolysis of the formed β-diketones by a specific hydrolase. Both enzymes are extracellular, and the polymer chains are cleaved by repeated enzyme-mediated reactions outside the cells into small fragments, which are further incorporated and assimilated inside the bacterial cytoplasm and metabolized up to carbon dioxide (Figure 5.3).[53]

The initial oxidation step of PVA macromolecules can also be promoted by ligninolytic enzymes (lignin peroxidase [Lip] and laccase) produced by white-rot fungal species such as *Phanerochaete crysosporium*[54,55] and *Pycnoporus cinnabarinus*.[56] The monoelectronic enzymatic oxidation reactions lead to formation of free radicals along with the formation of carbonyl groups as well as double bonds, thus increasing the macromolecule unsaturation.[54]

Similarly, the oxidative instability of polyolefins in the environment is due to physical-chemical radical reactions enhanced by the presence of sensitizing impurities in the polymer chain. To counter the mechanisms of physical aging (e.g., thermal and photolytic degradation) of poly(ethylene) (PE), heat and light stabilizers

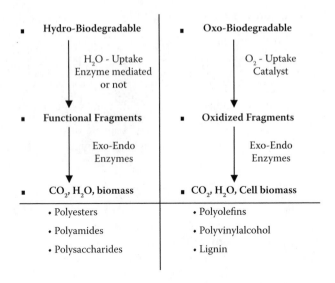

- **Hydro-Biodegradable**

  H_2O - Uptake
  Enzyme mediated
  or not

- **Functional Fragments**

  Exo-Endo
  Enzymes

- CO_2, H_2O, biomass

  • Polyesters
  • Polyamides
  • Polysaccharides

- **Oxo-Biodegradable**

  O_2 - Uptake
  Catalyst

- **Oxidized Fragments**

  Exo-Endo
  Enzymes

- CO_2, H_2O, Cell biomass

  • Polyolefins
  • Polyvinylalcohol
  • Lignin

FIGURE 5.2 General classification of environmentally degradable polymers.

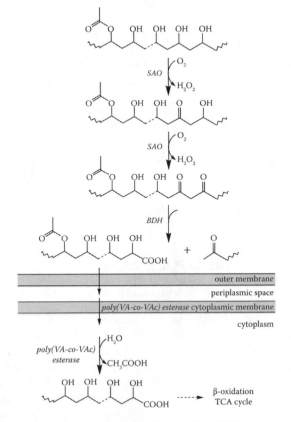

FIGURE 5.3 Biodegradation pathway of partially acetylated PVA.

have been used to improve the resistance of PE to environmental oxidation since the 1960s. Today, there is an opposite strategy aimed at accelerating the degradation rate of PE to overcome its intrinsic recalcitrance to biological attack. In this connection, copolymerization with a small amount of monomers containing carbonyl groups (carbon monoxide, methyl vinyl ketone) or the incorporation of transition metal compounds (dithiocarbamates) as photoinitiators or photosensitizers, and the addition of pro-oxidants (fatty acids and salts) constitute the major strategies for the introduction of functional groups and substances capable of promoting the degradation of macromolecules. When exposed to light and temperature, these additives generate free radicals that react with molecular oxygen to produce peroxides and hydroperoxides.[57]

Further, the hydroperoxides decompose in the presence of heat, light, and metallic ions, leading to the formation of macroalkoxy radicals, and the autooxidation of polyethylene proceeds through classical free radical chain reactions.[58,59] As a result, chain scission and cross-linking are the major consequences of thermal oxidation of polyolefins.[60,61] In the presence of oxygen, however, chain scission and oxidation of macromolecules are the predominant reactions,[62] leading to (a) a significant reduction of both Mn (Average Number Molecular Weight) and molecular weight of poly(ethylene) samples containing pro-oxidant after thermal degradation[63] and (b) to the formation of oxidation products, including carboxylic acids, ketones, lactones, and low-molecular-weight hydrocarbons.[62]

The rate and extent of radical-induced oxidation of polyolefins are also affected by structural parameters such as chain defeats and branching, these latter being representatives of weak links liable to bond cleavage and production of free radicals for succeeding chain fragmentation. The hierarchy in the oxidation susceptibility of polyolefins is: iPP (isotactic polypropylene) > LDPE (low-density PE) > LLDPE (linear low-density PE) > HDPE (high-density PE).[64,65]

By considering the same carbon skeleton and the identical requirement of an initial oxidation step, studies on the environmental fate of Environmental Degradable Polymers (EDPs) based on PVA and PE may represent an interesting tool for identifying similarities or differences between the degradation mechanisms and particularly the influence of different environmental parameters in accelerating the biodegradation of these two large-volume carbon-backbone synthetic polymers.

5.2 HYBRID POLYMER COMPOSITES

5.2.1 Formulations for Hydro-Biomulching Practice

Mulching practice controls radiation, soil temperature and humidity, weed growth, insect infestation, soil compaction, and the degree of carbon dioxide retention. Natural materials, such as straw and leaves, have always been used to provide an insulating layer around the roots of vegetables and protection of soft fruits. Today, the use of plastic sheets or films in mulching is the largest single application of plastics in agriculture. The techniques of "hydro-mulching" or "liquid mulching" also provide a conditioning effect on soil structure.[66,67] Water-soluble synthetic degradable polymers such as poly(acrylamide), poly(vinyl alcohol), carboxymethyl cellulose, and hydrolyzed starch-g-polyacrylonitrile (HSPAN) can be easily sprayed on the soil alone or in mixture with nutrients or other agronomically valuable fillers.[68–72] Polymer solution can also be applied as tackifiers, thus helping to hold the mulch in place once applied or aimed at forming a sort of thatch intended to protect seeds and soil against erosion.

Polyvinyl alcohol and gelatin were used as continuous matrices from synthetic and natural sources, respectively, for the formulation of hydro-biomulching mixtures with sugar cane bagasse and wheat flour as fillers.[73,74] The degree of hydrolysis for the PVA was 88% with a Mn of 67 kDa; the waste gelatin (WG) scraps, which contained several additives from its original formulation as pigments and glycerol, were provided by a pharmaceutical company (Rp Scherer, Egypt).[41] Sugar cane bagasse (SCB, Egypt) was in a powder form ($\phi < 0.212$ mm) and consisted of 42.6% crude fibers, 29,2% cellulose, 10.5% lignin, 9.1% crude protein, 2.6% fat, and 6.0% ash. The wheat flour (WF) was a thin white powder (Italy) with a composition of 45.4% starch, 2.0% ash, 2.2% cellulose, 34.8% hard fiber, and 15.5% moisture.

In countries where gelatin and sugar cane juice processing occupy an important position, the combined use of these by-products allows for low cost and *in situ* formulations of environmentally degradable hydro-biomulch. The addition of SCB fibers to WG hardened the resulting films and darkened the film color. Accordingly, films based on WG containing 10 to 30% SCB were rather flexible, whereas films containing 40 to 50% SCB were very tough and brittle. A WG/SCB 80/20 ratio (WGSCB20) appeared to be the most interesting composition as far as filler content and mechanical properties are concerned.

A series of experiments was performed by spraying PVA or gelatin-based water dispersions on pots (155 cm^2) containing loamy soil. Table 5.4 shows the compositions of the films sprayed on the loamy soil samples in the form of 10% (by weight) water suspensions prepared by dissolving PVA (and eventually glycerol and urea) and WG in water and then mixing with the selected filler. The pots were placed outdoors under open-air conditions. A pot containing the loamy soil without any treatment was used as a control. The evolution of film morphology formed on loamy soils was monitored for three weeks.

TABLE 5.4
Composition of Sprayed Films Based on Waste Gelatin, Polyvinyl Alcohol, Sugar Cane Bagasse, and Wheat Flour

Film Sample	WG (%)	PVA (%)	SCB (%)	WF (%)	Gly (%)	Urea (%)
WGSCB20	80	0	20	0	0	0
WGSCB50	50	0	50	0	0	0
WGP10SCB40	40	10	50	0	0	0
PSCBGU	0	33	33	0	17	17
PWFGU	0	33	0	33	17	17

Note: WG = waste gelatin; SCB = sugarcane bagasse; WF = wheat flour; Gly = glycerol.

Previous investigation on the applications of PVA-water solutions to soil outlined that, when sprayed on the soil surface, the solutions penetrated about 0.5 cm into the soil. When a PVA solution 5% by weight was applied, the soil appeared to be covered with a thin, wet, transparent, but poorly aggregated layer. With a PVA solution 10% by weight was applied, a thin plastic layer was formed on the soil surface; soil formed a very fragile crust, about 0.5 cm high, together with the polymer.[73]

The results of sprayed-film experiments showed that PVA and WG water suspensions were easy to spray; mixture-soil aggregates lasted for more than 3 weeks on the soil; and the soil appeared to be conditioned and in a better state when compared with the control sample. Formulations containing SCB conferred a marked brown color to the soil. The presence of WG, WF, and SCB enhanced PVA longevity on the soil, thus guaranteeing the resultant soil-structuring effect. In consideration of these positive effects, the hydro-biomulching formulations were further tested in the presence of seeds and plants.

Water suspensions based on PVA (P) and wheat flour (WF) and on PVA and sugar cane bagasse (SCB) were prepared at 15% by weight. Both suspensions also contained glycerol (G) and urea (U) (PWFGU and PSCBGU, respectively, in Table 5.4). To evaluate a possible influence of the amount of applied material, each suspension was applied at two different doses: 1 and 3 l/m^2. Eighty seeds of lettuce (*Lactuca sativa*) were planted in a 25-cm diameter pot. Selected water suspensions were sprayed on the soil surface. After 50 days, the average number (three replicates) of germinated seeds was recorded (Table 5.5).

In all cases, a homogeneous coverage of the pots was obtained. A strong negative effect on germination was detected, especially in pots treated with the WF-based formulations, which formed a hard crust on top of the soil, thus effectively inhibiting emergence through the film by the lettuce sprouts. Furthermore, possible direct contact of seeds with the composite formulation (lettuce seeds were buried only 0.5 cm deep) might have interfered with seed transpiration due to the viscosity of the water suspension. Also the large amount of urea applied could have had a negative effect on seed germination.

To verify whether the mixtures had a negative effect on growing plants or only on seed germination, the following test was set. Lettuce seedlings were transplanted into 14-cm-diam. pots (one plant in each pot), and then the soil was sprayed with the same formulations as above. For comparison, another test was prepared by directly spraying the liquid mixtures in the soil and then planting the seedling lettuce.

TABLE 5.5
Results of Lettuce Germination in Pots

Sample	Dose (l/m^2)	Germination (%)
Control	0	63
PWFGU	1	8
PWFGU	3	0
PSCBGU	1	33
PSCBGU	3	3

TABLE 5.6
Hydro-Biomulching Formulation Based on PVA and Lignocellulosic Filler

Treatment	PVA (g/m^2)	Organic Filler Type	g/m^2	Water (g/m^2)
PSCB	20	SCB	40	340
PWF	20	WF	40	340
PWS	20	WS	500	380
PSD	20	SD	500	380

Note: SCB = sugarcane bagasse; WF = wheat flour; WS = wheat straw (milled); SD = sawdust.

This procedure was carried out to promote a direct contact of the formulations with lettuce roots. With both these procedures no toxic effect was observed, and lettuce plants appeared to grow at a faster rate in the conditioned pots.

A further appraisal of the potential effect of hydro-biomulching based on PVA and natural fillers was conducted during a field trial at the University of Pisa in summer between June and September. Two conventional mulching materials — polyethylene film (PE) (70 g/m^2) and raw straw mulching (SM) (100 g/m^2) — were compared with the innovative PVA-based hydro-biomulching formulations represented by PSCBGU, PWFGU, and a PVA-water solution used as tackifiers on coarsely milled wheat straw (WS) with an average size of 20 mm, and on sawdust (SD), a commercial product from softwood sawmill (Table 5.6).

Sawdust and wheat straw were spread directly on the soil at a level of 500 g/m^2 prior to the spray application of the PVA solution. Corn (*Zea mays*) and lettuce (*Lactuca sativa*) were chosen to test the agronomic effect of the mulching treatments on seeded and transplanted crops. For each plot, three corn plants were established by sowing and three plants of lettuce were established by transplanting seedlings. Each trial was carried out on nine replicates. All the lettuce plants were harvested 50 days after transplanting, and fresh weight was measured and reported as an average weight (Figure 5.4).

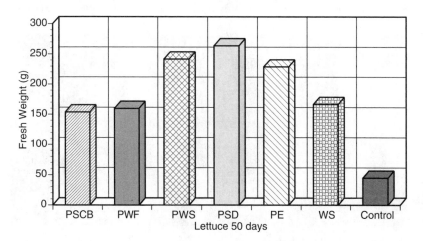

FIGURE 5.4 Average fresh weight of lettuce plants 50 days after transplanting.

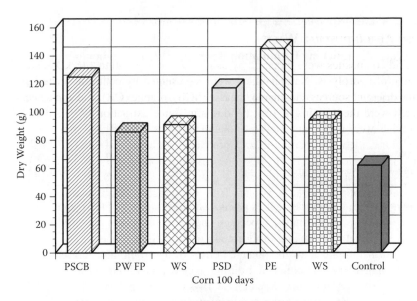

FIGURE 5.5 Average dry weight of corn plants after 100 days from seedling.

For lettuce, all of the PVA-based fluid mulching treatments produced results that were comparable with PE and significantly different from those obtained with the control and the semidry mulching (PWS and PSD). For corn, all of the fluid mulching treatments performed better than the untreated soil, although only the PSCB formulation approached the efficiency of the PSCB formulation (Figure 5.5).

Differences in the soil structure were evident from visual field observations at the end of the experiment. Figure 5.6 refers to the soil surface at the end of the trial in plots that had received PVA/SCB, PVA/WF, and the control not submitted to any mulching treatment. Soil aggregates (1 to 2 cm) were still present in treated soil, while they were completely disrupted in the control plot, thus confirming the positive effect on soil structure preservation previously observed in laboratory-scale tests.

FIGURE 5.6 Soil appearance at the end of the hydro field trial: a) control, b) PVA/SCB, c) PVA/WF.

5.2.2 Films and Laminates Based on PVA and Natural Fillers from Agroindustrial Waste

Hybrid composites were prepared by using PVA of Mn 100 to 146 kDa and a 96% degree of hydrolysis. Natural fillers were represented by cellulosic materials from three different sources: sugarcane bagasse (SCB), orange (OR), and apple peel (AP), which were the remains of fruit residue after juice extraction.[40,46,47] All cellulosic materials were milled and sieved to obtain particles sizes <0.188 mm.

Unmodified commercial-grade corn/starch with approximately 30% amylose and 70% amylopectin (U.S.) was added in some formulations to replace as much PVA as possible without compromising the film properties. Films were prepared by casting of water suspensions, 10% by weight. While maintaining a 1/1 PVA/fiber ratio, glycerol, urea, and starch were introduced in the formulation as reported in Table 5.7. Addition of glycerol and urea softened the films (Figure 5.7), as both glycerol and urea are known to act as a plasticizer for PVA in PVA/starch-based films.[75–77]

Effect of starch addition on mechanical properties is reported in Figure 5.8. In PVA/OR blends, starch addition caused significant reduction in elongation a break, with somewhat moderate variation in ultimate tensile strength (UTS). Excellent film-forming properties and flexibility were observed in films, even when starch concentration in formulations exceeded 25%. In PVA/SCB blends, elongation at break (EB) was mostly unaffected by starch content, while UTS increased with starch addition; thus the introduction of starch increased cohesiveness of PVA/SCB films. In contrast, in PVA/AP blends, addition of starch increased the presence of defects, and small holes were detected in the films, thus indicating a loss in mechanical properties due to increased starch content.

Given the positive results observed, particularly for the hybrid composites based on OR fibers prepared by casting, further tests were performed by compression

---

TABLE 5.7
Composition of the Composite Films Prepared by Casting from Water Suspensions of PVA and Natural Fillers (OR, SC, AP)

Sample	PVA (%)	Natural Fiber (%)	Glycerol (%)	Urea (%)	Starch (%)
POR50	50.0	50.0	0.0	0.0	0.0
PORGU1	40.0	40.0	10.0	10.0	0.0
PORGU2	33.3	33.3	16.6	16.6	0.0
PORGU3	25.0	25.0	25.0	25.0	0.0
PORSt1	31.0	31.0	15.0	15.0	8.0
PORSt2	28.6	28.6	14.3	14.3	14.3
PORSt3	25.0	25.0	12.5	12.5	25.0

Note: POR series based on orange peel (OR) as filler; PSC series based on sugarcane bagasse (SC) as filler; PAP series based on apple peel (AP) as filler. G = glycerol; U = urea; St = starch.

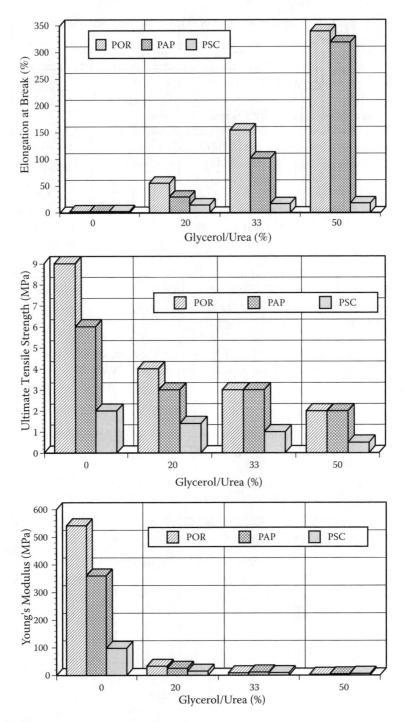

FIGURE 5.7 Mechanical properties as a function of plasticizers content; a) elongation at break, b) ultimate tensile strength, c) Young's modulus.

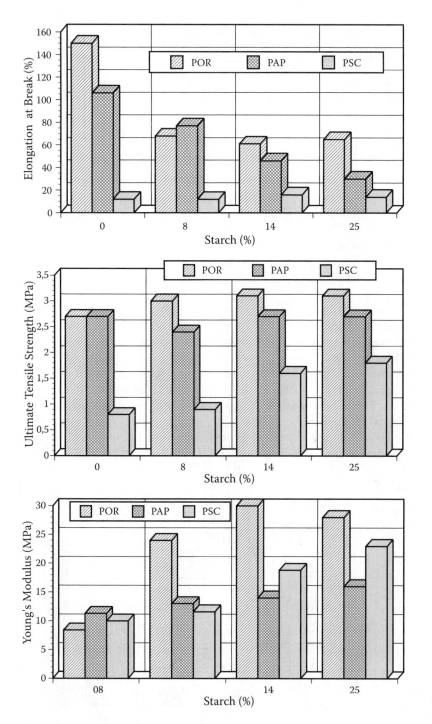

FIGURE 5.8 Mechanical properties as a function of starch content; a) elongation at break, b) ultimate tensile strength, c) Young's modulus.

TABLE 5.8
Composition of Biobased Mixtures in Compression Molding

Samples [a]	PVA (%)	Fibers Type	(%)	Starch (%)	Glycerol (%)
PStG	44	—	—	28	28
PORG	44	OR	28	—	28
PStSCB, PstOR, PStAP	34	SC, OR, AP	22	22	22
PStORU	25	OR	25	25	12.5 [b]

[a] P = poly(vinyl alcohol); St = starch; G = glycerol; SCB = sugar cane bagasse; OR = orange peel; AP = apple peel; U = urea.
[b] Containing 12.5 % of urea.

molding. Laminates based on PVA and natural fibers were prepared with compositions as reported in Table 5.8. About 20% water was added to each composition before processing.[11] Mechanical properties of the prepared laminates are shown in Figure 5.9.

OR was selected for the first attempts aimed at producing composites by compression molding, since films based on PVA and OR proved to be homogeneous, flexible, and more cohesive than composites based on PVA and SCB or AP. Thus the substitution of starch (PStG) with orange waste (PORG) in the hybrid composite based on PVA and glycerol presented a modest variation in mechanical properties. Comparison of PORG with PStOR mechanical properties suggested that the introduction of starch in the formulation moderately decreased mechanical properties.

Composites prepared with AP (PStAP) presented EB values (61%) similar to those of the composites prepared with OR (PStOR), but higher UTS (9 MPa) and

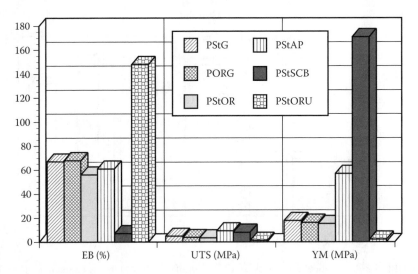

FIGURE 5.9 Mechanical properties of PVA-based composite laminates.

Young's modulus (57 MPa). PVA/SCB also resulted in a harder film with a reduced EB (7%) than PVA/OR and PVA/AP composites, but with a relatively high UTS (8 MPa) and Young's modulus (171 MPa). This behavior is related to the fibers' composition and type. Both AP and SCB have a fibrous shape that confers hardness to the composites when processing allows for orientation as in the Brabender device, while in casting films, the random distribution of the fibers was a source of fragility.

Composites containing urea presented a higher EB, as expected, because of the increased percentage of plasticizer additives. PStORU presented a high EB (148%), and a decreased UTS (1 MPa) and Young's modulus (2 MPa) in comparison with PStOR.

A 500-mg composite sample containing PVA, OR, starch and glycerol (PORSt), and the corresponding amounts of PVA (165 mg) and fillers composed by OR, starch, and glycerol (ORSt) (335 mg) were dried up to constant weight and ground. The resulting powders were mixed with 25 g of compost soil in the sample chamber of a closed-circuit Micro-Oximax respirometer system. Sample chambers were placed in the water bath thermostated at 25°C and connected to the Micro-Oxymax respirometer. The total CO_2 evolution was recorded every 6 h. Experiments were carried out over a period of 21 days. In this test, the large CO_2 production from the compost soil did not allow us to separate that produced by PVA; the limited amount of PVA used also was a factor. The CO_2 productions for one replicate of soil, PVA, PORSt, and ORSt are shown in Figure 5.10.

PVA and soil production are almost exactly overlapping, thus suggesting a very low mineralization of PVA. PVA resistance to degradation under soil burial conditions will be discussed in greater detail in the following sections of this chapter. In

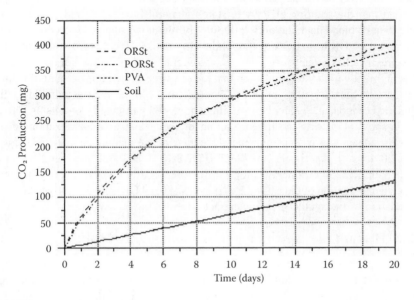

FIGURE 5.10 Time variation of the CO_2 production of PVA/OR blends in compost soil degradation test.

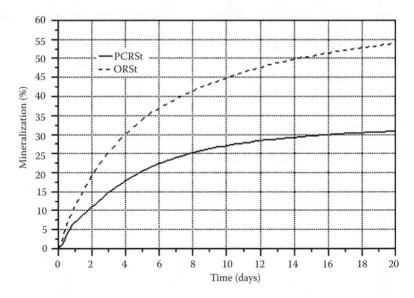

FIGURE 5.11 Time variation of the extent of mineralization of PVA/OR blends in compost soil degradation test.

a recent study on PVA/starch plastic degradation in activated sludge, it was observed that PVA degradation occurred only if PVA-degrading bacterium or enzyme were added to the sludge. This behavior was attributed both to the rare presence of PVA bacteria in the natural environment and to the long period of time necessary to enrich them to the level where they could exhibit PVA-degrading activity.[78] It has also been reported that the addition of PVA in starch-glycerol blends lowered both the rate and extent of blend degradation.[76] Interestingly, about the same CO_2 production was recorded for PORSt and ORSt, thus indicating that, in this hybrid composite, the PVA presence has no negative effect on the degradation of the natural components. Thus in investigating the supposition that the presence of a certain amount of natural polymers in blends with synthetic polymers can promote the degradation of the synthetic component, lignocellulosic fillers appeared particularly promising.[79]

Figure 5.11 reports the time variation of the extent of mineralization of the investigated samples evaluated as the percentage of the theoretical CO_2 production. After 20 days of experiment, the blend (PORSt) reached a 31% mineralization in comparison with 54% mineralization for the fillers (ORSt). These values confirmed the rapid degradation of the natural components (St and OR) and the not-yet-initiated degradation of the synthetic polymer fraction.

5.2.3 HYBRID COMPOSITES BY INJECTION MOLDING AND FOAMING

Hybrid composites with PVA and starch as continuous matrices were prepared with corn fiber (CF). CF is an industrial name given to the pericarp fraction of the corn kernel that is a coproduct of ethanol production by wet-milling technology. CF contains pericarp as well as starch and protein from the endosperm. The wet CF

TABLE 5.9

Composition of Biobased Mixtures Consisting of PVA and CF Processed by Injection Molding

Sample	PVA (%)	Corn Fiber (%)	Starch (%)	Glycerol (%)	Pentaerythritol (%)	PEG (%)
PCF1	42	26	0	21	11	0
PCF2	40	25	0	20	10	5
PCFSt1	36	23	9	18	9	5
PCFSt2	33	21	17	17	8	4
PCFSt3	29	32	14	14	7	4

Note: P = poly(vinyl alcohol) (PVA); CF = corn fibers; St = starch; PEG = poly(ethylene glycol).

(60% moisture) is sold at about \$15 per ton. Dried and ground at 10 mesh, CF is sold at about \$50 per ton, with its main use being in animal feeds. The CF used in our experiments had a composition of 1% fat, 14% protein, 25.5% starch, 59% lignocellulosic component, and 0.5% ash. Injection-molded specimens were prepared to test glycerol, pentaerythritol, and polyethylene glycol 2000 as suitable plasticizers.[48] Composition of the prepared samples is reported in Table 5.9. Compounding was performed at 160 to 170°C, thus avoiding fiber degradation during processing. PCF1 produced a composite that was cohesive and flexible. In the second mixture (PCF2), a limited amount of PEG (5%) was introduced to further lower the viscosity of the melt.

Mechanical properties of the produced items are reported in Table 5.10. Samples were tested after storing at 23°C and 50% relative humidity (RH) for 7 days and after 1 year.

Due to the greater amount of plasticizer in this composite, percent elongation at break (EB) increased in PCF2 compared with PCF1, with a concomitant decrease in ultimate tensile strength (UTS) and Young's modulus (YM). Only small changes in tensile properties were observed for composites after 1 year when compared with the composites stored for 7 days under the same conditions. In composites containing no starch (PCF1 and PCF2), changes in EB were not significant. UTS and YM

TABLE 5.10

Mechanical Properties of Injection-Molded Composites at Different Aging

Mechanical Property	Aging	PCF1	PCF2	PCFSt1	PCFSt2	PCFSt3
EB (%)	7 days	599	645	396	297	101
EB (%)	365 days	613	619	415	356	95
UTS (MPa)	7 days	11.7	7.1	8.3	8.0	7.8
UTS (MPa)	365 days	9.5	7.2	6.2	6.4	8.6
YM (MPa)	7 days	52.0	27.5	94.2	112.0	122.2
YM (MPa)	365 days	34.2	38.6	65.0	84.7	183.3

Note: EB = elongation at break; UTS = ultimate tensile strength; YM = Young's modulus.

decreased moderately for PCF1 and were almost the same (UTS) or moderately increased (YM) for PCF2.

The plasticizer proportion as in PCF2 was used for the rest of the composites. Corn starch was introduced in the formulation and the amount of CF was progressively increased up to a value of 32% in the mixture. As the starch and CF volume fraction increased, elongation at break decreased, while the modulus generally increased. Interestingly, UTS was not significantly affected by increasing CF. Also, composites made with 32% fibers and only 29% PVA resulted in cohesive extrusions.

For samples PCFSt1 and PCFSt2, prepared respectively with 9 and 17% starch, and approximately the same fiber:PVA weight ratio as in composite PCF2, EB was slightly increased with storage, and UTS and YM were decreased. These changes are probably an indication that water was absorbed during storage. Composite PCFSt3, which contained a higher ratio of fibers to PVA, had an increase in UTS and YM and a decrease in EB. This indicates that this composite was getting stiffer and less flexible with age. Composites tested after 1 year of storage had tensile properties similar to composites tested after 7 days of storage. Thus the prepared materials appear extremely promising for several practical applications.

In consideration of the positive results obtained by injection-molding processing of polymeric matrices with corn fibers, further investigations were done to evaluate the effect of using corn fibers in starch-based foam trays. Foam food containers contribute in large scale to the amount of plastics in municipal solid waste streams. These are mainly produced by expanded polystyrene (EPS) or coated paperboard. In recent years, several efforts have been devoted to the production of similar items based on polymers from renewable resources such as starch.[80,81]

Potato starch, CF, magnesium stearate, and PVA, respectively 88% (P88) and 98% (P98) hydrolysis degree, were first mixed using a kitchen mixer with a wire whisk attachment. For PVA-free batter and with less then 50-weight part of fibers, gum arabic (1% by weight of starch) was added to prevent starch settling (Table 5.11). Water was added to reach the required total solids content.[82]

TABLE 5.11
Composition of the Biobased Batter Used To Produce Trays by Foaming Technique

Batter	St (%)	PVA (%)	CF (%)	Arabic Gum (%)	Magnesium Stearate (%)
St	96	0	0	1	3
StCF50	65	0	31	1	3
StCF100	49	0	49	0	2
StCF150	39	0	59	0	2
StP	80	16	0	1	3
StPCF50	58	11	29	0	2
StPCF100	45	8	45	0	2
StPCF150	36	7	55	0	2

Note: St = starch, CF = corn fibers, P = poly(vinyl alcohol) (PVA).

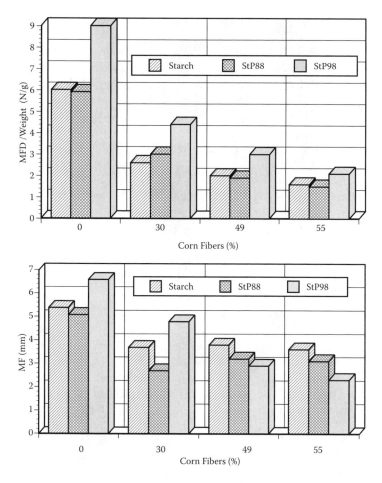

FIGURE 5.12 Dependence of a) Maximum Force (MF) and b) Deformation at Maximum Force (MFD) as a function of Corn Fibers (CF) content.

Foam trays were prepared using a lab model-baking machine essentially consisting of two heated steel molds, the top of which can be hydraulically lowered to mate with the bottom half for a set amount of time. Dimensions of the mold were 220-mm long, 135-mm wide, 20-mm deep, and 3-mm plate separation. Baking temperature was set at 200°C. Baking time was the minimum required to avoid soft or bubbled trays and varied in the range of about 120 to 180 sec.

Mechanical tests were performed on the trays, after conditioning for 1 week at 23°C and 50% RH, by using an Instron Model 4201 Universal Testing Machine equipped with a cylindrical probe (80-mm diameter). The probe was lowered onto the tray until a load of 0.5 N was reached and then lowered at 30 mm/min. Parameters calculated were the maximum force (MF) and deformation to MF (MFD).

Increasing fiber content resulted in lowered MFD and MF values, indicating that the fibers' irregular shape was not allowing a strengthening effect (Figure 5.12). PVA presence improved MFD and MF values, especially for P98, and the resulting

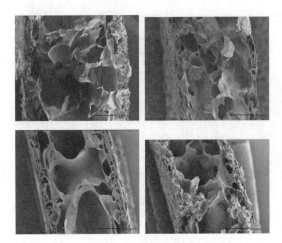

FIGURE 5.13 Scanning electron micrographs (SEM) of a) foams based on potato starch (St), b) foams based on potato starch/Corn fibers (StCF100), c) foams based on potato starch/PVA (SW), d) foams based on potato starch/PVA/corn fibers (StPCF100).

trays were still cohesive, even with a high content of fibers in the composition (45%). Moreover, if 88% hydrolyzed PVA is added as a powder grade, its addition can be performed in the dry mixture, thus avoiding the time- and energy-consuming step of PVA dissolution in hot water that was adopted in previous procedures.[80]

Distilled water (100 ml) was poured in trays with fibers and trays without fibers. After 30 min, trays based on potato starch or strengthened by PVA easily softened because of water addition. In contrast, trays containing a high percentage of fibers (45%) and PVA (8%), such as StPCF100, softened but remained cohesive. This effect is attributed to the disposition of the fibers in the external part of the trays. Thus the foam trays have a dense outer skin with large, thin-walled channels comprising the core as a result of the faster drying of the paste placed nearer to the mold, as shown in the SEM (scanning electron microscope) micrographs reported in Figure 5.13. We can conclude that the introduction of corn fibers improves resistance to moisture and water effects due to the fibers' disposition on the external sides of the tray.

5.2.4 BIODEGRADATION OF POLY(VINYL ALCOHOL)

The environmental fate of water-soluble poly(vinyl alcohol) (PVA) has been primarily investigated due to its large utilization in textile and paper industries that generate considerable amounts of wastewaters contaminated by PVA. In 1936, it was observed that PVA was susceptible of sustaining ultimate biodegradation when submitted to the action of *Fusarium lini*.[83] Afterward, the nature of PVA as a truly biodegradable synthetic polymer was repeatedly and intensively assessed.

Most of the PVA-degrading microorganisms have been identified as aerobic bacteria belonging to *Pseudomonas*, *Alcaligenes*, and *Bacillus* genera. Some species degrade and assimilate PVA axenically, although symbiotic association exhibiting complex cross-feeding processes is a rather common feature of PVA biodegradation.[84-87]

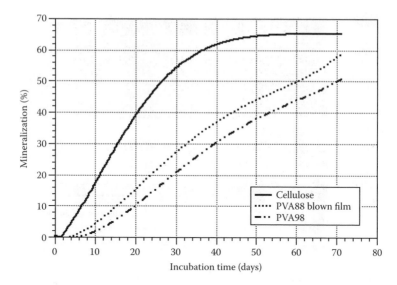

FIGURE 5.14 Time profiles of mineralization of PVA-based blow films, 98% hydrolized PVA (PVA98), and cellulose in an aqueous medium in the presence of paper mill sewage sludge.

Nevertheless, the extensive biodegradation of PVA is accomplished almost exclusively by specific degrading microorganisms whose occurrence in the environment appears to be uncommon and, in most cases, strictly associated with PVA-contaminated environments.[88,89] Correspondingly, limited mineralization of PVA-based blown film was recorded during respirometric biodegradation tests in an aqueous medium inoculated with municipal sewage sludge.[90] In contrast, in the presence of the sewage sludge collected from a paper mill wastewater treatment plant, the mineralization extent of PVA and PVA-based blown film was comparable with that of cellulose, although in a longer incubation time (Figure 5.14).[90] This behavior demonstrates that the selective pressure exerted by the constant presence of large amounts of PVA in wastewater from paper factories is effective in establishing a microbial consortia able to degrade and assimilate PVA. The enrichment procedure of paper mill sewage sludge, as obtained by repeated sequential transfers of this microbial inoculum in the presence of PVA as sole carbon and energy source, induced a significant acceleration in the degradation rate, with a substantial increase of the extent of PVA mineralization. On the other hand, the acclimation of the microbial strains to PVA led to an almost total abatement of cellulose assimilation (Figure 5.15), thus confirming the high specificity of PVA-degrading microorganisms.[90]

Molecular weight, degree of hydrolysis (HD), at least in the 80 to 100% range, and content of head-to-head units does not appear to greatly affect the enzymatic random endocleavage of PVA macromolecules in aqueous media.[91–93] However, earlier investigations are suggesting that some structural characteristics of PVA, such as the hydrophobic character (associated with the residual acetyl group content), may considerably influence the activity of different PVA-degrading enzymes.[90,94,95] Accordingly, no major differences in the extent of mineralization in the presence of an acclimated microbial inoculum have been observed among three commercial PVA

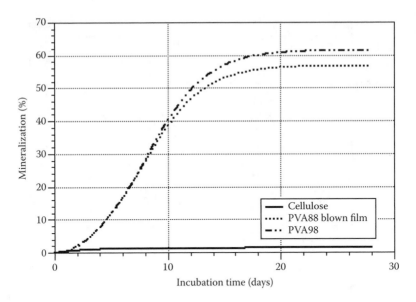

FIGURE 5.15 Time profiles of mineralization of PVA-based blown films, 98% hydrolized PVA (PVA98), and cellulose in an aqueous medium in the presence of selected PVA-degrading microorganisms.

samples (PVA72, PVA88, PVA98) having HD 72, 88, and 98%, respectively.[90] Nevertheless, the kinetics of the mineralization process of the sample having the lowest HD was affected by an appreciably longer lag phase with respect to the other two samples (Figure 5.16).

In contrast, it has been observed that both the degree of polymerization (DPn) and HD strongly affect the biodegradation process of PVA in solid state incubation tests aimed at reproducing natural environments such as soil or compost. In particular, the preferential assimilation of low-molecular-weight fractions has been ascertained in soil burial tests,[96] whereas a very limited propensity of high-molecular-weight and fully hydrolyzed PVA to be biodegraded in soil and compost environments was repeatedly observed.[78,97–99]

This behavior can be attributed in a first instance to the very rare distribution of specific PVA-degrading microorganisms in soil and compost media.[100] However, the strong complexing propensity of free hydroxyl groups to polar mineral components of the soil that render the PVA impervious to even exoenzyme attack also has to be taken into account.[101] An investigation of the adsorption of PVA by soil components provided much information on the influence of both DPn and HD on the adsorption process.[101] For instance, the specific PVA adsorption on montmorillonite at equilibrium increased from 31 to 37 mg/g with increasing HD from 72.5 to 98%, while in the case of an 88% hydrolyzed samples, the specific PVA adsorption decreased from 34 to 28 mg/g when the number average molecular weight of the sample was increased from 36 to 88 kDa.[101]

PVA adsorption on whole soil was shown to behave similarly, thus providing further evidence for the preferential adsorption of low-molecular-weight fractions.

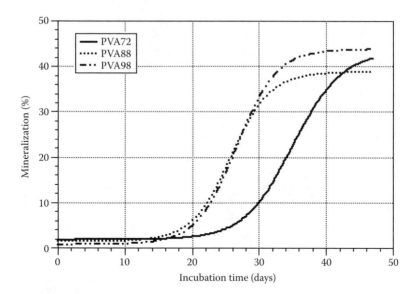

FIGURE 5.16 Time profiles of mineralization of PVA samples having different degree of hydrolysis in an aqueous medium in the presence of acclimated PVA-degrading microorganisms.

No detectable amount of PVA was released from montmorillonite samples when suspended in water, suggesting that the adsorption process is almost irreversible.[101]

The influence of the adsorption by montmorillonite on the biodegradation of PVA was investigated in the presence of an acclimated microbial inoculum[90] in liquid cultures containing either soluble PVA or PVA adsorbed on montmorillonite. The recorded mineralization profiles clearly showed that PVA in solution was extensively mineralized (34%), whereas when adsorbed on montmorillonite, only 4% mineralization was achieved within 1 month of incubation (Figure 5.17), thus indicating that PVA adsorption on inorganic substrates effectively inhibits the biodegradation processes.[101]

Based on these results, the dependence of mineralization rate on the HD and DPn has been further evaluated in soil-burial respirometric experiments by using three different commercial PVA samples having 72, 88, and 98% HD, respectively, and corresponding molecular weight of 25.0, 7.2, and 93.5 kDa. Very limited mineralization was observed in the case of the PVA sample having the highest molecular weight (MW) and HD, whereas lower HD and in particular DPn have been shown to promote the biodegradation propensity of PVA in soil (Figure 5.18).[102,103] The results achieved with these studies indicate different biodegradation behaviors of PVA in aqueous and in solid-state tests reproducing different environmental conditions, as well as the role exerted by PVA structural parameters such as molecular weight and degree of hydrolysis.

To clarify this point, different PVA samples having similar degree of polymerization and noticeably different hydrolysis degrees (ranging between 11 and 75%) were prepared by controlled acetylation of a commercial-grade sample (HD = 98%)

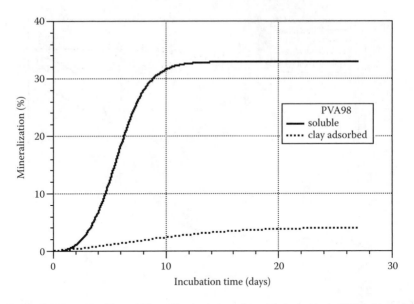

FIGURE 5.17 Time profiles of mineralization of soluble and clay-adsorbed PVA in an aqueous medium in the presence of acclimated PVA-degrading microorganisms.

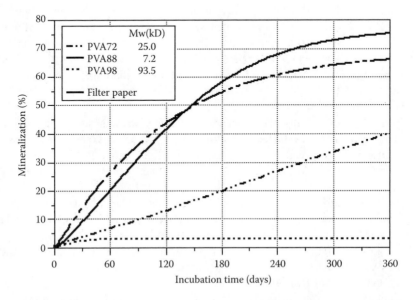

FIGURE 5.18 Time profiles of mineralization of PVA samples having different degree of hydrolysis in an aqueous medium in the presence of acclimated PVA-degrading microorganisms.

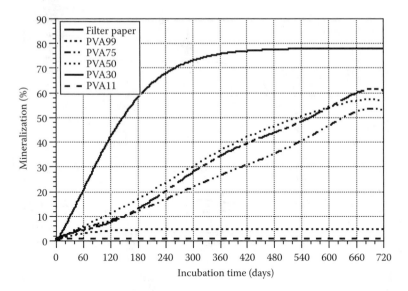

FIGURE 5.19 Time profiles of mineralization of re-acetylated PVA samples in a soil burial respirometric test.

and submitted to biodegradation respirometric experiments under aqueous, mature compost, as well as soil incubation media.[103]

In a soil-burial respirometric test, the biodegradation extents of four different PVA (PVA11, PVA30, PVA50, PVA75) samples, obtained by the controlled reacetylation of PVA99, in comparison with the biodegradation behavior of PVA99 (representing the polymeric substrate utilized in the reacetylation reactions) were assessed. Reacetylated PVA samples having a molar content of vinyl acetate units ranging between 24 and 73%, PVA75, PVA50, and PVA30 showed comparable biodegradation profiles, reaching fairly high mineralization extents (53 to 61%) at the end of the test (Figure 5.19). On the contrary, the almost completely hydrolyzed sample (PVA99), as well as the sample having the highest content of vinyl acetate units (PVA11) (Figure 5.19) did not undergo any significant microbial degradation after approximately 2 years of incubation time.[103] A similar behavior, characterized however by lower mineralization extents, was observed in the presence of the same PVA samples when tested in a series of respirometric trials in mature compost (Figure 5.20).

For comparison, reacetylated PVA samples were also submitted to a biodegradation test in aqueous medium in the presence of selected microorganisms. As expected, they dissolved in the aqueous medium at a rate depending upon the degree of hydrolysis. Accordingly, PVA75 was found to be readily soluble in cool water. PVA50 was shown to disintegrate and partially solubilize, whereas PVA30 proved to be swellable and PVA11 was almost completely insoluble.

For comparison PVA99 was also tested both as an insoluble film and as a water solution (attained after heating at 90°C). Under these conditions, soluble PVA75 and PVA99 samples were promptly mineralized to a similar extent by the selected microbial inoculum (Figure 5.21). Lower but appreciable biodegradation degrees

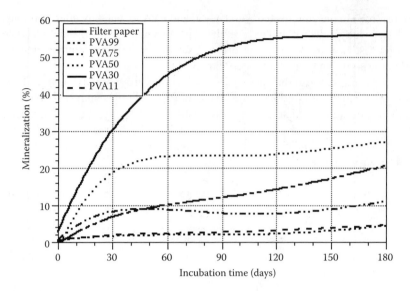

FIGURE 5.20 Time profiles of mineralization of re-acetylated PVA samples in a mature compost respirometric test.

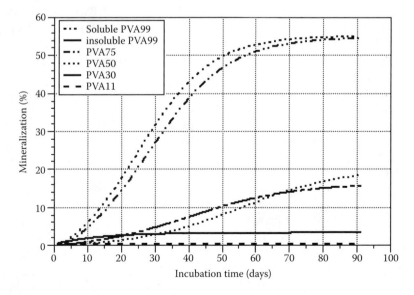

FIGURE 5.21 Time profiles of mineralization of re-acetylated PVA samples in an aqueous medium in the presence of acclimated PVA-degrading microorganisms.

were recorded in the cultures fed with PVA30 and PVA50, but only after a prolonged lag phase, whereas the insoluble reacetylated PVA sample having the highest content of acetyl residues (PVA11) and insoluble PVA99 sample were almost completely recalcitrant to the microbial attack (Figure 5.21).

These data suggest that a certain grade of hydrophobicity achieved by a dominant content of vinyl acetate units in PVA samples may be taken as a practical parameter indicating the propensity of a PVA to be biodegraded in solid media. A similar influence was recognized for the degree of polymerization, resulting, as expected, in a higher mineralization degree for the sample with lower molecular weight.

An opposite trend was instead observed in the biodegradation tests carried out in aqueous medium in the presence of PVA-acclimated microorganisms. In these conditions, the driving force in the biodegradation of PVA appear to be its solubility, which makes the solvated chains vulnerable to the endocleavage of the polymer chains by specific enzymatic systems.

5.2.5 OXO-BIODEGRADATION OF POLYETHYLENE

Free-radical oxidation, as induced by thermal or photolytic preabiotic treatment, constitutes the first step for promoting the eventual biodegradation of both LDPE and LDPE-containing pro-oxidant additives (Figure 5.22). This can be accomplished by monitoring the initial variation of sample weight, molecular weight, and other structural parameters (tensile strength, degree of crystallinity, spectroscopic characteristics). Biodegradation is then observed when degraded oxidized polymer fragments are exposed to biotic environments.[104–106]

Additionally, the required degree of macromolecular breakdown for the microbial assimilation of polyethylene to occur is substantiated by the observation of the propensity to biodegradation of lower-molecular-weight hydrocarbon molecules. It has been demonstrated that linear hydrocarbon having molecular weight below 500[107] or n-alkanes up to tetratetracontane ($C_{44}H_{90}$, MW = 618)[108] can be utilized as carbon source by microorganisms. The degradation of higher-molecular-weight, untreated high-density poly(ethylene) with molecular weights up to 28,000 by a *Penicillium simplicissimum* isolate has also been reported,[109] although the extent of fungal attack of polyethylene matrix was monitored only by physicochemical tools (FT-IR, HT-GPC) as well as by growth-proliferation assays in agar plates containing the polyolefin sample. No respirometric data have been reported.

Therefore, it has been generally accepted that the initial abiotic degradation step represents the major prerequisite in the induction of potential microbial assimilation of poly(ethylene). Many studies have been undertaken to obtain an insight into the mechanisms of radical oxidation of (PE).[57,58,62,63] On the other hand, relatively little information on the influence of physical parameters such as humidity, oxygen pressure, as well as the whole biological environmental conditions on the propensity of thermal oxidation of degradable PE films are available.[110]

Accordingly, the oxidation propensity of PE samples from EPI Inc. (Canada) containing TDPA™ pro-oxidant additives, as induced by heat, has been assayed in

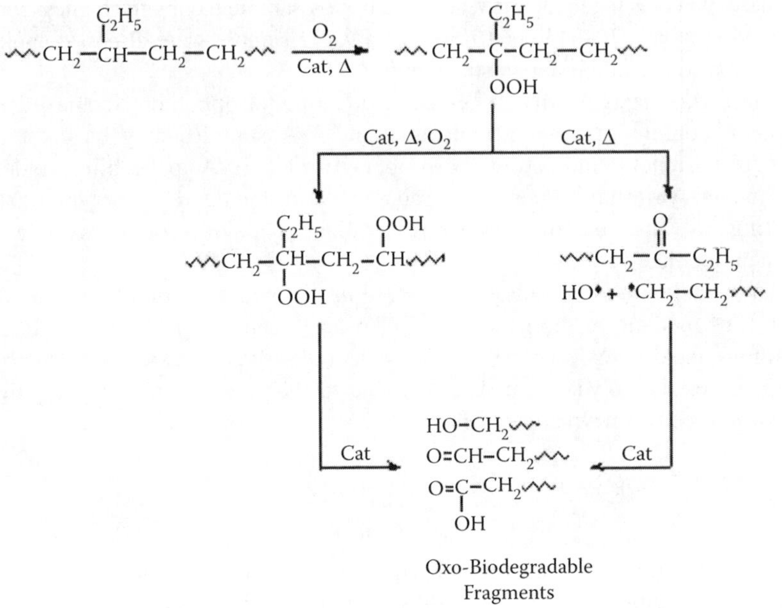

$$\text{www}CH_2-\underset{\underset{H}{|}}{\overset{\overset{C_2H_5}{|}}{C}}-CH_2-CH_2\text{www} \xrightarrow[\text{Cat, }\Delta]{O_2} \text{www}CH_2-\underset{\underset{OOH}{|}}{\overset{\overset{C_2H_5}{|}}{C}}-CH_2-CH_2\text{www}$$

Cat, Δ, O₂ | Cat, Δ

$$\text{www}CH_2-\underset{\underset{OOH}{|}}{\overset{\overset{C_2H_5}{|}}{C}}H-CH_2-\overset{\overset{OOH}{|}}{C}H\text{www}$$

$$\text{www}CH_2-\overset{\overset{O}{||}}{C}-C_2H_5$$
$$HO^\bullet + {}^\bullet CH_2-CH_2\text{www}$$

Cat ← HO-CH₂www
O=CH-CH₂www
O=C-CH₂www
  |
  OH
→ Cat

Oxo-Biodegradable
Fragments

Parameters to be monitored: 1. Weight increase;
2. Carbonyl index;
3. Wettability;
4. Molecular weight;
5. Fractionation by solvent extraction

FIGURE 5.22 Mechanism of radical oxidation of polyethylene.

oven at different temperatures (55 and 70°C) mimicking the thermophylic conditions of the composting process. The influence of the vapor pressure of water has been also investigated by comparing the thermal aging under dry conditions and in atmosphere conditioned at approximately 75% RH.

As the carbonyl group is representative of most of the oxidation products (carboxylic acids, ketones, aldehydes, and lactones)[105] of the oxidative degradation of polyethylene, the concentration of carbonyl groups, as determined by the carbonyl index (COi), can be used to monitor the progress of oxygen uptake and degradation.[111] COi determinations have been therefore utilized to compare the thermal oxidation behavior of test samples under dry and 75% RH conditions at 55 and 70°C.

The oxidation of the poly(ethylene) matrix in the thermally aged samples was clearly confirmed by FT-IR (Fourier-transform infrared) spectroscopy. The increasing absorption and broadening within time of the band in the carbonyl region was recorded in all samples aged at 70°C (Figure 5.23). Overlapping bands corresponding to acids (1712 cm⁻¹), ketones (1723 cm⁻¹), aldehydes (1730 cm⁻¹), and lactones (1780 cm⁻¹)[112] were also observed, thus indicating the presence of different oxidized species (Figure 5.23). Among these, carboxylic acids and ester groups have been

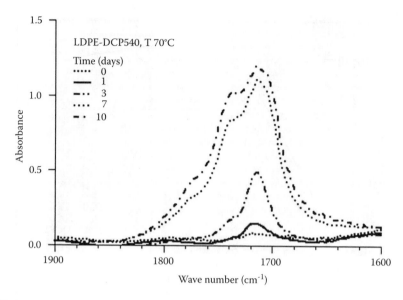

FIGURE 5.23 Time variation of carbonyl absorption band of LDPE sample containing pro-oxidant additives thermally treated in air in oven at 70°C.

shown to be produced in the early and later stages of polymer matrix oxidation, respectively.[113] The increase of specimen weight up to 4 to 5% of the original value, as well as the increase of wettability of the thermally treated films, as determined by contact-angle measurements, further demonstrated the polymer oxidation (Figure 5.24 and Figure 5.25).

COi profiles recorded at 55 and 70°C under dry and 75% RH condition showed that the predominant effect on the oxidation kinetics is the test temperature (Figure 5.26 and Figure 5.27). The high humidity level, comparable with that occurring under real environmental conditions (e.g., composting), was only influencing, in some cases, the rate but not the overall extent of oxidation (Figure 5.26 and Figure 5.27).[114]

Thermally treated samples were also characterized by high-temperature gel permeation chromatography (HT-GPC) , and the recorded molecular weight and molecular weight distribution (ID) were compared with the COi values detectable at the same time of thermal aging (Table 5.12, Table 5.13, and Table 5.14).[114] A drastic decrease of molecular weight below 5 kDa and significant reduction of the ID were recorded after a few days of thermal degradation under both dry and 75% RH conditions. In HT-GPC chromatograms relevant to LDPE samples retrieved at the longest thermal degradation period, and thus having the highest level of oxidation, the elution peaks showed a bimodal shape, and very low (1.7 kDa) molecular-weight fractions (Figure 5.28) were detected.

In addition, the relationship between the molecular weight and COi has been found to fit a mono-exponential trend (Figure 5.29). Accordingly, COi values can be used to predict the rate of MW decrease as a function of the oxidation extent. Moreover, the recorded trend is in agreement with a statistical chain-scission mechanism repeatedly suggested in the photophysical and thermal degradation of polyolefin.[57, 58, 62]

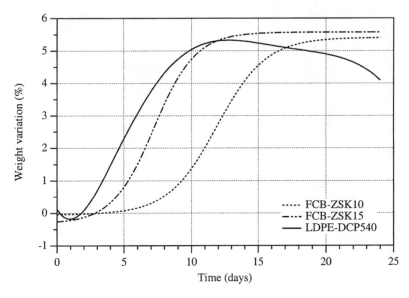

FIGURE 5.24 Weight variation profile of LDPE samples containing pro-oxidant additives thermally treated in air in oven at 70°C.

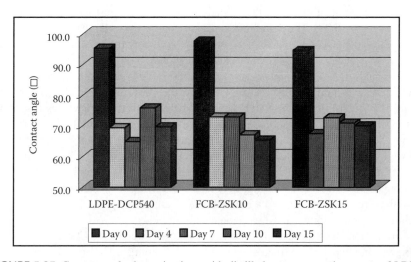

FIGURE 5.25 Contact angle determinations with distilled water as wetting agent of LDPE samples containing different EPI-TDPA pro-oxidant additives thermally treated in air in oven at 70°C.

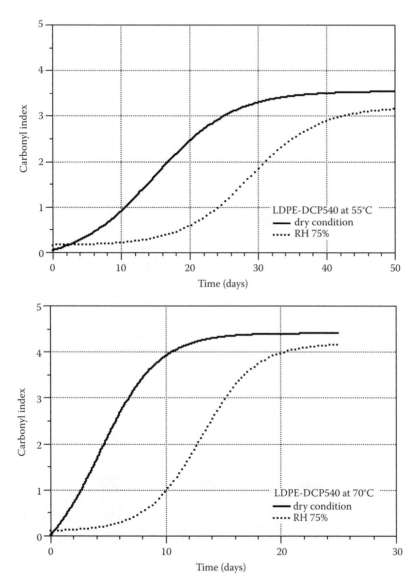

FIGURE 5.26 Carbonyl index (COi) variation of LDPE-DCP54O film sample aged in oven at 55 (a) and 70°C (b), under both dry and 75% RH atmosphere.

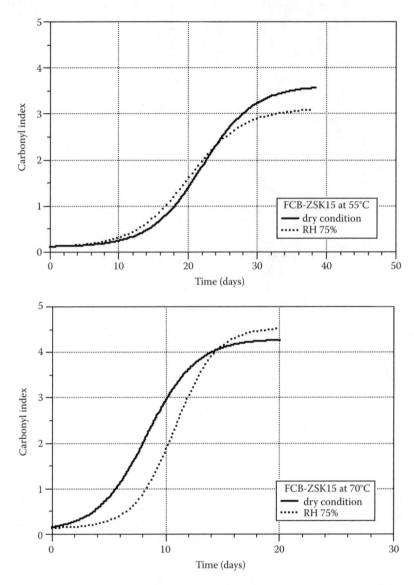

FIGURE 5.27 Carbonyl index (COi) variation of FCB-ZSK film sample aged in oven at 55 (a) and 70°C (b), under both dry and 75% RH atmosphere.

TABLE 5.12
Molecular Weight Analysis by HT-GPC of LDPE-DCP540 Sample Thermally Treated at 70°C

				Test Condition			
Dry (open air)				RH ≈ 75%			
Time (days)	COi [a]	MW (kDa)	ID [b]	Time (days)	COi [a]	MW (kDa)	ID [b]
0	0.61	39.4	4.24	2	1.60	10.7	2.92
1	1.14	19.5	2.96	3	2.23	10.4	2.88
2	2.32	9.7	2.59	9	5.37	4.8	1.37
9	5.44	4.5	1.27	N/A	N/A	N/A	N/A

[a] Carbonyl index as $D_{B\ 1720}/D_{B\ 1435}$.
[b] Dispersity index.

TABLE 5.13
Molecular Weight Analysis by HT-GPC of FCB-ZSK10 Sample Thermally Treated at 70°C

				Test Condition			
Dry (open air)				RH ≈ 75%			
Time (days)	COi [a]	MW (kDa)	ID [b]	Time (days)	COi [a]	MW (kDa)	ID [b]
1	0.63	45.7	3.94	2	0.57	28.6	3.63
2	1.40	16.4	2.75	6	2.54	9.9	2.59
3	2.33	10.1	2.47	11	6.20	4.9	1.33
9	5.35	4.4	1.25	N/A	N/A	N/A	N/A

[a] Carbonyl index as $D_{B\ 1720}/D_{B\ 1435}$.
[b] Dispersity index.

TABLE 5.14
Molecular Weight Analysis by HT-GPC of FCB-ZSK15 Sample Thermally Treated at 70°C

				Test Condition			
Dry (open air)				RH ≈ 75%			
Time (days)	COi [a]	MW (kDa)	ID [b]	Time (days)	COi [a]	MW (kDa)	ID [b]
1	0.67	34.4	3.66	3	1.49	18.1	3.53
3	1.59	14.9	2.91	5	2.89	8.1	2.68
5	2.83	7.6	2.44	11	6.20	4.2	1.44
6	4.53	5.1	1.32	N/A	N/A	N/A	N/A

[a] Carbonyl index as $D_{B\ 1720}/D_{B\ 1435}$.
[b] Dispersity index.

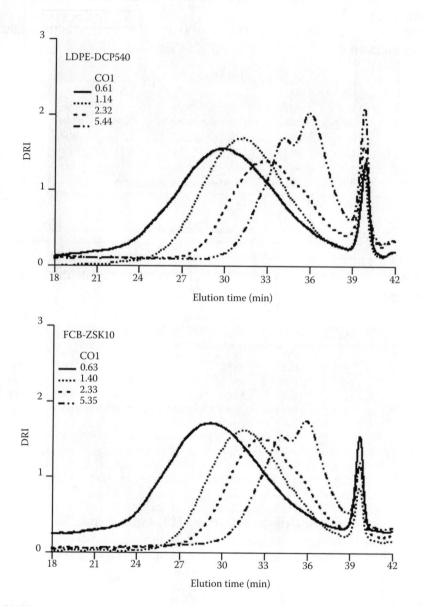

FIGURE 5.28 HT-GPC chromatograms of LDPE samples containing pro-oxidant additives thermally treated in air in oven at 70°C.

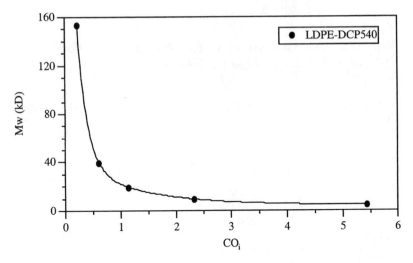

FIGURE 5.29 Molecular weight vs Carbonyl index (COi) relationship in LDPE-DCP54O film sample thermally treated in air in oven at 70°C.

The amount of oxidized degradation intermediates extractable by acetone, which can be considered to be more likely to be diffused in the environment, has also been shown to be positively correlated to the level of oxidation. The amount of degradation intermediates is fairly high, corresponding to 25 to 30% of specimen weight (Table 5.15).

TABLE 5.15
Acetone-Extractable Fractions from Original and Thermally Treated LDPE Samples

Sample	Aging Time (days) 70°C	55°C	COi [a]	Extract (%)	Residue (%)
FCB-ZSK10	0 [b]	0 [b]	0.534	7.7	91.9
	4	—	0.453	6.5	92.6
	—	42	3.583	17.9	82.2
	24		6.816	27.1	72.7
FCB-ZSK15	0 [b]	0 [b]	0.212	5.9	93.9
	5	—	2.864	9.2	88.4
	—	42	5.193	23.8	75.6
	24	—	7.256	22.6	74.4
LDPE-DCP540	0 [c]	0 [c]	0.627	5.5	94.2
	4	—	2.243	11.3	88.5
	—	42	4.818	21.1	78.2
	24	—	5.441	27.7	71.9

[a] Carbonyl index as $D_{B\ 1720}/D_{B\ 1435}$.
[b] Six months storage at room temperature.
[c] Seven months storage at room temperature.

TABLE 5.16
Molecular-Weight Analysis of Acetone-Extractable Fractions from Original and Thermally Aged LDPE Samples

Sample	Treatment	COi[a]	Extract (% weight)	MW (kDa)	ID
FCB-ZSK10	None [b]	0.534	7.7	1.52	1.49
	4 days at 70°C	0.453	6.5	1.47	1.46
	42 days at 55°C	3.583	17.9	1.30	1.39
	24 days at 70°C	6.816	27.1	0.92	1.32
FCB-ZSK15	None [b]	0.212	5.9	1.58	1.46
	5 days at 70°C	2.864	9.2	1.67	1.52
	42 days at 55°C	5.193	23.8	1.27	1.43
	24 days at 70°C	7.256	22.6	1.03	1.36
LDPE-DCP540	None [c]	0.627	5.5	1.08	1.27
	4 days at 70°C	2.243	11.3	1.49	1.41
	42 days at 55°C	4.818	21.1	1.08	1.37
	24 days at 70°C	5.441	27.7	0.89	1.33

[a] Carbonyl index as $D_{B\ 1720}/D_{B\ 1435}$.
[b] Six months storage at room temperature.
[c] Seven months storage at room temperature.

It is also presumed that the acetone extraction selected almost exclusively for polar (e.g., oxidized) degradation intermediates, whereas hydrophobic low-molecular-weight compounds might not be extracted by this polar solvent. Therefore the recorded data have to be considered as partially representative of the overall amount of low-molecular-weight fractions produced during the thermal oxidation of the test material.

Molecular weights as low as 0.8 to 1.6 kDa for the acetone extracts of thermally treated samples were recorded by SEC (Table 5.16). It is also worth noting that at higher levels of oxidation, increasing amounts of the extractable fractions were recorded, and these fractions are characterized by very low molecular weight (Table 5.16).[114] Therefore, the recorded data suggest that cross-linking reactions do not notably affect the oxo-degradation behavior of the analyzed samples, which seems to proceed with cleavage of the macromolecules up to low-molecular-weight fractions capable of being assimilated by microorganisms.[104,107,115] Indeed, earlier studies have demonstrated the microbial utilization, as a carbon source, of the oxidation products formed during the thermal and photo-oxidation of polyethylene.[116,117] Accordingly, in a respirometric experiment, designed to evaluate the biodegradation behavior of low-molar-mass aliphatic hydrocarbons in soil, including docosane (C22), branched hexamethyltetracosane (C30), and α,ω docosandioic acid (C22), a fairly high rate and extent of mineralization (70%) of the acetone extractable fraction from a thermally oxidized LDPE-TDPA sample was recorded (Figure 5.30).[118] This suggests that, similarly to low-molar-mass hydrocarbons, the oxidized fragments from LDPE can be rapidly biodegraded in natural soil.

Nevertheless, the ultimate biodegradation propensity of the whole thermally treated LDPE-TDPA samples is a fundamental issue that has to be assessed.

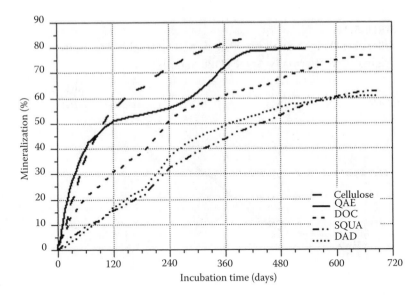

FIGURE 5.30 Mineralization profile of carbon substrates in soil burial respirometric tests. Cellulose = pure Whatman cellulose; DOC = docosane; SQUA = squalane; DAD = cçw docosandioic acid; QAE = acetone extractable fraction from a thermally oxidized LDPE-TDPA sample.

Thermally treated LDPE samples containing TDPA pro-oxidant additives from EPI Environmental Products Inc. (Canada) were therefore assayed in respirometric tests aimed at simulating soil and mature compost incubation media.

Several studies that have evaluated the biodegradation of LDPE samples containing pro-oxidant and natural fillers have shown only limited and slow conversion to carbon dioxide. The mineralization rate from long-term biodegradation experiments of UV-irradiated samples,[119] nonpretreated samples, and additive-free LDPE samples in natural soils indicates more than 100 years for the ultimate mineralization of polyethylene.[120]

In a soil-burial test, carried out in the presence of forest soil as incubation media, the biodegradation profile of a thermally pretreated LDPE-TDPA sample was monitored during 830 days of incubation.[118] The recorded mineralization kinetics (Figure 5.31) were characterized by the presence of a first exponential step that approached a 4% degree of biodegradation within 30 days of incubation. Afterward, a prolonged (120 days) stasis in the microbial respiration was recorded. At approximately 5 months of incubation it was noted that a further and marked exponential increase in the biodegradation profile took place, thus reaching the highest extent of biodegradation (63.0%) after 85 weeks of incubation (Figure 5.31). This two-step mineralization profile of thermally fragmented LDPE-TDPA samples has been repeatedly observed in both soil and mature compost biodegradation tests (Figure 5.32). In contrast to previous studies,[119, 120] very large degrees of mineralization (70 to 80%) have been then recorded, although these were obtained over a relatively long time frame (more than 800 days).[114\]

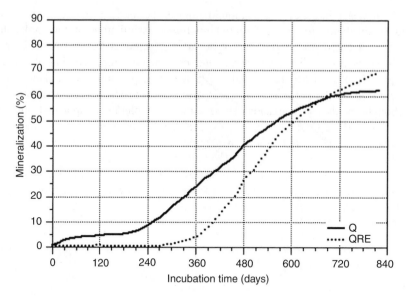

FIGURE 5.31 Mineralization profile LDPE-TDPA samples in soil burial respirometric tests. Q = LDPE sample containing a pro-oxidant additive, aged in an air oven at 70°C for 14 days prior to testing; Q-RE = residue of the aged Q-LDPE sample after extraction of the fraction (ca 25 wt%) soluble in refluxing acetone.

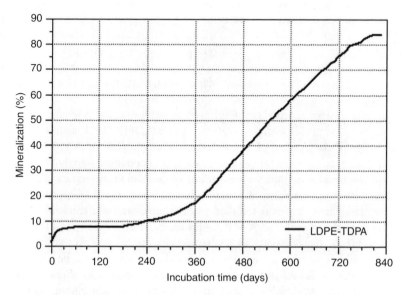

FIGURE 5.32 Mineralization profile LDPE-TDPA thermally oxidized sample in mature compost respirometric tests.

The first step in the biodegradation kinetics of thermally degraded LDPE in soil could be attributed to the preferential microbial assimilation of low-molecular-weight oxidized fragments present on the surface of the LDPE specimens, as previously suggested.[117,121–123] Albertsson and coworkers[105] reported that abiotically aged pure LDPE, LDPE/starch, and LDPE-additives promoting degradation were characterized by the presence of several degradation products such as mono- and dicarboxylic acids and ketoacids. These almost completely disappeared after the incubation of the polymer samples in the presence of *Arthrobacter paraffineus* as a consequence of the assimilation of the degradation products by the bacterial strain.

This hypothesis appears to be confirmed by the FT-IR characterization of soil-buried specimens retrieved after a few months of incubation. Indeed, a significant reduction of the intensity of the carbonyl absorption band with respect to the original value has been repeatedly recorded. This result has also been validated by SEM microanalysis that revealed an appreciable reduction of the oxygen from 20.6 to 16.9% by weight in the surface elemental composition of a soil-incubated sample.[118]

In contrast, a substantial increase in the carbon-carbon unsaturation, as revealed by the increase of the double-bond index from 0.55 at the beginning to 0.88 after 62 weeks of incubation, was recorded. This can be attributed to enzymatic dehydrogenation.[105,120] Hence it is likely that the macromolecular cleavage of thermally oxidized LDPE-TDPA can be achieved concomitantly by abiotic oxidation and biotic, enzymatic scission. Furthermore, a dramatic change in the fingerprint region of the IR spectrum between 1300 and 950 cm^{-1} was observed with increasing incubation time, probably attributable to the presence of lower-molecular-weight fragments.

Therefore, it seems that the ongoing abiotic and biotic degradation of thermally degraded LDPE-TDPA polymer bulk occurs during the incubation in forest soil, with the production of a large amount of degradation intermediates capable of being assimilated by the soil microflora. This might explain the prolonged period of stasis as well as the second, more pronounced exponential phase repeatedly observed in the biodegradation kinetics of these materials in soil-burial respirometric tests.

5.2.6 MICROBIAL CELL BIOMASS PRODUCTION DURING THE BIODEGRADATION OF POLYVINYLICS

The increasing development of biodegradable plastic items introduced in the late 1980s has prompted government authorities and decision makers to issue — through technical standardization bodies — standard norms and laboratory test methods to assess the ultimate environmental behavior of plastics. All of these indicated procedures are stating that the biodegradation of a test material, under aerobic conditions, has to be determined as the extent of carbon substrate evolved to carbon dioxide as a consequence of microbial attack. The amount of carbon dioxide evolved is quantitatively compared with the corresponding value achieved by cellulose taken as positive control, under the same operative conditions and time frame. It must be taken into consideration, however, that under aerobic conditions, heterotrophic soil microbes metabolize carbon substrates both as carbon dioxide and for the production of new biomass, while some of it is converted, through chemo-enzymatic reactions, to humic substances. It is generally accepted that over a relatively short term, 50% of the carbon content of most organic substrates is converted to CO_2, the remaining part being assimilated as biomass or humified.

The relationship between the structural features of low-molar-mass carbon substrates and the growth efficiency of microorganisms incubated on them has also been thoroughly investigated by thermodynamic studies. In particular, relatively simple, closed systems, such as those represented by liquid cultures of certain microorganisms fed with specific carbon sources, have been assessed. In aqueous media, heterotrophic bacteria utilize the organic substrate both as carbon and energy source, i.e., the input carbon is driven into the biosynthetic pathway by anabolism and can as well be utilized for energetic demand through catabolism.

Metabolic efficiency is therefore strictly correlated to the cellular metabolic network that includes interrelated catabolic and anabolic pathways, especially under substrate-limited conditions,[124–126] with the carbon being converted to biomass or the growth yield proportional to the anabolic activity. For instance, it has been repeatedly established that anabolic processes, and consequently growth efficiency, are positively correlated with the standard free energy of oxidation ($\Delta G°$) of a particular carbon substrate.[127–131] This correlation has recently been confirmed by studying the ratio between organic carbon channeled into biomass and that converted to CO_2 under substrate-limited cultures when fed with different types of organic substrates.[132]

In the presence of carbon substrate characterized by a relatively low free-energy content, a large portion of carbon can be converted to CO_2, and the energy liberated from catabolic reactions is not substantially utilized for cell growth and proliferation, but it is rather used to sustain the living activities of bacterial cells. Therefore, in the presence of low-free-energy (e.g., oxidized substrates) carbon sources, large amounts of carbon appear to be utilized for the maintenance functions, thus providing great production of carbon dioxide (e.g., high SCO_2/S_B ratio) rather than being consumed for growth purposes (e.g., low Ys and hence limited cell growth) (see Table 5.17).

The role of free-energy content of carbon substrate, which is repeatedly evidenced in the metabolic efficiency of heterotrophic microorganisms in aqueous media and under substrate-limiting conditions, could also be reasonably applied to the transformation of organic substances by microbes in soil. The substrate carbon channeling between the anabolic and catabolic pathways was evaluated by determining the gross cell growth yield (GY) in the aqueous medium, as well as the net soil microbial biomass (SMB-C), as a function of the substrate addition in soil-burial respirometric tests. The collected data were then analyzed against the amounts of carbon converted to CO_2 (mineralization).

The GY was determined by gravimetric analysis of dry cell biomass divided by the amount of organic carbon in the substrate submitted to biodegradation test. Similarly, when the liquid cultures approached the plateau phase, they were submitted to centrifugation, and the resulting pellets washed with distilled water and dried at 105°C up to constant weight. In soil burial respirometric tests, the organic C fixed in soil microbial biomass (SMB-C) was estimated at the end of the tests by applying chlorofumigation, followed by a K_2SO_4 extraction procedure reported by Turner et al.[132]

TABLE 5.17
Growth Yield (Ys) and SCO$_2$/S$_B$ Ratio in Substrate-Limited Cultures of Activated Sludge

Substrate	Formal C-Oxidation Number	$\Delta H^{\circ}_{c}{}^{a}$ (kJ·g^{-1})	Y$_s{}^{b}$	SCO$_2$/S$_B{}^{c}$
Carbohydrates				
Glucose	−0.16	−15.6	0.80	0.25
Fructose	−0.16	−15.6	0.71	0.41
Sucrose	−0.16	−16.5	0.72	0.40
Lactose	−0.16	−16.5	0.71	0.41
Xylose	−0.16	−15.6	0.63	0.59
Amino acids				
Glycine	+1	−12.9	0.38	1.67
Alanine	0	−18.2	0.53	0.88
Glutamic acid	+0.4	−13.5 ^{d}	0.48	1.08
Phenylalanine	−0.25	−28.1	0.55	0.82
Fatty acids				
Acetic acid	0	−14.6	0.53	0.90
Propionic acid	−0.6	−20.6	0.63	0.58
Butyric acid	−1	−24.8	0.84	0.19
Miscellaneous				
Butanol	−1.75	−36.1	0.78	0.29
Benzoic acid	+0.3	−26.4	0.46	1.19

[a] Values taken from Handbook of Chemistry and Physics, 85th ed., CRC Press, Boca Raton, FL, 2004.
[b] Determined as weight of carbon incorporated in cell biomass/overall weight carbon in substrate (g C_{MB}/mg C_S).
[c] Ratio of carbon evolved as CO_2 (SCO$_2$) to that converted to biomass (S$_B$) (mg C/mg C).
[d] ΔG^0 of oxidation.

The net CO_2 emission (e.g., mineralization) and the GY associated with the metabolization of a water-soluble polymeric material, such as PVA grade 99 and various low-molecular-weight compounds, were evaluated in respirometric biodegradation experiments carried out in a mineral liquid medium (0.4 to 0.6 g substrate/l) under substrate-limited condition in the presence of a selected microbial inoculum.[133] In particular, the metabolization patterns of aliphatic alcohols such as 1-decanol and 2-octanol were compared with the carbon utilization profiles of compounds with lower free-energy content, i.e., with carbon in an average higher oxidation state, such as 3-hydroxybutyric acid and 2,4-pentanediol. The percentage of theoretical cumulative net CO_2 emissions of tested substances are reported in Figure 5.33. 3-Hydroxybutyric acid, with a 0.5 formal average oxidation number of carbon, was shown to be easily metabolized, without a significant lag-phase, throughout an extensive catabolic conversion to CO_2. In contrast, the other substrates required a longer time (6 to 10 days) before the onset of the mineralization process.

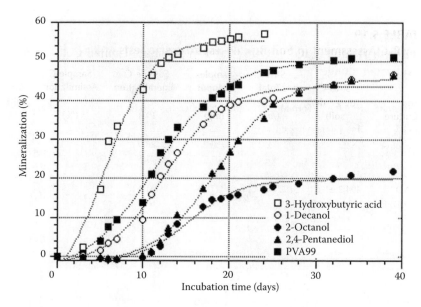

FIGURE 5.33 Mineralization profiles of different carbon substrates in the aqueous respirometric test.

The overall extension of mineralization was also significantly different for the various substrates tested, with the highest value recorded in the culture fed with 3-hydroxybutyric acid (57%) and the lowest (22%) in those supplemented with 2-octanol (Figure 5.33, Table 5.18). Interestingly, complementary GY values were recorded in the case of 2-octanol, whose assimilation amounted to 0.61 GY per gram of added organic carbon despite a fairly low extent of mineralization (22.0%) (Table 5.18). It is notable that a positive correlation was detected between the recorded GY value and the enthalpy of combustion of the selected organic substrates (Table 5.18),

TABLE 5.18
Growth Yield, Maximum Mineralization Extent, and Heat of Combustion of Carbon Substrate Used in Aqueous Respirometric Test

Carbon Substrate	$\Delta H^\circ C$ [a] (kJ·g⁻¹)	Th.CO$_2$ (%)	Growth Yield (mg MB/g C) [b]
3-Hydroxybutyric acid	−21.5	57.1	0.39
1-Decanol	−41.5	46.6	0.56
2-Octanol	−40.6	22.0	0.61
2,4-Pentanediol	−31.0	46.5	0.41
PVA99	−19.0	51.0	0.34

[a] Values taken from Handbook of Chemistry and Physics, 85th ed., CRC Press, Boca Raton, FL, 2004.
[b] MB : dry microbial cell biomass

TABLE 5.19
SMB-C Assessment in Soil Burial Respirometric Tests

Soil Culture	SMB-C ($\mu g \cdot g^{-1}$ dry soil)	SMB-C Total (mg)	Sample-C Input (mg)	Sample-C Mineralization (%)	Sample-C Assimilation (%)	ΔH°_c (kJ·g^{-1})
Blank	131.1 ± 11.0	3.3 ± 0.3	—	—	—	—
Cellulose	83.2 ± 3.0	2.1 ± 0.1	150.7	85.9	−0.7	−14.7
DOC	780.0 ± 14.0	19.5 ± 0.4	307.2	77.5	5.3	−47.2
SQUA	594.4 ± 13.0	14.9 ± 0.3	294.3	63.1	3.9	−47.3
DAD	494.2 ± 13.0	12.4 ± 0.3	277.3	61.1	3.3	−37.0
Q-LDPE	1034.1 ± 46.6	25.8 ± 1.2	278.1	63.8	8.5	−46.5
Q-RE	1105.5 ± 54.7	27.6 ± 1.4	312.7	69.6	7.8	−46.5

thus confirming the existence of a correlation between the metabolic efficiency and the free-energy content of the carbon substrate.[127–131]

Table 5.19 shows the amount of SMB-C in different soil cultures along with the relevant overall fate of substrate carbon input. The recorded data indicate that the observed considerable differences in both mineralization extent and carbon substrate conversion to biomass are also in this case correlated to the free-energy content of the substrate.

As previously reported, the soil metabolization of a glucosidiclike material such as cellulose does not allow the fixation of appreciable amounts of carbon as microbial biomass, the major part being converted to carbon dioxide. In particular, in the present test, the SMB-C associated with the soil cultures supplemented with cellulose was slightly lower than the blank, thus substantiating the promotion of the *priming effect* involving at least the increased turnover of soil microbial biomass.

In contrast, hydrocarbon substrates appeared to be metabolized throughout a substantial conversion to SMB-C, thus accounting for 10 to 15% carbon assimilation in the presence of relatively high free-energy content in high-molecular-weight samples, whereas the carbon conversion to CO_2 was noticeably lower than for cellulose. As a supporting element to this behavior, a picture of the extensive microbial colonization of an oxidized polyethylene fragment is shown in Figure 5.34.

There is some evidence that biodegradation in soil of carbon substrates with low free-energy content (i.e., high level of oxidation) as measured only by net CO_2 evolution could be, to some extent, overestimated. Actually, these compounds have been found to provoke the soil microbial consortia to shift toward fast-growing species (e.g., zymogeneous) with low metabolic efficiency (i.e., high rate of carbon substrate conversion to CO_2). In these microorganisms, carbon substrate utilization patterns are correlated with relatively simple enzymatic tools, which accounts for the almost exclusive metabolization of easily assimilable compounds, especially carbohydrates and amino acids, whereas more complex carbon substrates cannot be utilized. As a consequence, once the substrate (e.g., glucose) is consumed, zymogeneous microbial cells die, thus initiating a swift microbial biomass turnover, accompanied by the release of additional CO_2 deriving from the cellular component.

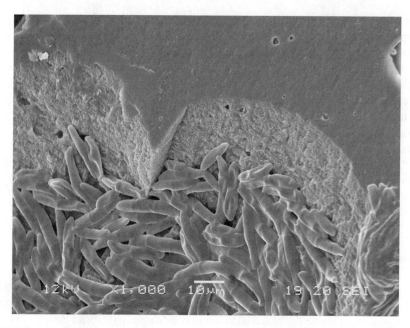

FIGURE 5.34 SEM micrograph of thermally oxidized LDPE fragment from soil burial test.

As a result, in a relatively short time, the addition of glucosidic material to soil is characterized by a great extent of respiration accompanied by very little conversion to biomass or humic substance. Furthermore, the induction of the extramineralization of soil organic matter (e.g., *priming effect*) by glucosidic materials may also increase the level of soil respiration, with resultant depletion of carbon in the soil.

These effects, which may act synergistically, could be crucial in biodegradation tests where cellulose or starch are used as the reference standard materials. Indeed, this could account for the potential underestimation of the biodegradation level of test compounds, especially when they consist of carbon atoms in fairly low formal oxidation state and hence in relatively high free-energy content.

Given these considerations, the use of different reference materials that are representative of the different classes of carbon substrates in standardized soil bio-degradation tests is recommended. Moreover , the various interacting factors affecting the microbial transformation of carbon substrates in soil — the carbon balance, including both biomass and carbon dioxide evolution — should also be taken into account in standardized test procedures.

5.3 CONCLUSIONS

The accumulation of plastic waste and problems connected with waste collection and eventual recycling are promoting interest in the use of plastic materials suitable for biorecycling, particularly in compost plants. The worldwide available biomass could largely supply low-cost polymeric matrices and fillers for bioplastics production without competing with food production. Thus the use of by-products and waste from agricultural, ethanol, dairy, and pharmaceutical industries, among others, is gaining increasing attention in the bioplastics field.

Gelatin, starch, and lignocellulosic fibers derived from fruit-juice production or fermentation to ethanol represent a valuable, low-cost, and largely available source of natural materials for production of biocomposites. Given the modest percentage of fossil fuels devoted to plastics production (4% of worldwide oil usage), the utilization of degradable petro-derived polymeric matrices such as poly(vinyl alcohol) (PVA) or polyethylene (PE) appears feasible for the production of ecocompatible materials. The effect on greenhouse gas emissions attributable to plastics production using fossil fuels is insignificant compared with the impact of fossil fuels used for energy production (93% of total oil usage).

In field trials using conventional machinery, hydro-biomulching formulations were produced using water suspensions of either gelatin or PVA and natural fillers. These materials showed a positive mulching and soil-structuring effect in both laboratory-scale tests and field trials.

Hybrid biocomposites were produced by blending PVA with waste gelatin from the pharmaceuticals industry, with sugarcane bagasse and orange and apple waste (by-products of juice extraction), and with corn starch and corn fiber (by-products of ethanol production). Compression-molded, injection-molded, and foamed items were prepared, and these items showed good thermal and mechanical properties along with good resistance to aging. For example, injection-molded samples maintained their consistency and mechanical properties after a year of storage at room conditions. Improvements in the processing and the water resistance of starch-based foamed trays were achieved by adding corn fibers to the formulations.

Biodegradation in soil of the prepared items was promoted by the presence of natural components in the composites. Studies on the biodegradation of PVA-based materials suggested that a threshold grade of hydrophobicity achieved by a discrete content of vinyl acetate units can be taken as a practical parameter for estimating the propensity of PVA samples to be biodegraded in solid media. Similarly, biodegradation was influenced by the degree of polymerization, with low-molecular-weight materials showing a higher degree of mineralization. In contrast, the biodegradation tests carried out in aqueous medium (in the presence of PVA-acclimated microorganisms) suggested that the driving force in PVA biodegradation appears to be its solubility, which promotes endocleavage of the polymer chains by specific enzymatic systems.

Thermally promoted oxidation of reengineered polyethylene samples (as introduced by EPI Inc. [Canada]) containing TDPA pro-oxidant additives were assayed in oven at different temperatures (55 and 70°C) to reproduce the thermophylic conditions of the composting process. Oxidation of the polyethylene matrix in the thermally aged samples was clearly established by FT-IR spectroscopy. The relationship between the molecular weight and carbonyl index (COi) was found to fit a mono-exponential trend, thus indicating that COi values can be conveniently used to predict the rate of molecular-weight decrease as a function of the extent of oxidation. Moreover, the recorded trend is in agreement with a statistical chain-scission mechanism that is repeatedly suggested in the photodegradation and thermal degradation of polyolefin. Therefore, it appears that the ongoing abiotic degradation followed by biotic breakdown of the thermally degraded LDPE-TDPA polymer bulk occurs during incubation in forest soil, with the production of a large amount of degradation intermediates capable of being assimilated by the soil microflora.

Finally, it is inferred that in assessing the level of biodegradation of different polymeric as well as low-molar-mass components, it is important to perform a mass balance taking into account the amount of carbon that is converted to biomass. This approach would put all materials submitted to biodegradation tests on the same level for a fair comparison consisting of carbon atoms in different states of oxidation. Therefore, in standardized soil-biodegradation tests, it is recommended that (a) different reference materials representative of the different classes of carbon substrates should be used and (b) the carbon balance should be taken into account, including both biomass and carbon dioxide evolution.

REFERENCES

1. Van Os, G., The European plastic industry: a sunset industry, *EPF*, special issue, July 2001, 19.
2. Packtech D4.1, Assessment of the Extent of Uptake and Use of Recyclable and Biodegradable Materials for use in the Food Industry; Packtech D4.4 Technical Appraisal and Recommendation of the Use of Alternative Recyclable Technologies and Biodegradable Material Alternatives for Specific Applications of this Project; available on-line at http://www.pack-tech.org, 2003.
3. Plastics Resource; available on-line at http://plasticsresource.com/s_plasticsresource/docs/900/840.pdf, and http://www.plasticseurope.org, 2006.
4. Chiellini, E., Chiellini, F., and Cinelli, P., Polymers from renewable sources, in *Degradable Polymers: Principles and Applications,* 2nd ed., Scott, G., Ed., Kluwer Academic, New York, 2002, chap. 7, pp. 163–233.
5. Sperling, L.H. and Carraher, C.E., Polymers from renewable resources, in *Encyclopedia of Polymer Science and Engineering*, Vol. 12, Mark H.F., Bikales, N.M., Overberger, C.G., and Menges G., Eds., Wiley, New York, 1988, pp. 658–690.
6. Bozell, J.J., Ed., Chemicals and Materials from Renewable Resources, in *Proc. 218th ACS National Meeting*, New Orleans, 22–26 Aug. 1999, ACS Symp. Ser. 784, ACS, Washington, DC, 2001.
7. Van Wyk, J.P.H. Biotechnology and the utilization of biowaste as a resource for bioproduct development, *Trends Biotechnol.*, 19, 172, 2001.
8. Rangasprad, R. and Vasudeo, Y.B., *Green Plastics: Plastics from Renewable Resources*; available on-line at www.ril.com/downloads/D128Biopolymer_presentation_at_Hyderabad.ppt, 2001.
9. Various Web sites: http://www.ibaw.org; http://www.mli.kvl.dk/foodchem/special/biopack; http://www.ienica.net/bioplastics/bioplasticsindex.htm; http://www.biopro.de/en/region/ulm/magazin/01391/Diepenbrock/Pelzer/Radtke: Energiebilanz im Ackerbaubetrieb 1995, 2006.
10. Young, R.A., Utilization of natural fibres: characterization, modification, and applications, in *Lignocellulosic-Plastic Composites*, Leao, A., Carvalho, F.X., and Frollini, E., Eds., USP-UNESP Sao Paulo, Brazil, 1997, pp. 1–22.
11. Chiellini, E. et al., Environmentally degradable biobased polymeric blends and composites, *Macromolecular Biosci.*, 4, 218, 2004.

12. Doane, W.M., Biodegradable plastics, *J. Polym. Mater.*, 11, 229, 1994.
13. Heyde, M., Ecological consideration on the use production of biosynthetic and synthetic biodegradable polymers, *Polym. Degrad. Stab.*, 59, 3, 1998.
14. Mayer, J.M. and Kaplan, D.L., Biodegradable materials: balancing degradability and performance, *TRIP/ Rev.*, 2, 227, 1994.
15. Reijnders, L., A normative strategy for sustainable resource choice and recycling, *Resour. Conserv. Recycl.,* 28, 121, 2000.
16. Van Wyk, J.P.H., Biotechnology and the utilization of biowaste as a resource for bioproduct development, *Trends Biotechnol.*, 19, 172, 2001.
17. Weber, C.J., KVL Denmark: *Biobased Packaging Materials for the Food Industry*, report of Project PL98 4046 funded by the EU, 1995; available on-line at http://www.ibaw.
18. E. Würdinger (BIfA), U. Roth (BIfA), G. A. Reinhart, A. Detzel: LCA for loose-fill packaging material from starch and polystyrene, resp. (Specific results; in German). Proceedings of the 8th Symposium Nachwachsendce Rohstoffe für die Chemie (Biobased materials for chemistry), Tübingen (D), 26–27 March 2003, BMVEL, FNR, VCI, 2003.
19. Scott, G., Green polymers, *Polym. Degrad. Stab.*, 68, 1, 2000.
20. Roper, H., Renewable raw materials in Europe: industrial utilisation of starch and sugar, *NutraCos*, May–June, 37, 2002.
21. Patel, M., Closing Carbon Cycles, Ph.D. thesis, Utrecht University, Utrecht, Netherlands, 1999.
22. El Bassam, N., Renewable energy for rural communities, *Renewable Energy*, 24, 40, 2001.
23. Rowell, R.M. et al., Utilisation of natural fibres in plastic composites: problems and opportunities, in *Lignocellulosic-Plastic Composites*, Leao, A.L., Carvalho, F.X., and Frollini, E., Eds., USP-UNESP, Sao Paulo, Brazil, 1997, pp. 23–52.
24. Warwel, S. et al., Polymers and surfactants on the basis of renewable resources, *Chemosphere*, 43, 39, 2001.
25. Narayan, R., Biomass (renewable) resources for production of materials, chemicals, and fuels: a paradigm shift, in *Emerging Technologies for Materials and Chemicals from Biomass*, Rowell, R.M., Schultz, T.P., and Narayan, R., Eds., ACS Symp. Ser. 476, ACS, Washington, DC, 1992, pp. 1–10.
26. Hocking, P.J. and Marchessault, R.H., Biopolyesters, in *Chemistry and Technology of Biodegradable Polymers*, Griffin, G.J.L., Ed., Chapman & Hall, New York, 1994, pp. 48–96.
27. Steinbüchel, A. and Fuchtenbusch, B., Bacteria and other biological systems for polyester production, *TIBTECH*, 16, 419, 1998.
28. Steinbüchel, A. et al., Molecular basis for biosynthesis and accumulation of polyhydroxyalkanoic acid in bacteria, *FEMS Microbiol. Rev.*, 103, 217, 1992.
29. Lenz, R.W. et al., Bacteria synthesis of poly(-β-)hydroxyalkanoates with functionalized side chains, in *Biodegradable Plastics and Polymers*, Doi, Y. and Fukuda, K., Eds., Elsevier Science, Amsterdam, 1994, pp. 109–119.
30. Mergaert J. et al., Biodegradation of polyhydroxyalkanoates, *FEMS Microbiol. Rev.*, 103, 317, 1992.
31. Lee, S.Y., Bacterial polyhydroxyalkanoates, *Biotechnol. Bioeng.*, 49, 1, 1996.
32. Choi, J. and Lee, S.Y., Process analysis and economic evaluation for poly(3-hydroxybutyrate) production by fermentation, *Bioproc. Eng.*, 17, 335, 1997.
33. Yamane, T., Cultivation engineering of microbial bioplastics production, *FEMS Microbiol. Rev.*, 103, 257, 1992.

34. Reddy, C.S.K. et al., Polyhydroxyalkanoates: an overview, *Bioresource Technol.*, 87, 137, 2003.
35. Madison, L.L. and Huisman, G.W., Metabolic engineering of poly(3-hydroxyal-kanoates): from DNA to plastic, *Microbiol. Mol. Biol. Rev.*, 63, 21, 1999.
36. Bohlmann, G.M., Process economics of biodegradable polymers from plants, in *227th ACS National Meeting Proceedings*, Anaheim, CA, March 2004.
37. Snell, K.D. and Peoples, O.P. Polyhydroxyalkanoate polymers and their production in transgenic plants, *Metab. Eng.*, 29, 4, 29–40, 2002.
38. Coffin, R., Fishman, M.L., and Ly, T.V. Thermomechanical properties of blends of pectin and poly(vinyl alcohol), *J. Appl. Polym. Sci.*, 57, 71, 1996.
39. Lahalih, S.M., Akashah, S.A., and Al-Hajjar, F.H., Development of degradable slow release multinutritional agricultural mulch film, *Ind. Eng. Chem. Res.*, 26, 2366, 1987.
40. Chiellini, E. et al., Composite films based on biorelated agro-industrial waste and poly(vinyl alcohol): preparation and mechanical properties characterization, *Biomacromolecules*, 2, 1029, 2001.
41. Chiellini, E. et al., Gelatin based blends and composites: morphological and thermal mechanical characterization, *Biomacromolecules*, 2, 806, 2001.
42. Chiellini, E. et al., Biodegradable hybrid polymer films based on poly(vinyl alcohol) and collagen hydrolyzate, *Macromol. Symp.*, 197, 125, 2003.
43. Cinelli, P. et al., Characteristics and degradation of hybrid composite films prepared from PVA, starch and lignocellulosics, *Macromol. Symp.*, 197, 143, 2003.
44. Grillo Fernandes, E., Cinelli, P., and Chiellini, E., Thermal behavior of composites based on poly(vinyl alcohol) and sugar cane bagasse, *Macromol. Symp.*, 218, 231, 2004.
45. Chiellini, E. et al., Environmentally sound blends and composites based on water-soluble polymer matrices, *Macromol. Symp.*, 152, 83, 2000.
46. Chiellini, E. et al., Biobased polymeric materials for agriculture applications, in *Biodegradable Polymers and Plastics*, Chiellini, E. and Solaro, R., Eds., Kluwer Academic-Plenum, New York, 2003, chap. 13, p. 185.
47. Imam, S.H. et al., Characterization of biodegradable composite films prepared from blends of poly(vinyl alcohol), cornstarch, and lignocellulosic fiber, *J. Polym. Environ.*, 13, 47, 2005.
48. Cinelli, P. et al., Injection molded hybrid composites based on corn fibers and poly(vinyl alcohol), *Macromol. Symp.*, 197, 115, 2003.
49. *Kunstoff Plastics Business Data and Chart*, 2004; available on-line at http://www.vke.de.
50. Scott, G. and Wiles, D.M., Programmed-life plastics from polyolefins: a new look at sustainability, *Biomacromolecules*, 2, 615, 2001.
51. Sakai, K., Hamada, N., and Watanabe, Y., Studies on the poly(vinyl alcohol)-degrading enzyme; part VI: degradation mechanism of poly(vinyl alcohol) by successive reactions of secondary alcohol oxidase and β-diketone hydrolase from *Pseudomonas* sp., *Agric. Biol. Chem.*, 50, 989, 1986.
52. Suzuki, T., Degradation of poly(vinyl alcohol) by microorganisms, *J. Appl. Polym. Sci., Appl. Polym. Symp.*, 35, 431, 1979.
53. Sakai, K. et al., Purification and characterization of an esterase involved in poly(vinyl alcohol) degradation by *Pseudomonas vesicularis* PD, *Biosci. Biotech. Biochem.*, 62, 2000, 1998.
54. Inés Mejía, A.G., Lucy López, B.O., and Mulet, A.P., Biodegradation of poly(vinyl alcohol) with enzymatic extracts of *Phanerochaete chrysosporium*, *Macromol. Symp.*, 148, 131, 1999.

55. Mori, T. et al., Secondary alcohol dehydrogenase from a vinyl alcohol oligomer-degrading *Geotrichum fermentans*: stabilization with Triton X-100 and activity toward polymers with polymerization degrees less than 20, *World J. Microbiol. Biotechnol.*, 14, 349, 1998.

56. Larking, D.M. et al., Enhanced degradation of polyvinyl alcohol by *Pycnoporus cinnabarinus* after pretreatment with Fenton's reagent, *Appl. Environ. Microbiol.*, 65, 1798, 1999.

57. Scott, G., Environmental biodegradation of hydrocarbon polymers: initiation and control, in *Biodegradable Plastic and Polymers*, Doi, Y. and Fukuda, K., Eds., Elsevier, Amsterdam, 1994, Session 2, pp. 79–91.

58. Gugumus, F., Re-examination of the thermal oxidation reactions of polymers; 1: new views of an old reaction, *Polym. Degrad. Stab.*, 74, 327, 2001.

59. Gugumus, F., Re-examination of the thermal oxidation reactions of polymers; 3: various reactions in polyethylene and polypropylene, *Polym. Degrad. Stab.*, 77, 147, 2002.

60. Erlandsson, B., Karlsson, S., and Albertsson, A.-C., The mode of action of corn starch and a pro-oxidant system in LDPE: influence of thermooxidation and UV-irradiation on the molecular weight changes, *Polym. Degrad. Stab.*, 55, 237, 1997.

61. Burman, L. and Albertsson, A.-C., Chromatographic fingerprint: a tool for classification and for predicting the degradation state of degradable polyethylene, *Polym. Degrad. Stab.*, 89, 50, 2005.

62. Khabbaz, F., Albertsson, A.-C., and Karlsson, S., Chemical and morphological changes of environmentally degradable polyethylene films exposed to thermo-oxidation, *Polym. Degrad. Stab.*, 63, 127, 1999.

63. Karlsson, S., Hakkarainen, M., and Albertsson, A.-C., Dicarboxylic acids and ketoacids formed in degradable polyethylenes by zip depolymerization through a cyclic transition state, *Macromolecules*, 30, 7721, 1997.

64. Iring, M. et al., Thermal oxidation of linear low density polyethylene, *Polym. Degrad. Stab.*, 14, 319, 1986.

65. Winslow, F.H., Photooxidation of high polymers, *Pure Appl. Chem.*, 49, 495, 1977.

66. Kay, B.L., Evans, R.A., and Young, J.A., Effect of hydroseeding on bermuda grass, *Agronomy J.*, 69, 555, 1977.

67. Young, J.A., Kay, B.L., and Evans, R.A., Accelerating the germination of common bermuda grass seed for hydroseeding, *Agronomy J.*, 69, 115, 1977.

68. Stefanson, R.C., Soil stabilization by poly(vinyl alcohol) and its effect on the growth of wheat, *Aust. J. Soil Res.*, 12, 59, 1974.

69. Oades, J.M., Prevention of crust formation in soils by poly(vinyl alcohol), *Aust. J. Soil. Res.*, 12, 139, 1976.

70. Painuli, D.K. and Pagliai, M., Effect of polyvinyl alcohol, dextran and humic acid on some physical properties of a clay and loam soil; I: cracking and aggregate stability, *Agrochimica*, 34, 117, 1990.

71. Painuli, D.K. et al., Effect of polyvinyl alcohol, dextran and humic acid on some physical properties of a clay and loam soil; II: hydraulic conductivity and porosity, *Agrochimica*, 34, 131, 1990.

72. Orts, W.J., Sojka, R.E., and Glenn, G.M., Biopolymer additives to reduce erosion-induced soil losses during irrigation, *Ind. Crops Prod.*, 11, 19, 2000.

73. Cinelli P., Formulation and characterization of environmentally compatible polymeric materials for agriculture applications, Ph.D. thesis, Pisa University, Pisa, 1996.

74. Kenawy, E.R. et al., Biodegradable composite films based on waste gelatin, *Macromol. Symp.*, 144, 351, 1999.

75. Lawton, W. and Fanta, G.F., Glycerol-plasticized films prepared from starch—poly(vinyl alcohol) mixtures: effect of poly(ethylene-co-acrylic acid), *Carbohydrate Polym.*, 23, 275, 1994.

76. Mao, L. et al., Extruded cornstarch–glycerol–polyvinyl alcohol blends: mechanical properties, morphology, and biodegradability, *J. Polym. Environ.*, 8, 205, 2000.

77. Chiellini, E. et al., Thermo-mechanical behaviour of poly(vinyl alcohol) and sugarcane bagasse blends, *J. Appl. Polym. Sci.*, 92, 426, 2004.

78. Sawada, H., Field testing of biodegradable plastics, in *Biodegradable Plastics and Polymers*, Doi, Y. and Fukuda, K., Eds., Elsevier, Amsterdam, 1994, pp. 298–310.

79. Chapman, G.M., Status of technology and applications of degradable products, in *Polymers from Agricultural Coproducts*, Fishman, M.L., Friedman, R.B., and Huang, S.J., Eds., ACS Symp. Ser. 575, ACS, Washington, DC, 1994, pp. 29–47.

80. Shogren, R.L. et al., Starch-poly(vinyl alcohol) foamed articles prepared by a baking process, *J. Appl. Polym. Sci.*, 68, 2129, 1998.

81. Lawton, J.W., Shogren, R.L., and Tiefenbacher, K.F., Effect of batter solids and starch type on the structure of baked starch foams, *Cereal Chem.*, 76, 682, 1999.

82. Cinelli, P. et al., Foamed articles based on potato starch, corn fibers and poly(vinyl alcohol), *Polym. Degrad. Stab.*, 91, 11947, 2006.

83. Nord, F.F., Dehydrogenation activity of *Fusarium lini* B., *Naturwiss*, 24, 763, 1936.

84. Shimao, M. et al., Properties and roles of bacterial symbionts of polyvinyl alcohol-utilizing mixed cultures, *Appl. Environ. Microbiol.*, 46, 605, 1983.

85. Mori, T. et al., Isolation and characterization of a strain of *Bacillus megaterium* that degrades poly(vinyl alcohol), *Biosci. Biotech. Biochem.*, 60, 330, 1996.

86. Kawagoshi, Y. and Fujita, M., Purification and properties of the polyvinyl alcohol-degrading enzyme 2,4-pentanedione hydrolase obtained from *Pseudomonas vesicularis* var. *povalolyticus* PH, *World J. Microbiol. Biotechnol.*, 14, 95, 1998.

87. Matsumura, S. et al., Effects of molecular weight and stereoregularity on biodegradation of poly(vinyl alcohol) by *Alcaligenes faecalis*, *Biotechnol. Lett.*, 16, 1205, 1994.

88. Porter, J.J. and Snider, E.H., Long term biodegradability of textile chemicals, *J. Water Pollut. Control Fed.*, 48, 2198, 1976.

89. Solaro, R., Corti, A., and Chiellini, E., Biodegradation of poly(vinyl alcohol) with different molecular weights and degree of hydrolysis, *Polym. Adv. Technol.*, 11, 873, 2000.

90. Chiellini, E., Corti, A., and Solaro, R., Biodegradation of poly(vinyl alcohol) based blown films under different environmental conditions, *Polym. Degrad. Stab.*, 64, 305, 1999.

91. Solaro, R., Corti, A., and Chiellini, E., Biodegradation of poly(vinyl alcohol) with different molecular weights and degree of hydrolysis, *Polym. Adv. Technol.*, 11, 873, 2000.

92. Suzuki, T. et al., Some characteristics of *Pseudomonas* O-3 which utilize polyvinyl alcohol, *Agric. Biol. Chem.*, 37, 747, 1973.

93. Watanabe, Y. et al., Purification and properties of a polyvinyl alcohol-degrading enzyme produced by a strain of *Pseudomonas Arch. Biochem. Biophys.*, 174, 575, 1976.

94. Matsumura, S. and Toshima, K., Biodegradation of poly(vinyl alcohol) and vinyl alcohol block as biodegradable segments, in *Hydrogels and Biodegradable Polymers for Bioapplications*, Ottenbrite, R.M., Huang, S.J., and Park, K., Eds., John Wiley and Sons, New York, 1992, pp. 767–772.

95. Hatanaka, T. et al., Effects of the structure of poly(vinyl alcohol) on the dehydrogenation reaction by poly(vinyl alcohol) dehydrogenase from *Pseudomonas* sp. 113P3, *Biosci. Biotech. Biochem.*, 59, 1229, 1995.

96. Takasu, A. et al., New chitin-based polymer hybrids, 4: soil burial degradation behavior of poly(vinyl alcohol)/chitin derivative miscible blends, *J. Appl. Polym. Sci.*, 73, 1171, 1999.

97. Chen, L. et al., Starch-polyvinyl alcohol crosslinked film: performance and biodegradation, *J. Environ. Polym. Degrad.*, 5, 111, 1997.

98. Krupp, L.R. and Jewell, W.J., Biodegradability of modified plastic films in controlled biological environments, *Environ. Sci. Technol.*, 26, 193, 1992.

99. Bloembergen, S. et al., Biodegradation and composting studies of polymeric materials, in *Biodegradable Plastics and Polymers*, Doi, Y. and Fukuda, K., Eds., Elsevier, Amsterdam, 1994, pp. 601–609.

100. Tokiwa, Y. and Iwamoto, A., Establishment of biodegradability evaluation for plastics of PVA series, in *Proceedings of the Annual Meeting '92 of the Fermentation and Bioengineering Society*, 1992, p. 158.

101. Chiellini, E. et al., Adsorption/desorption of poly(vinyl alcohol) on solid substrates and relevant biodegradation, *J. Polym. Environ.*, 8, 67, 2000.

102. Chiellini, E. et al., Environmentally degradable plastics: poly(vinyl alcohol) — a case study, in *International Workshop on Environmentally Degradable Polymers: Polymeric Materials*, Said, Z.F. and Chiellini, E., Eds., Doha , Qatar, 1999, pp. 103–118.

103. Chiellini, E. et al., Synthesis and biodegradation behavior of poly(vinyl alcohol) with different degree of hydrolysis, *Polym. Degrad. Stab.*, in press.

104. Volke-Sepulveda, T. et al., Thermally treated low density polyethylene biodegradation by *Penicillium pinophilum* and *Aspergillus niger*, *J. Appl. Polym. Sci.*, 83, 305, 2002.

105. Albertsson, A.-C. et al., Molecular weight changes and polymeric matrix changes correlated with the formation of degradation products in biodegraded polyethylene, *J. Environ. Polym. Degrad.*, 6, 187, 1998.

106. Contat-Rodrigo, L. and Ribes Greus, A., Biodegradation studies of LDPE filled with biodegradable additives: morphological changes: I, *J. Appl. Polym. Sci.*, 83, 1683, 2002.

107. Potts, J.E., Clendinning, R.A., and Ackart, W.B., EPA-R2-72-046, U.S. Environmental Protection Agency, Washington, DC, 1972.

108. Haines, J.R. and Alexander, M., Microbial degradation of high-molecular-weight alkanes, *Appl. Microbiol.*, 28, 1084, 1975.

109. Yamada-Onodera, K. et al., Degradation of polyethylene by a fungus, *Penicillium simplicissimum* YK, *Polym. Degrad. Stab.*, 72, 323, 2001.

110. Weiland, M. and David, C., Thermal oxidation of polyethylene in compost environment, *Polym. Degrad. Stab.*, 45, 371, 1994.

111. Gugumus, F., Thermooxidative degradation of polyolefins in the solid state; part 1: experimental kinetics of functional group formation, *Polym. Degrad. Stab.*, 52, 131, 1996.

112. Hinsken, H. et al., Degradation of polyolefins during melt processing, *Polym. Degrad. Stab.*, 34, 279, 1991.

113. Khabbaz, F. and Albertsson, A.-C., Great advantages in using a natural rubber instead of asynthetic SBR in a pro-oxidant system for degradable LDPE, *Biomacromolecules*, 1, 665, 2000.

114. Chiellini, E. et al., Oxobiodegradable full carbon backbone polymers: totally degradable polyethylene under different environmental conditions, *Polym. Degrad. Stab.*, submitted.

115. Kawai, F., Shibata, M., Yokoyama, S., Maeda, S., Tada, K., and Hayashi, S., *Macromol. Symp.*, 144, 73, 1999.
116. Arnaud, R. et al., Photooxidation and biodegradation of commercial photodegradable polyethylenes, *Polym. Degrad. Stab.*, 46, 211, 1994.
117. Albertsson, A.-C. et al., Degradation product pattern and morphology changes as means to differentiate abiotically and biotically aged degradable polyethylene, *Polymer*, 36, 3075, 1995.
118. Chiellini, E., Corti, A., and Swift, G., Biodegradation of thermally oxidized, fragmented, low-density polyethylenes, *Polym. Degrad. Stab.*, 81, 341, 2003.
119. Albertsson, A.-C. and Karlsson, S., The three stages in degradation of polymers: polyethylene as a model substance, *J. Appl. Polym. Sci.*, 35, 1289, 1988.
120. Ohtake, Y. et al., Studies on biodegradation of LDPE-observation of LDPE films scattered in agricultural fields or in garden soil, *Polym. Degrad. Stab.*, 60, 79, 1998.
121. Albertsson, A.-C., Andersson, S.O., and Karlsson, S., The mechanism of biodegradation of polyethylene, *Polym. Degrad. Stab.*, 18, 73, 1987.
122. Volke-Sepulveda, T. et al., Microbial degradation of thermo-oxidized low-density polyethylene, *J. Appl. Polym. Sci.*, 73, 1435, 1999.
123. Bonhomme, S. et al., Environmental biodegradation of polyethylene, *Polym. Degrad. Stab.*, 81, 441, 2003.
124. Forrest, W.W., Energetic aspects of microbial growth, in *Microbial Energetics*, Cambridge University Press, London, 1969, pp. 65–86.
125. Lehninger, A.L., *Biochemistry*, Worth, New York, 1975.
126. Bitton, G., *Wastewater Microbiology*, John Wiley & Sons, New York, 1994.
127. Schroeder, E.D. and Busch, A.W., Validity of energy change as a growth parameter, *J. Sanitary Eng.*, 94, 193, 1968.
128. Burkhead, C.E. and McKinney, R.E., Energy concepts of aerobic microbial metabolism, *J. Sanitary Eng.*, 95, 253, 1969.
129. Sykes, R.M., Theoretical heterotrophic yields, *J. Water Pollut. Control Fed.*, 47, 591, 1975.
130. Heijnen, J.J. and van Dijken, J.P., In search of a thermodynamic description of biomass yields for the chemotrophic growth of microorganisms, *Biotechnol. Bioeng.*, 39, 833, 1992.
131. Schill, N.A., Liu, J.S., and von Stockar, U., Thermodynamic analysis of growth of *Methanobacterium thermoautotrophicum*, *Biotechnol. Bioeng.*, 64, 74, 1999.
132. Turner, B.L., Bristow, A.W., and Hygarth, P.M., Rapid estimation of microbial biomass in grassland soils by ultra-violet absorbance, *Soil Biol. Biochem.*, 33, 913, 2001.
133. Chiellini, E., Corti, A., and Solaro, R., Biodegradation of poly(vinyl alcohol) based blown films under different environmental conditions, *Polym. Degrad. Stab.*, 64, 305, 1999.

6 Production of Plastics from Waste Derived from Agrofood Industry

Gerhart Braunegg, Martin Koller, Paula Varila, Christoph Kutschera, Rodolfo Bona, Carmen Hermann, Predvag Horvat, Jose Neto, and Luis Pereira

CONTENTS

6.1 INTRODUCTION

The literature contains a number of articles about microbially mediated production of polyhydroxyalkanoates (PHAs). These polyesters constitute biodegradable polymers possessing physical properties of thermoplastics or elastomers. Therefore, they can be applied as alternatives to common plastic materials originating from mineral oils [1–4].

The most common representative of these biopolymers is poly-3-hydroxybutyrate (PHB). PHB constitutes a rather stiff and brittle material with a low extension to break [5–7]. These mechanical properties can be enormously enhanced by incorporation of different comonomers into the PHB matrix. This results in co- and terpolyesters possessing mechanical properties similar to high-tech materials [7, 8].

PHAs are synthesized in the cytoplasm of various prokaryotic strains, usually from carbohydrates, but also from other renewable resources [9–11]. Under unfavorable growth conditions (surplus of carbon source plus limitation of an essential substrate, i.e. nitrogen, phosphate, or oxygen), some strains are able to divert the usual carbon flux (acetyl-CoA synthesized in the central metabolic pathways of the microorganism) from biosynthesis of protein (biomass) constituents to the formation of compounds acting as precursors for the production of PHA, mainly PHB. This typical feature is valid for most PHA producers, e.g., *Wautersia eutropha*, the production strain of BIOPOL® (Metabolix Inc., Cambridge, MA) that was commercialized in the 1990s by Zeneca Bio Products and later by Monsanto [8]. Some organisms exist that accumulate PHA during balanced growth at substantial rates. This was first found for strains of *Alcaligenes latus* [12]. The outstanding properties of these growth-associated PHA producers enabled the development of an industrial polyester production process by

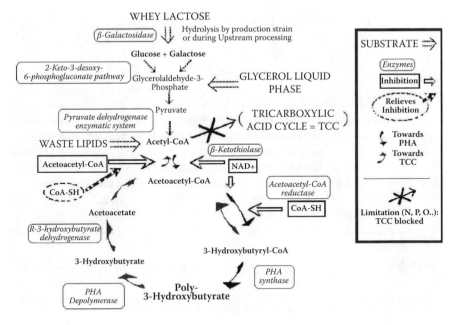

FIGURE 6.1 The cyclic nature of PHA metabolism: pathway for anabolism and catabolism of polyhydroxyalkanoates (PHAs) from whey lactose, glycerol liquid phase, and waste lipids. If the depletion of an essential nutritional component is blocking the biosynthesis of biomass components (proteins and others), carbon flux is directed toward PHA formation. (From Koller, M. et al., Production of Poly-3(hydroxybutyrate-*co*-hydroxyvalerate) from Agricultural Surplus Materials, presented as a poster at the annual meeting of ÖGBT, Innsbruck, Austria, 20–22 Sept. 2004. With permission.)

the Austrian company Chemie Linz in the late 1980s [8]. Figure 6.1 illustrates the metabolic pathway for production and reutilization of PHB from three cheap carbon sources: whey lactose, liquid-phase glycerol, and waste lipids.

Due to a high degree of polymerization, the average molecular weights of PHAs can reach several millions [13, 14]. These polymers are organized in granules and serve as intracellular reserve carbon and energy sources that normally are degraded when external carbon sources are depleted. Because of their outstanding property of complete degradability to water and CO_2, PHAs are embedded in nature's closed cycle of carbon. If PHAs are applied instead of fossil-oil-originated polymers, the mass balance of carbon from biomass will be closed, but its durability will be prolonged if compared with the usual biological life cycle of PHA-carbon. The fossil fuels energy demand during the span of life of PHAs will not exceed the amount necessary for the industrial production and processing of PHAs itself. The negative effects of CO_2 accumulation in the atmosphere, i.e., the greenhouse effect, will be related merely to the amount originated from energy production and not to the carbon in the biopolymer itself. Therefore, substituting common plastic materials by PHAs significantly lowers the amount of fossil-oil-originated carbon in the cycle of nature. Hence, it can be concluded that there are ecologically sound reasons for using PHAs instead of mineral-oil-originated plastics.

6.2 ENHANCEMENTS IN THE PROFITABILITY OF PHA PRODUCTION

6.2.1 FACTORS INFLUENCING THE PROFITABILITY OF BIOPLASTIC PRODUCTION

Conventional plastics like polyethylene (PE) and polypropylene (PP) are produced at a price of less than US $1/kg. In 1995, the Monsanto Company sold PHA products at about 17 times the price of petrol-based plastics.

It is clear that to achieve a truly cost-effective process, all production steps must be taken into account [15]. Hänggi pointed out that the raw materials constitute the major part of the production cost for biopolymers [16]. Recent studies explicitly show that PHA production from purified substrates such as glucose or sucrose has more or less been optimized [17]. Therefore, it is of importance to concentrate further process development on cheaper carbon sources as basic feedstocks.

Beyond the expenses for raw materials, improving the efficiency of biotechnological production in bioreactors (optimization of process parameters such as dissolved oxygen concentration that can strongly influence product yields, for details see Section 6.2.6) is decisive for cost effectiveness.

After biosynthesis of the polyester, the needed recovery process (typically a solid-liquid extraction procedure) can constitute another important cost factor, especially in large-scale production. Here extraction solvents that can easily be recycled will be of interest [18]. To avoid leaving the patterns of sustainability solely in biopolymer production, it will be indispensable to concentrate the development of new extraction processes on recyclable solvents that are also of an environmentally benign nature, such as lactic acid esters [19]. Typical precarious solvents like chloroform will have to be avoided.

Product quality of PHAs is very much dependent on the polyester composition. PHB homopolymer, a bioplastic that is easy to produce from a variety of cheap feedstocks, constitutes a material with restricted possibilities of commercial application due to its brittleness and high glass-transition point. Varying the intramolecular composition of PHAs, polyesters with properties ranging from crystalline thermoplasticity to characteristics of typical elastomers can be produced. This can be achieved by supplementing the cultivation medium with cosubstrates that can act as precursors for co- and terpolyester production. In the polymer chains of these co- and terpolyesters, building blocks such as 3-hydroxyvalerate (3HV) or 4-hydroxybutyrate (4HB) are incorporated. These building blocks possess a number of carbon atoms in their side chains or backbones that are different from 3HB and that disturb the crystalline PHB lattice. In 1987, Byrom discovered that poly-(3HB-co-3HV) could be produced on a large scale by supplementing a fed-batch culture of a glucose-utilizing mutant of *Alcaligenes eutrophus* (recently known as *Wautersia eutropha*) with glucose and propionic acid (precursor for 3HV formation) [20]. Lefebvre et al. demonstrated that the dissolved oxygen concentration not only affects the conversion rate of the main carbon source toward PHB, but is also of high importance for the rates and especially yields of the formation of 3HV building blocks from propionic acid (see also Section 6.2.6) [21]. In addition, it has been shown that the utilization of valeric acid instead of propionic acid results in a higher proportion of 3HV units [22].

The use of these often expensive precursors to obtain a higher product quality of co- and terpolyesters translates directly to higher costs for the entire process. A possible backdoor solution is the direct conversion of 3HV-unrelated substrates such as sugars or glycerol toward 3HV by some selected organisms. This rare feature can be found among a few wild-type strains belonging to the genera of *Norcardia* and *Rhodococcus* [8, 23–25]. Koller et al. described the formation of poly-(3HB-co-8-10%-3HV) from whey sugars and crude glycerol phase without the cofeeding of precursors using an osmophilic wild-type organism (see also Sections 6.2.4 and 6.2.5) [26].

It is estimated that, at an annual production scale of 100,000 t of PHB, the production costs of PHA will decrease from US \$4.91/kg to US \$3.72/kg if hydrolyzed corn starch (US \$0.22/kg) is chosen as carbon source instead of glucose (US \$0.5/kg) [15]. But this is still far beyond the cost for conventional polymers, which in 1995 was less than US \$1/kg [8]. In 1999, Lee et al. estimated that P(3HB) and medium-chain-length PHA (*mcl*-PHA) could be produced at a cost of approximately US \$2/kg [27], the preconditions being a highly efficient production process and the use of inexpensive carbon sources. Among these substrates, molasses, starch and hydrolyzed starch, whey from the dairy industry (see also Section 6.2.4), surplus glycerol from biodiesel production (see also Section 6.2.3), xylose, and various lipids (see also Section 6.2.2) are available [26, 28–36].

An additional cost factor in normally phosphate-limited production processes for PHAs is the expense for complex nitrogen sources. It was previously found that supplementation of a small amount of complex nitrogen source such as tryptone could enhance 3-PHB production by recombinant *Escherichia coli* in a defined nutrition medium containing glucose as the sole carbon source [37]. Examples for complex nitrogen sources are fish peptone, meat extract, casamino acids, corn steep liquor, soybean hydrolysate, and cotton seed hydrolysate [33, 38, 40]. Instead of expensive complex nitrogen sources such as yeast extract or casamino acids, cheaper products like silage juice or meat and bone meal can successfully be applied in PHA production processes [26, 39].

6.2.2 WASTE LIPIDS: A VERSATILE FEEDSTOCK

Several waste lipids of different origin can be applied in a variety of sustainable, future-oriented technologies:

Used cooking oil is a waste product that is available in huge amounts [40].
Tallow is an inexpensive source of triacylglyceride (TAG) [41].
From the meat and bone meal (MBM) hydrolysis process that can provide a useful nitrogen source for PHA-producing organisms (see Section 6.2.5), about 11% of lipids remain as surplus material after the degreasing step [42].
In PHA production, biomass has to be degreased before isolation of PHA. Typically 2 to 4% of lipids are isolated from the cells [43].

All of these materials would have to be expensively disposed of unless they can be further used, e.g., in one of the following two ways:

6.2.2.1 Transesterification

Alkaline methanolysis of TAGs results in a mixture of fatty acid methyl esters (FAME) and glycerol. When phosphoric acid is used for the necessary subsequent neutralization, cheap green fertilizers are generated (sodium phosphates). The FAME fraction is mainly used as an ecologically benign fuel (biodiesel) that is gaining increased interest because of its better CO_2-emission qualities than diesel from petroleum. For example, the public transit system in Graz (Austria) is expected to complete its switch from petrol to 100% biodiesel fuel by the end of 2005 [44]. The current worldwide annual production is estimated to be 350 million gallons [32].

A smaller share of FAME is converted to fine chemicals such as surfactants [40]. Additionally, biodiesel is directly converted by several bacterial strains toward PHA [45].

6.2.2.2 Direct Utilization of TAGs as a Carbon Source for PHA Production

Tallow is one of the cheapest fats. Production of PHA from tallow has been achieved using *Pseudomonas resinovorans*. Although the raw material is inexpensive, the process is not profitable due to the low amounts of PHA produced (ca. 15% of cell dry mass) [41].

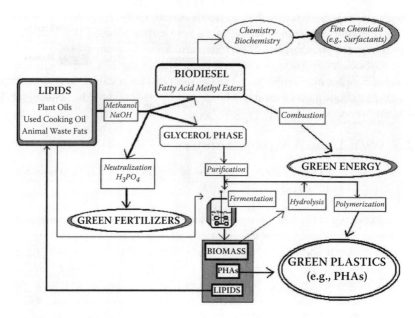

FIGURE 6.2 Different possibilities of utilization of lipids toward higher-value products. (From Braunegg, G. et al., *Production of Plastics from Waste Derived from Agrofood Industry*, paper presented at International Conference Renewable Resources and Renewable Energy: A Global Challenge, Trieste, Italy, 10–12 June 2004. With permission.)

The production of poly-(3HB-*co*-3HV) from olive oils by *Aeromonas caviae* was described by Doi in 1995. However, the polyester content in cells was rather low (6 to 12%), and therefore this process also was not profitable [46].

Higher amounts of pure PHB (up to 80% of cell dry mass) from different plant oils were produced by *Wautersia eutropha* [35]. Crude palm oil is a substrate of interests for *Erwinia* sp. USMI-20. Studies done by Majid et al. show that, using this strain, 46% (PHB in cell mass) was achieved after 48 h of cultivation [36].

Figure 6.2 depicts the different possibilities of utilization of lipids toward production of green plastics (PHAs), green fuels (biodiesel), and other by-products.

6.2.3 PHAs from Raw Glycerol Deriving from Biodiesel Production

Glycerol liquid phase (GLP), the major side stream of biodiesel production from triacylglycerides (see Section 6.2.2), contains about 70% (w/w) glycerol. In all Europe, the total production of biodiesel for 2005 is estimated at 1,925,000 tonnes (t), and the estimate for 2008 is 2,649,000 t, corresponding to 192,500 and 264,900 t of glycerol, respectively [47]. GLP currently constitutes a surplus material. Its utilization leads to an enormous cost advantage compared with commercially available pure glycerol, possessing a market value of 900/t (2002)

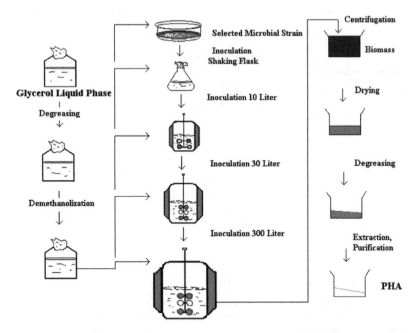

FIGURE 6.3 Scheme of PHA production from the glycerol liquid phase. (From Braunegg, G. et al., *Production of Plastics from Waste Derived from Agrofood Industry*, paper presented at International Conference Renewable Resources and Renewable Energy: A Global Challenge, Trieste, Italy, 10–12 June 2004. With permission.)

[48]. From the substrate GLP, PHA is produced by different organisms [27, 33]. Figure 6.3 shows the production cycle for GLP from lipid waste and its subsequent utilization as a carbon source in PHA biosynthesis. Starting from crude GLP, and depending on the strain used, a degreasing and a demethanolization step might be needed before the substrate can be applied as a carbon source. The scale-up starting from single colonies of the production strain on solid medium until PHA production on a 300:l scale is demonstrated in Figure 6.3, which also shows the needed downstream processing.

Figure 6.4 depicts the process pattern of production of poly-(3HB-*co*-3HV) from GLP. Here, GLP was supplied as the sole carbon source without additional precursor feeding. Therefore, an osmophilic wild-type strain was used that converts a broad spectrum of cheap substrates toward PHA production [26]. It is apparent that the organism is accumulating PHA already in parallel with the formation of active biomass (expressed as protein, see Figure 6.4a). Until the end of the cultivation, the cells accumulated PHA up to 70% of their total mass (see Figure 6.4b) at a final copolyester concentration of 16.2 g/l (see Figure 6.4a and Figure 6.4c). During the whole process, 3HV was incorporated into the PHB matrix in a constant amount (8 to 10%). The polymer isolated at the end of the fermentation showed a molecular mass of MW = 253,000 [26].

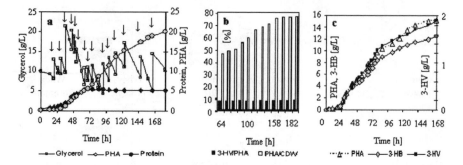

FIGURE 6.4 Production of poly-(3HB-*co*-3HV) from the glycerol liquid phase (GLP). Arrows in (a) indicate the refeeding of GLP. (From Braunegg, G. et al., *Production of Plastics from Waste Derived from Agrofood Industry*, paper presented at International Conference Renewable Resources and Renewable Energy: A Global Challenge, Trieste, Italy, 10–12 June 2004. With permission.)

1 kg P(HB-co-15%HV) can be produced from	3.567 kg Lactate
3 567 kg Lactate can be produced from	3.963 kg Lactose
3.963 kg Lactose correspond to	80.878 kg Whey

Annual Whey Production:

a) For one Italian Region:

     6×10^5 t Whey can potentially be converted to $\boxed{7418 \text{ t P(HB-co-15\%HV)}}$

b) For Europe:

     5×10^7 t Whey can potentially be converted to $\boxed{618000 \text{ t P(HB-co-15\%HV)}}$

FIGURE 6.5 Possible amounts of PHA from whey in an Italian region and the whole EU from whey by a two-step production process via lactic acid. (From Braunegg, G. et al., *Production of Plastics from Waste Derived from Agrofood Industry*, paper presented at International Conference Renewable Resources and Renewable Energy: A Global Challenge, Trieste, Italy, 10–12 June 2004. With permission.)

6.2.4 PHA PRODUCTION FROM SURPLUS WHEY

Whey is the major by-product from cheese and casein production. Annually, 13,500,000 t of whey containing 620,000 t of lactose constitute a surplus product in the EU [49].

From the feedstock milk, casein is precipitated enzymatically or by acidification. This so-called transformation results in the generation of curd cheese (casein fraction) and full fat whey (liquid fraction). By subsequent skimming, most of the lipids are removed, leaving overskimmed whey. The sweet skimmed whey undergoes a concentration step, where 80% of the water is removed. This whey concentrate is separated via ultrafiltration into a whey permeate (lactose fraction) and a whey retentate (protein fraction with considerable lactose residues). Whereas the retentate

TABLE 6.1
Composition of Different Types of Whey

Compound (% w/w)	Sweet Whey	Fermented Whey	Whey Permeate	Whey Retentate
Lactose	4.7–4.9	4.5–4.9	23	14
Lactic acid	traces	0.5	—	—
Proteins (nitrogen compounds)	0.75–1.1	0.45	0.75	13
Lipids	0.15–0.2	traces	—	3–4
Inorganic compounds	ca. 7	6–7	ca. 27	ca. 7

Source: Braunegg, G. et al., *Production of Plastics from Waste Derived from Agrofood Industry*, paper presented at International Conference Renewable Resources and Renewable Energy: A Global Challenge, Trieste, Italy, 10–12 June 2004. With permission.

fraction is of interest due to the importance of lactalbumin and lactoferrin for the pharmaceutical industry [50], the whey permeate (containing 81% of the total lactose originally included in the feedstock milk) can be used as a carbon source for biotechnological production of PHAs. Table 6.1 summarizes the composition of sweet whey, fermented whey, whey permeate, and retentate.

Biotechnological production of PHAs from different sugars via condensation of acetyl-CoA units stemming from hexose catabolism is well described, but only a limited number of microorganisms directly convert lactose into PHAs [51]. Depending on the production strain, three possible ways of applying whey for PHA production are viable:

Metabolization of lactose to lactic acid and subsequent conversion to PHA
Direct conversion of lactose to PHA
Hydrolysis of lactose to glucose and galactose prior to conversion to PHA

6.2.4.1 Metabolizing Lactose to Lactic Acid and Subsequent Conversion to PHA

The anaerobic conversion of whey lactose by different *Lactobacilli* strains provides lactic acid with high yields. Lactic acid is a substrate that can be metabolized in a following aerobic fermentation toward PHA by numerous strains, e.g., by *Paracoccus denitrificans* [40]. Figure 6.5 shows the amounts of poly-(3HB-*co*-3HV) that can be produced by this two-step process from surplus whey of one Italian region and for the whole of Europe (data refer to the entirely produced whey).

6.2.4.2 Direct Conversion of Lactose to PHA

The principal possibility of direct conversion of whey lactose toward PHA using different wild-type bacterial strains is reported in the literature [30, 31, 52, 53]. The cultivation of recombinant *E. coli* strains harboring PHA synthesis genes for direct utilization of purified lactose has already been studied and is of interest due to high volumetric productivities [54–57]. The application of whey permeate by recombinant *E. coli* as carbon source for PHA production has also been well investigated [30, 58, 59]. Ahn et al. published the results from a fed-batch fermentation using a cell-recycling system for recombinant *E. coli* on whey permeate. The determined volumetric productivity for PHB (4.6 g/l·h) is the highest reported for PHA production from whey lactose [59].

6.2.4.3 Hydrolysis of Lactose to Glucose and Galactose Prior to Conversion to PHA

Hydrolysis of lactose by β-galactosidase is the first step in biological PHA production from whey lactose (see Figure 6.1). PHA-producing strains exist that show no or insufficient activity of lactose utilization, but they accept the hydrolysis products of lactose, namely glucose and galactose, as substrates [60]. Due to insufficient β-galactosidase activity of the production strain, the hydrolysis rate can be so low that this step is decisive for the total duration of the process. An extension of the process increases costs because of a higher demand of energy and labor. To overcome this problem, lactose can be hydrolyzed chemically or enzymatically prior to the cultivation. This way, glucose and galactose are provided to the microorganisms as carbon substrates [26, 60].

Recently it was shown that the same osmophilic wild-type strain that was used for PHA production from raw glycerol (see Section 6.2.3) is also able to accumulate poly-(3HB-*co*-3HV) from hydrolyzed whey lactose. Therefore, crude whey was separated from proteins and concentrated directly at a dairy company. The concentrated whey had a lactose concentration of ca. 20% (w/v), which constitutes the upper limit of lactose solubility in water. Lactose included in whey permeate was hydrolyzed enzymatically and used as carbon source for the investigation of poly-hydroxyalkanoate synthesis. Figure 6.6 and Figure 6.7 depict the process pattern of substrate utilization and production of poly-(3HB-*co*-3HV) from hydrolyzed whey lactose. Also in this case, supplementing of 3HV-precursors was not needed for the

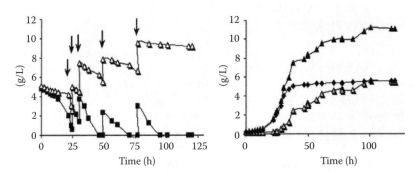

FIGURE 6.6 PHA production on whey: time curves of substrates (a) and products (b). Arrows indicate the refeeding with hydrolyzed whey permeate. ■ Glucose; X Galactose; ▲ Cell dry mass; X PHA; ◆ Protein.

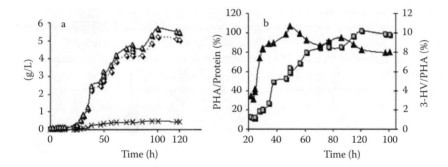

FIGURE 6.7 Time curves of hydroxyalkanoates (a) and percentages of 3-HV in polymer and ratio PHA/protein (b). (a) ◊ 3-HB; × 3-HV, X PHA; (b) ■ PHA/Protein; ▲ Protein.

generation of 8 to 10% (mol/mol) of 3HV. A final copolyester concentration of 5.5 g/l was obtained until the end of the fermentation. The average molecular weight was 696,000, with a polydispersity index of 2.2 [61].

6.2.5 HYDROLYZED MEAT AND BONE MEAL: A CHEAP NITROGEN SOURCE

Severe problems have arisen during the last couple of years in the EU from the emergence of BSE (bovine spongiform encephalopathy). At its peak, the disease had infected 3500 head of cattle weekly in Great Britain [62]. In 2001, this encouraged several scientists at Graz University of Technology (Austria) to research the development of technologies for safe utilization of meat and bone meal (MBM) [63]. One of these novel technologies is the production of alternative cheap nitrogen sources for biotechnological purposes from MBM that was proven to be free of prions. For this purpose, MBM was degreased and further chemically or enzymatically hydrolyzed. Figure 6.8 illustrates the procedure of acidic hydrolysis of MBM and the subsequent neutralization and separation of (predominantly inorganic) precipitates.

The remaining hydrolyzate contains 70% of the original organic nitrogen from MBM (see Figure 6.8) and can therefore be successfully applied as a complex nitrogen source in biotechnological PHA production processes. A fermentation on the 10:1 scale using the same osmophilic wild-type organism as described in Sections 6.2.3 and 6.2.4 was reported on hydrolyzed MBM as the sole nitrogen source; crude glycerol from biodiesel production was used as a carbon source. Under these conditions, 1.75 g/l of cell protein and approximately 6 g/l poly-(3HB-*co*-3HV) were produced [61].

6.2.6 INFLUENCE OF DOC ON THE COST-EFFECTIVENESS OF PHA PRODUCTION

Restricted availability of nitrogen, oxygen, or phosphate as initiator for PHA formation was reported several years ago [64]. As a novelty, the effects of a double-limitation by depletion of nitrogen together with low concentration of dissolved oxygen (DOC) was investigated by Lefebvre et al. in the late 1990s using the strain *Wautersia eutropha* [21]. In this study, glucose and propionic acid were cosupplemented in the PHA production phase to enable the biosynthesis of poly-(3HB-*co*-

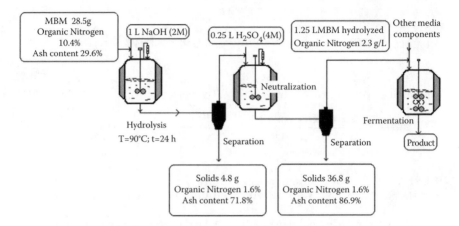

FIGURE 6.8 Hydrolyzed meat and bone meal: production steps and application for fermentation purposes. (From Braunegg, G. et al., *Production of Plastics from Waste Derived from Agrofood Industry*, paper presented at International Conference Renewable Resources and Renewable Energy: a Global Challenge, Trieste, Italy, 10–12 June 2004. With permission.)

3HV). In direct comparison to a control fermentation with sufficient oxygen supply (50 to 70% of air saturation during accumulation phase), low-DOC experiments (carried out at a DOC between 1 and 4% of air saturation) resulted in a slower production rate for 3HB, probably due to lower glucose uptake rates. On the other hand, the production rate for 3HV increased significantly, although the same enzymes are involved in 3HB and 3HV synthesis. Prior to the experiments, it had to be expected that the slowdown in enzyme activities would also negatively influence 3HV formation! Additionally, it turned out that higher yields of 3HV from propionate are achieved at low DOC.

The described findings can be explained as follows: propionate is typically converted to propionyl-CoA, which undergoes a further condensation with acetyl-CoA to form C5-building blocks. By releasing CO_2, propionyl-CoA can easily lose its carbonyl atom, thus building acetyl-CoA [65]. The condensation of two acetyl-CoA units generates 3HB. The experiments done by Lefebvre et al. [21] indicate that this unwanted oxidative loss of CO_2 from propionyl-CoA can obviously be avoided by restricting the oxygen supply, resulting in higher yields of 3HV from propionate at the expense of 3HB and, due to the lower glucose uptake, the total copolyester formation. On an industrial scale, one would have to decide, as the case arises, whether the increased 3HV yield from propionate, together with the minimized need for aeration, can economically compensate the lower overall PHA production. This will very much be dependent on the desired end composition of the copolyester. Table 6.2 collects the detailed data for the discussed findings.

It is known that, due to the lack of depolymerase, recombinant *E. coli* does not require nutrient limitation for PHA synthesis and produces PHA during growth [66]. Because it was noticed that PHA production was increased in recombinant *E. coli*

TABLE 6.2
Influence of DOC on Production of Poly-(3HB-co-3HV)

Experiment	PHA (g/l) [a]	Mol% 3HV in PHA [a]	PHA in CDM (%) [a]	Y 3HV/Propionate	rPHA (g/l·h)	r3HB (g/l·h)	r3HV (g/l·h)
Control (DOC 50–70%) [b]	11.7	21.5	ca. 80	0.25	0.44	0.33	0.07
Low DOC (DOC 1–4%) [c]	9.1	31.5	ca. 80	0.73	0.36	0.23	0.14

[a] Data from end of cultivation.
[b] A total of 16.0 g·l^{-1} of sodium propionate was added; all of it was used by the cells.
[c] A total of 5.65 g·l^{-1} of sodium propionate was added; 4.7 g·l^{-1} was consumed by the cells.

Source: Lefebvre, G., Rocher, M., and Braunegg, G., *Appl. Environ. Microbiol.*, 63, 827–833, 1997. With permission.

when biomass production was gradually retarded, Kim [30] compared PHA production from whey lactose at sufficient and insufficient oxygen supply by changing the maximum agitation speed during different fed-batch fermentations in stirred-tank fermentors. It turned out that cell concentration as well as PHB homopolyester productivity and total concentration of PHB (g/l reactor volume) increased with higher agitation speed (better oxygen input), whereas the PHB content in cells was highest at an agitation speed of 500 rpm (80% PHB in cells compared with approximately 60% at 700, 900, and 1300 rpm). A further lowering of the agitation speed (300 rpm) did not further positively influence the PHB content (70% PHB in cells) [30]. The increased content of PHA at 500 rpm is most probably due to the reduced formation of non-PHA biomass in combination with a specific PHA production that was not influenced; the decrease in PHA content at 300 rpm might again be due to a restricted specific sugar uptake rate of the cells (see previously described experiments of Lefebvre et al. [21]). In addition to the previously described findings for *Wautersia eutropha*, the results achieved by Kim [30] clearly indicate the importance of optimizing the oxygen supply during PHA synthesis.

6.3 CONCLUSION

The presented studies clearly demonstrate that a variety of surplus materials deriving from the agrofood industry are of interest for the production of biopolymers. Some of these inexpensive substrates act directly as a carbon source for industrial production of future-oriented bioplastics such as poly-(3HB-co-3HV). To illustrate this point, waste materials like whey lactose from cheese production, crude glycerol liquid phase from biodiesel production, and several waste lipids are discussed. Meat and bone meal are presented as a suitable raw material for complex nitrogen sources that are needed to achieve sufficient concentrations of PHA-producing biomass, a precondition for economical, profitable biopolyester production. The application of surplus materials enhances the cost efficiency of PHA production while reducing the amounts of waste originating from these surplus materials.

The costs derived from the addition of precursors for copolyester production can be mitigated by using microbial strains that convert cheap materials directly into poly-(3HB-*co*-3HV). Production of these precursors can also be enhanced using new fermentation strategies based on limiting the oxygen supply.

The results obtained look promising in regard to future production of economically and ecologically profitable high-quality bioplastics.

REFERENCES

1. Brandl, H. et al., Plastics from bacteria and for bacteria: poly(beta-hydroxyalkanoates) as natural, biocompatible, and biodegradable polyesters, *Adv. Biochem. Eng. Biotechnol.*, 41, 77–93, 1990.

2. Steinbüchel, A. and Valentin, H.E., Diversity of bacterial polyhydroxyalkanoic acids, *FEMS Microbiol. Lett.*, 128, 219–228, 1995.

3. Holmes, P.A., Biologically produced PHA polymers and copolymers, in *Developments in Crystalline Polymers,* Vol. 2, Basset, D.C., Ed., Elsevier, London, 1988, pp. 1–65.

4. Lee, S.Y., Choi, J., and Chang, H.N., Process development and economic evaluation for the production of polyhydroxyalkanoates by *Alcaligenes eutrophus*, in *Proceedings of the 1996 International Symposium on Bacterial Polyhydroxyalkanoates,* Eggink, G. et al., Eds., NRC Research Press, Ottawa, 1996, pp. 127–136.

5. Barham, P.J. et al., Crystallization and morphology of a bacterial thermoplastic: poly-3-hydroxybutyrate, *J. Mater. Sci.*, 19, 2781–2794, 1984.

6. Luzier, W.D., Materials derived from biomass/biodegradable materials, *Proc. Natl. Acad. Sci. USA*, 89, 839–842, 1992.

7. Sudesh, K., Abe, H., and Doi, Y., Synthesis, structure and properties of polyhydroxyalkanoates: biological polyesters, *Prog. Pol. Sci.*, 25, 1503–1555, 2000.

8. Braunegg, G., Lefebvre, G., and Genser, K.F., Polyhydroxyalkanoates, biopolyesters from renewable resources: physiological and engineering aspects, *J. Biotechn.*, 65, 127–161, 1998.

9. Hocking, P.J. and Marchessault, R.H., *Chemistry and Technology of Biodegradable Polymers*, Chapman & Hall, New York, 1994, pp. 48–96.

10. Bourque, P.Y., Pomerleau, Y., and Groleau, D., High-cell-density production of poly-β-hydroxybutyrate (PHB) from methanol by *Methylobacterium extorquens*: production of high-molecular-mass PHB, *Appl. Microbiol. Biot.*, 44, 367–376, 1995.

11. Ramsay, J.A., Aly Hassan, M.-C., and Ramsay, B.A., Hemicellulose as a potential substrate for production of PHA, *Can. J. Microbiol.*, 41, 262–266, 1995.

12. Lafferty, R.M. and Braunegg, G., Verfahren zur biotechnologischen Herstellung von Poly-D-(-)-3-Hydroxybuttersäure, German Patent No. 379,613, 1983.

13. Wang, F. and Lee, S.Y., Poly(3-hydroxybutyrate) production with high polymer content by fed-batch culture of *Alcaligenes latus* under nitrogen limitation, *Appl. Environ. Microbiol.*, 63, 3703–3706, 1997.

14. Steinbüchel, A. and Lütke-Eversloh, T., Metabolic engineering and pathway construction for biotechnological production of relevant polyhydroxyalkanoates in microorganisms, *Biochem. Eng. J.*, 3734, 1–16, 2003.

15. Choi, J. and Lee, S.Y., Factors affecting the economics of polyhydroxyalkanoate production by bacterial fermentation, *Appl. Microbiol. Biotechnol.*, 51, 13–21, 1999.

16. Hänggi, U.J., Requirements on bacterial polyesters as future substitute for conventional plastics for consumer goods, *FEMS Microb. Rev.*, 16, 213–220, 1995.

17. Tsuge, T., Metabolic improvements and use of inexpensive carbon sources in microbial production of polyhydroxyalkanoates, *J. Biosci. Bioeng.*, 94, 579–584, 2002.

18. Chen, G.Q. et al., Industrial scale production of poly(3-hydroxybutyrate-co-3-hydroxyhexanoate), *Appl. Microbiol. Biotechnol.*, 57, 50–55, 2001.

19. Metzner, K., Sela, M., and Schaffer, J., Agents for extracting polyhydroxyalkane acids, U.S. Patent No. 9,708,931, 1997.

20. Byrom, D., Polymer synthesis by microorganisms: technology and economics, *Tibtech.*, 5, 246–250, 1987.

21. Lefebvre, G., Rocher, M., and Braunegg, G., Effects of low dissolved-oxygen concentrations on poly-(3-hxdroxybutyrate-co-3-hydroxyvalerate) production by *Alcaligenes eutrophus*, *Appl. Environ. Microbiol.*, 63, 827–833, 1997.

22. Haywood, G.W., Anderson, A.J., and Dawes, E.A., The importance of PHB synthase substrate specificity in polyhydroxyalkanoate synthesis by *Alcaligenes eutrophus*, *FEMS Microbiol. Lett.*, 57, 1–6, 1989.

23. Alvarez, H.M., Kalscheuer, A., and Steinbüchel, A., Accumulation of storage lipids in species of *Rhodococcus* and *Norcardia* and effect of inhibitors and polyethylene glycol, *Fett-Lipid*, 99, 239–246, 1997.

24. Williams, R. et al., Production of a co-polyester of 3-hydroxybutyric acid and 3-hydroxyvaleric acid from succinic acid by *Rhodococcus rubber*: biosynthetic considerations, *Appl. Microbiol. Biotechnol.*, 62, 717–723, 1994.

25. Valentin, R. and Dennis, D., Metabolic pathway for poly(3-hydroxybutyrate-co-3-hydroxyvalerate) formation in *Norcardia corallina*: inactivation of *mutB* by chromosomal integration of a kanamycin resistance gene, *Appl. Environ. Microbiol.*, 62, 372–379, 1996.

26. Koller, M. et al., Production of polyhydroxyalkanoates from agricultural waste and surplus materials, *Biomacromol.*, 6, 561–565, 2005.

27. Lee, S.Y., Choi, J., and Wong, H.H., Recent advances in polyhydroxyalkanoate production by bacterial fermentation: mini-review, *Int. J. Biol. Macrom.*, 25, 31–36, 1999.

28. Braunegg, G. et al., Production of PHAs from agricultural waste material, *Macromol. Symp.*, 144, 375–383, 1999.

29. Zhang, H. et al., Production of polyhydroxyalkanoates in sucrose-utilizing recombinant *Escherichia coli* and *Klebsiella* strains, *Appl. Env. Microbiol.*, 60, 1198–1205, 1994.

30. Kim, B.S., Production of poly(3-hydroxybutyrate) from inexpensive substrates, *Enzyme Microb. Tech.*, 27, 774–777, 2000.

31. Povolo, S. and Casella, S., Bacterial production of PHA from lactose and cheese whey permeate, *Macromol. Symp.*, 197, 1–9, 2003.

32. Ashby, R.D., Solaiman, D.K.Y., and Foglia, T.A., Bacterial poly(hydroxyalkanoate) polymer production from the biodiesel co-product stream, *J. Polym. Environ.*, 12, 105–112, 2004.

33. Lee, S.Y., Poly(3-hydroxybutyrate) production from xylose by recombinant *Escherichia coli*, *Bioprocess Eng.*, 18, 397–399, 1998.

34. Silva, L.F. et al., Poly-3-hydroxybutyrate (P3HB) production by bacteria from xylose, glucose and sugarcane bagasse hydrolysate, *J. Ind. Microbiol. Biot.*, 31, 245–254, 2004.

35. Fukui, T. and Doi, Y., Efficient production of polyhydroxyalkanoates from plant oils by *Alcaligenes eutrophus* and its recombinant strain, *Appl. Microbiol. Biot.*, 49, 333–336, 1998.
36. Majid, M.I.A. et al., Production of poly(3-hydroxybutyrate) and its copolymer poly(3-hydroxybutyrate-co-3-hydroxyvalerate) by *Erwinia* sp. USMI-20, *Int. J. Biol. Macrom.*, 25, 95–104, 1999.
37. Lee, S.Y. and Chang, H.N., Effect of complex nitrogen source on the synthesis and accumulation of poly(3-hydroxybutyric acid) by recombinant *Escherichia coli* in flask and fed-batch cultures, *J. Environ. Polym. Degr.*, 2, 169–176, 1994.
38. Page, W.J. and Cornish, A., Growth of *Azotobacter vinelandii* UWD in fish peptone medium and simplified extraction of poly-(beta)-hydroxybutyrate, *Appl. Environ. Microb.*, 59, 4236–4244, 1993.
39. Koller, M. et al., Biotechnological production of poly(3-hydroxybutyrate) with *Wautersia eutropha* by application of green grass juice and silage juice as additional complex substrates, *Biocat. Biotrans.*, 223, 561–565, 2005.
40. Braunegg, G. et al., Production of plastics from waste derived from agrofood industry, paper presented at *International Conference Renewable Resources and Renewable Energy: a Global Challenge*, Trieste, Italy, 10–12 June 2004.
41. Cromwick, A.M., Foglia, T., and Lenz, R.W., The microbial production of poly(hydroxyalkanoates) from tallow, *Appl. Microbiol. Biotechnol.*, 46, 464–469, 1996.
42. Neto, J., personal communication, 2005.
43. Koller, M., unpublished data, 2005.
44. Available on-line at: http://taten.municipia.at/alle/f0002084.html.
45. Steinbüchel, A., Voss, I., and Gorenflo, V., Interesting carbon sources for biotechnological production of biodegradable polyesters: the use of rape seed oil methyl ester (biodiesel), in *Polymers from Renewable Resources: Biopolyesters and Biocatalysis*, Scholz, C. and Gross, R.A., Eds., ACS Symp. Ser. 764, ACS, Washington, DC, 2000, pp. 14–24.
46. Doi, Y., Kitamura S., and Abe, H., Microbial synthesis and characterization of poly(3-hydroxybutyrate-co-3-hydroxyhexanoate), *Macromol.*, 28, 4822–4828, 1995.
47. Oleoline® HBI Glycerine Market Report, Dec. 2003.
48. Biodiesel BDI Anlagenbau Ges.m.b.H., Graz, Austria, personal communication, 2003.
49. Latterie Vicentine (a dairy company), Italy, and Bundesamt für Statistik (statistical office), Neuchâtel, Switzerland, 2000.
50. Daniel, H.J. et al., Production of sophorolipids from whey: development of a two-stage process with *Cryptococcus curvatus* ATCC 20509 and *Candida bombicola* ATCC 22214 using deprotonized whey concentrates as substrates, *Appl. Microbiol. Biotechnol.*, 51, 40–45, 1999.
51. Sudesh, K., Abe, H., and Doi, Y., Synthesis, structure and properties of polyhydroxyalkanoates: Biological polyesters, *Prog. Pol. Sci.*, 25, 1503–1555, 2000.
52. Yellore, V. and Desai, A., Production of poly-3-hydroxybutyrate from lactose and whey by *Methylobacterium* sp. ZP24, *Lett. Appl. Microbiol.*, 26, 391–394, 1998.
53. Young, F.K., Kastner, J.R., and May, S.W., Microbial production of poly--hydroxybutyric acid from D-xylose and lactose by *Pseudomonas cepacia*, *Appl. Environ. Microbiol.*, 60, 4195–4198, 1994.
54. Fiedler, S. and Dennis, D., Polyhydroxyalkanoate production in recombinant *Escherichia coli*, *FEMS Microbiol. Rev.*, 103, 231–236, 1992.
55. Lee, S.Y., *E. coli* moves into the plastic age, *Nat. Biotechnol.*, 15, 17, 18, 1997.

56. Schubert, P., Steinbüchel, A., and Schlegel, H.G., Cloning of the *Alcaligenes eutrophus* poly-β-hydroxybutyrate synthetic pathway and synthesis of PHB in *Escherichia coli*, *J. Bacteriol.*, 170, 5837–5847, 1988.

57. Wong, H.H. and Lee, S.Y., Poly(3-hydroxybutyrate) production from whey by high density cultivation of recombinant *Escherichia coli*, *Appl. Microbiol. Biotechnol.*, 50, 30–33, 1998.

58. Ahn, W.S., Park, S.J., and Lee, S.Y., Production of poly(3-hydroxybutyrate) by fed-batch culture of recombinant *Escherichia coli* with a highly concentrated whey solution, *Appl. Environ. Microbiol.*, 66, 3624–3627, 2000.

59. Ahn, W.S., Park, S.J., and Lee, S.Y., Production of poly(3-hydroxybutyrate) from whey by cell recycle fed-batch culture of recombinant *Escherichia coli*, *Biotechnol. Lett.*, 23, 235–240, 2001.

60. Koller, M. et al., unpublished data, article submitted in 2005.

61. Koller, M. et al., Production of poly-3(hydroxybutyrate-*co*-hydroxyvalerate) from agricultural surplus materials, presented as a poster at annual meeting of ÖGBT, Innsbruck, 20–22 Sept. 2004.

62. Available on-line at: www.med4you.at/bse/bse_rind.htm.

63. Graz Technical University: *Task Force, Safe Technical Utilization of Meat and Bone Meal*, press release, Graz Technical University, Austria, 2001.

64. Vollbrecht, D. et al., Excretion of metabolites by hydrogen bacteria, IV: respiration rate-dependent formation of primary metabolites and of poly-3-hydroxybutyrate, *Eur. J. Appl. Microbiol. Biotechnol.*, 7, 267–276, 1979.

65. Doi, Y. et al., Biosynthesis of copolyesters in *Alcaligenes eutrophus* H16 from ^{13}C-labeled acetate and propionate, *Macromolecules*, 20, 2988–2991, 1987.

66. Lee, S.Y. and Chang, H.N., Production of poly(3-hydroxybutyric acid) by recombinant *Escherichia coli*: genetic and fermentation studies, *Can. J. Microbiol.*, 41, 207–215, 1995.

7 On the Environmental Performance of Biobased Energy, Fuels, and Materials: A Comparative Analysis of Life-Cycle Assessment Studies

Martin Weiss and Martin Patel

CONTENTS

7.1 INTRODUCTION

The interest in biomass applications to produce energy, fuels, and materials has grown rapidly in recent years. Among other aspects such as the replacement of scarce fossil resources or the generation of alternative markets for agricultural products, environmental considerations are key drivers for this development.

Several studies have shown the great potential of biobased products for contributing to savings of nonrenewable energy resources and greenhouse gas emissions [1, 2, 3]. A recent survey among stakeholders revealed that manufacturers, researchers, and technologists expect biobased materials to reduce both negative environmental impacts and the consumption of nonrenewable resources relative to their fossil-based counterparts [4].

On the other hand, options for using biomass to produce energy, fuels, and materials may be limited due to requirements concerning product quality. Moreover, cultivation, extraction, and processing of biomass are by no means environmentally neutral and involve the consumption of fossil fuels as well as the application of fertilizers and hazardous chemicals. Large-scale biomass production requires agricultural land and might cause adverse environmental effects such as degradation of soils, eutrophication of ground and surface waters, or fragmentation of ecosystems. Especially in Western Europe, biomass cultivation for nonfood applications might be limited due to demand of land for other purposes such as infrastructure, housing, or recreation. A thorough analysis of all environmental impacts is, therefore, of fundamental importance to quantify and evaluate the advantages and disadvantages associated with the production of biobased energy, fuels, and materials. To this end, life-cycle assessment (LCA) can be applied as a method to assess environmental impacts throughout the entire life cycle of products and services.

The main objective of this chapter is to review and compare the results of life-cycle assessment studies on biobased energy, fuels, and materials. A total of 11 publications were reviewed, out of which three address energy or fuels and another eight the environmental performance of biobased materials such as loose fills, hemp-fiber composites, and disposable dishes. Most life-cycle assessment studies compare the environmental impacts of biobased fuels and materials with their fossil counterparts based on a functional unit of, e.g., 1 MJ energy or 1 kg polymer. Only a few studies in the scientific literature address the aspect of land use, either as a separate environmental impact category [5] or as a functional unit in LCA comparisons [1, 6, 7]. The long-term availability of agricultural land is likely to become scarce due to land requirements for other purposes. It will, hence, become increasingly important to maximize environmental benefits from biomass utilization per unit of agricultural land. The inclusion of land demand in LCA studies of biobased energy, fuels, and materials is especially desirable in comparing the environmental performance of alternative crops used to manufacture the same product (e.g., starch for TPS [thermoplastic starch] loose fills can be produced from corn, wheat, or potato). For this reason, we recalculated LCA results for biobased products and their conventional counterparts and consistently reference all environmental impacts to the functional unit of 1 hectare (Ha). This comparison ultimately aims at giving a comprehensive and comparative insight into the environmental performance of the various options of biomass utilization.

Following the introduction, this chapter presents a short overview of LCA methodology. We then explain the scope and assumptions used in the LCA studies analyzed in this chapter, and we briefly state the methods chosen for our comparison. We finally present results and discuss the implications of our findings for future research and development on biobased energy, fuels, and materials.

7.2 METHODOLOGY OF LIFE-CYCLE ASSESSMENT

The increasing awareness of environmental impacts associated with the production, consumption, and disposal of commodities has led to the development of life-cycle assessment (LCA), which is a widely accepted and internationally standardized method to analyze and quantify environmental impacts of products and services.

A typical life-cycle assessment (LCA) analysis consists of four independent elements [8]:

Definition of goal and scope
Life-cycle inventory analysis
Life-cycle impact assessment
Life-cycle interpretation

The four phases of LCA are illustrated in Figure 7.1.

The starting point of each LCA is to define the *goal and scope* of the analysis. This includes a decision about the functional unit to which the analysis should refer, the definition of the product system and system boundaries, as well as a choice of allocation procedures, types of impact categories to be studied, and the methodology of impact assessment. The functional unit can either be a certain service or a product, with the latter being the choice of the studies reviewed in this chapter (e.g., 1 kg of polylactic acid [PLA], 1 m^3 of loose-fill packaging material, or 1 Ha of agricultural land required for the production of biomass). Critical issues for a comparative environmental analysis of biobased versus fossil-based products are typically (a) the cultivation of biomass in agriculture and forestry (intensive vs. extensive practices), (b) the choice of the conventional product serving as a reference, and (c) the waste-

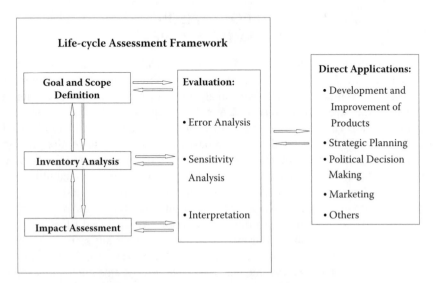

FIGURE 7.1 General scheme of LCA. (Modified from CEN [Comité Européen de Normalisation], ISO 14040.)

management option assumed for biobased and petrochemical materials (e.g., land-filling, incineration, recycling) [2].

The second step of LCA consists of life-cycle *inventory analysis*. This involves data collection and calculation procedures to quantify all environmentally relevant inputs and outputs of the system, with the most prominent ones being resource use; gaseous, liquid, and particulate emissions; solid waste; and land use.

The third step involves life-cycle *impact assessment* to (a) evaluate the relevance of system inputs and outputs for the environment and to (b) aggregate parameters belonging to the same environmental impact category into a single value by use of so-called characterization or equivalence factors. For example, to determine the greenhouse potential of substances emitted along a product's life cycle, all relevant emissions determined during the inventory analysis are multiplied with their specific greenhouse gas potential (characterization factor) and summed up to one single value. This value represents the specific impact (expressed in CO_2 equivalents) in a particular environmental impact category (greenhouse gas potential). As an optional step, the results per impact category can be divided by a reference value (e.g., total greenhouse gas emissions of a country or total greenhouse gas emissions per capita and year) to better understand and illustrate the relative importance of the various environmental impacts. This step is generally referred to as normalization.

Life-cycle interpretation is the last step of the LCA. Here, final conclusions are drawn from both life-cycle inventory analysis and life-cycle impact assessment. This includes a check for completeness and consistency of inventory data, a sensitivity analysis, as well as a report of all results. As an outcome of this step, recommendations for producers, consumers, policy makers, or environmental stakeholders can be formulated according to the original goal of the LCA.

7.3 ANALYSIS AND COMPARISON OF LIFE-CYCLE ASSESSMENT STUDIES

In our assessment, we analyze 11 LCA studies for biobased energy, fuels, and materials [3, 5, 9–17]. These studies compare biobased energy, fuels, or materials with their conventional counterparts made from fossil resources. The criteria that led to the selection of these LCA publications were that (a) the studies should compare environmental impacts of biobased products with fossil-based product alternatives on a cradle-to-grave basis, i.e., including the entire process chain from the extraction of raw materials, via processing of intermediate and final products, up to the final disposal and that (b) the publications should cover several environmental impacts and not only energy consumption and CO_2 emissions. As a consequence of these criteria, prominent studies on biopolymers (e.g., [18–20]) are not included in our comparison. All analyzed LCA publications refer to the situation in Europe and to the current state of technologies. The selected publications differ considerably with respect to the amount of published background data and the degree of detail regarding explanations about methodology and results. Also, studies with a comparatively limited level of detail are included in this analysis if (a) they contribute to a better understanding of environmental impacts from biobased materials not studied from this perspective before and (b) they provide an indication about uncertainty and sensitivity of LCA results [2].

Furthermore, the LCA publications vary regarding the assumptions made for agricultural yields, treatment of agricultural residues, allocation procedures for by-products, and waste-treatment scenarios. We did not correct for all of these differences; instead, we compared biobased and fossil products based on the assumptions made in each respective LCA study.* The most important assumptions made in the LCA publications analyzed in this chapter are given in Table 7.1.

The life-cycle assessment studies consider a wide range of different environmental impact categories. We select four categories for our analysis here, based on data availability and quality. In total, the environmental performance of 45 product pairs made from renewable and fossil resources is evaluated by use of the following four environmental impact categories:

Consumption of non–renewable primary energy resources[†]
Global warming potential
Eutrophication potential
Acidification potential

Several studies differentiate between terrestrial and aquatic eutrophication potential (e.g., [17]), others, such as [5, 9], choose a different method and give total eutrophication potential. We follow the latter approach and uniformly calculate eutrophication potential as the sum of terrestrial and aquatic eutrophication, expressed in phosphate equivalents. We further assume that the carbon dioxide originating from biomass is equivalent to the amount that was previously withdrawn from the atmosphere during the growth period of crops and, therefore, does not contribute to global warming. Fossil fuels required for transport and processing of biomass as well as the production of auxiliaries (e.g., application of mineral fertilizers) are, however, taken into account. For the comparison of biobased energy, fuels, and materials with their fossil counterparts, land is only used for the production of biobased products. This means that no reference land use is defined for fossil-based products because quantities required to produce the latter are negligible in comparison with biobased products.

We obtain the data for this LCA comparison from the relevant publications and recalculate the environmental impacts consistently for 1 hectare (1 Ha) of agricultural land. In the case of biopolymers, this means that the results refer to the amount of a particular biopolymer that can be produced using the agricultural crop yield available from 1 Ha (e.g., 38 t potatoes to produce starch-

---

* This approach differs somewhat from the analysis of Dornburg et al. [1]. There, the regional- and producer-specific LCA data are harmonized by (a) assuming identical average yields for agricultural crops, by (b) adopting waste incineration without energy recovery as the single only waste treatment option, and by (c) taking the life-cycle data for fossil-based polymers from one single source, e.g., the Association of Plastics Manufacturers Europe [21].
[†] In the following, we refer to the consumption of nonrenewable primary energy resources as "nonrenewable energy consumption."

TABLE 7.1
Overview of Core Assumptions Made in LCA Publications

Publication	Product	Scope of Analysis	Crop Yield per Hectare	Allocation Method	Waste-Treatment Scenario	Utilization of Agricultural Residues
Corbiére-Nicollier et al. 2001 [9]	Transport-pallets	Cradle to grave	17.2 t china reed	Substitution	Incineration	No
Diener and Siehler 1999 [10]	Fiber composites for car construction	Cradle to grave	5.5 t flax	—	Recycling	—
Dinkel et al. 1996 [11]	Polymer films, cups (single serving)	Cradle to grave[a]	37.7 t potato; 12.5 t corn	Allocation of environmental impacts	20% landfilling, 80% incineration[b]	No
Dinkel and Waldeck 1999 [12]	Plates (single serving)	Cradle to grave	38.9 t potato	—	Incineration, composting, production of biogas	No
Gärtner et al. 2002 [13]	Packaging material	Cradle to grave	n.a.	Substitution	Incineration	—
Müller-Sämann et al. 2002 [5]	Lubricants, hydraulic oils, packaging material, fiber composites, insulation material	Cradle to grave	n.a.	Substitution	Incineration	Substitution of oils[c]
Reinhardt et al. 2000 [14]	Heat, electricity, fuels	Cradle to grave	12.1 t triticale[d]; 3.3 t rapeseed; 2.7 t sunflower; 20.5 t china reed; 17.3 t willow; 8.4 t various woods; 5.7 t wheat straw	Substitution, allocation of environmental impacts	Incineration	—

Reinhardt and Gärtner 2003 [15]	Fuels and rapeseed oil	Cradle to grave	3.3 t rapeseed	Substitution	Incineration	Substitution of mineral fertilizers
Reinhardt and Zemanek 2000 [3]	Heat, fuels	Cradle to grave	11.5 t wheat **; 5.7 t orchard grass; 9.9 t cottonwood; 1.8 t wheat straw; 2.8 t pasture; 14.8 t sugar-beet; 10.1 t china reed; 2.8 t corn	Substitution	Incineration	Substitution of mineral fertilizers
Wötzel et al. 1999 [16]	Fiber composites for car construction	—	9.0 t hemp	—	—	No
Würdinger et al. 2002 [17]	Loose-fill packaging material	Cradle to grave	38.9 t potato; 7.5 t corn; 6.8 t wheat, 4.4 t wheat (ext.[e])	Substitution, allocation of environmental impacts	Various waste-treatment scenarios	Substitution of mineral fertilizers

Note: — = no information available.

[a] Without use phase.

[b] Without energy recovery.

[c] For the consumption of flax and hemp as fiber material, it is assumed that the oil contained in the seed is used to replace sunflower and primrose oil.

[d] Total plant.

[e] Wheat produced by extensive farming.

Adapted from: Weiss, M., Bringezu, S., and Heilmeier, H., *Zeitschrift Angewandte Umweltforschung*, 15/16, 3–5, 2003/2004. With permission.

based loose fills). For the petrochemical counterparts, the same amount of final product was then chosen as reference. Crop yields and data on biomass processing are obtained from the respective LCA studies. If agricultural yields are not reported, we use yield data from [22] as estimates. The differences between biobased and conventional product alternatives are calculated for each environmental impact category separately:

$$D_{(i)} = EI_{bio-based(i)} - EI_{fossil(i)} \qquad (7.1)$$

where

$D_{(i)}$ = difference between biobased and fossil product alternative
$EI_{biobased(i)}$ = environmental impact of the biobased product alternative
$EI_{fossil(i)}$ = environmental impact of the fossil product alternative
i = index for the environmental impact category

As the result of this approach, negative values indicate advantages of biobased energy, fuels, and materials, whereas positive values represent environmental advantages of their fossil counterparts. For final characterization, we divide the various options for biomass use into three categories: energy (power and heat), fuels, and materials. For the three groups, median values are calculated separately for each of the four environmental impact categories.

7.3.1 Results

The relative differences between biobased and conventional products per hectare and year with respect to the four environmental impact categories are presented in Figure 7.2[‡]. The results show large variations, both between single product alternatives and among the different environmental impact categories.

7.3.1.1 Nonrenewable Energy Consumption

All biobased product alternatives consume less nonrenewable energy throughout their life cycle than their fossil counterparts. Energy savings range from around 1300 GJ/(Ha·a) for the china reed/PP transport pallet to 8 GJ/(Ha·a) for PLA packaging material produced from cornstarch. The outstandingly high value of the china reed/PP transport pallet is mainly a consequence of (a) the relatively high yield of china reed per hectare of agricultural land, (b) the large proportion of the crop (70%) that is usable for fiber-composite production, and (c) the high resource and energy requirements to produce conventional glass fiber/PP pallets. Glass fiber/PP pallets are, however, not the most common fossile-based alternative for transportation

---

[‡] It is important to note that the units for all environmental impact categories are chosen to visualize the performance of biobased and fossil product alternatives. The size of the bars representing the different environmental impact categories in Figure 7.2 is, therefore, not an indication for the relative importance of one environmental impact category compared with another.

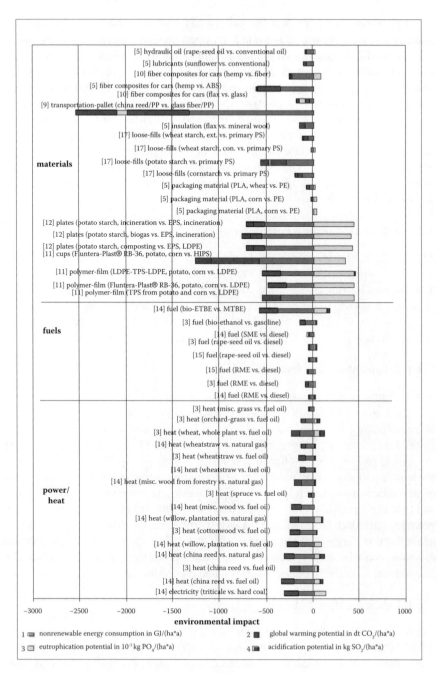

FIGURE 7.2 Relative environmental impacts of biobased and conventional products.

pallets. Instead, mainly wood pallets are used. It is very likely that the environmental advantages of the china reed pallet would decrease if this pallet were compared with other conventional transport pallets.

High savings of nonrenewable energy can also be achieved with disposable plates, cups, and polymer films made from potato and potato/cornstarch (580 to 290 GJ/[Ha·a]). ETBE (ethyl t-butyl ether) from sugar beet, which can be used as a substitute for the fuel additive MTBE (methyl t-butyl ether), reduces nonrenewable energy consumption by approximately 380 GJ/(Ha·a). The substitution of conventional packaging materials by corn- or wheat-based biomaterials and the replacement of conventional diesel by RME, SME, or rapeseed oil offers only minor reductions of nonrenewable energy consumption (8 to 54 GJ/[Ha·a]). This is mainly caused by (a) the relatively low hectare specific yield of biobased raw material (i.e., corn and wheat) and (b), especially in the case of RME (rapeseed methyl ester) and SME (sunflower methyl ester), by the energy-intensive chemical conversions of raw oil to final fuel products.

7.3.1.2 Global Warming Potential

The results in this category vary from around −520 dt CO_2/(Ha·a) for the china reed/PP transport pallet to +12 dt CO_2/(Ha·a) for loose fills made from wheat starch. The findings for this category are strongly related to the results in the category of nonrenewable energy consumption because energy production is the main source of CO_2 emissions. Consequently, products that score best with respect to nonrenewable energy use also generally offer the greatest reduction of CO_2 emissions.

7.3.1.3 Eutrophication Potential

The total results in this category range between −10 kg PO_4/(Ha·a) for the transport pallet and around +45 kg PO_4/(Ha·a) for polymer films studied by Dinkel et al. [11]. Biobased energy, fuels, and materials score in general worse than their fossil counterparts. Exceptions are, among others, the china reed/PP transport pallet (−10 kg PO_4/(Ha·a) and the fiber-reinforced composite for cars (−5 kg PO_4/[Ha·a]). The relatively poor performance of biobased products in this category is mainly a consequence of atmospheric ammonia emissions as well as nitrate and phosphate leaching to surface and ground waters from fertilizer application in agriculture [17]. As improved farming practices can reduce these fertilizer losses by (a) reducing the total amount of fertilizers applied and (b) improving the management of fertilizer application, a high potential to reduce eutrophication associated with biomass production exists. This potential is independent of further biomass utilization and processing options.

7.3.1.4 Acidification Potential

The results in this category range from −450 kg SO_2 equivalents/(Ha·a) for the china reed/PP transport pallet to 45 kg SO_2 equivalents/(Ha·a) for heat produced from wheat versus fuel oil. Relative to their fossil-based counterparts, disposable plates and cups made from potato/corn starch as well as loose fills made from potato starch also reduce acidification considerably. In contrast, the production of heat from wheat or china reed and the generation of RME and especially ETBE from sugar beet (+32 kg SO_2/[Ha·a]) increase acidification compared with their fossil product alternatives. The use of biomass to manufacture materials (mainly polymers and fiber-composite products) generally decreases acidification. In contrast, acidification is increased if biomass is used to replace fossil fuels to produce energy (power/heat) and transportation fuels. The specific acidification potential for energy, fuels, and materials is mainly caused by sulfur and nitrogen oxide emissions from incineration processes. This holds for biobased and conventional products. In addition, ammonia emissions from fertilizer application are a typical source for acidification caused by biobased products. Energy and fuels from biomass score worse than their fossil counterparts because, for these products, the process chains from raw biomass to incineration are typically short, and the production of biomass contributes significantly to the total product-specific acidification potential. In contrast, these emission sources constitute only a smaller portion of the total acidifying emissions for biobased materials because process chains are longer. As a result, there is a higher chance that biobased materials (typically characterized by long process chains) are advantageous over their fossil product alternatives.

7.3.2 SPECIFIC RESULTS BY CATEGORIES OF BIOMASS UTILIZATION

In Figure 7.3[**], the product alternatives analyzed in this chapter are grouped according to three different options of biomass utilization: power and heat production, fuel production, and the manufacture of materials. The data shown represent medians as well as value ranges.

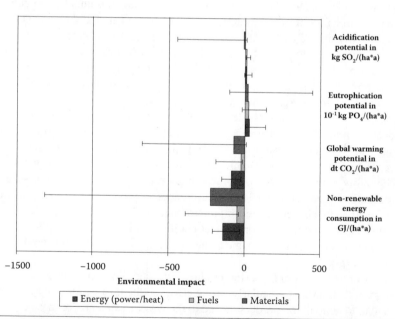

FIGURE 7.3 Medians and value ranges of different biomass utilization options.

[**] For this comparison, we chose the median instead of the mean because the former is more resistant against statistical outlier. This is important here as our sample sizes are comparatively small and the data do not conform to a Gausian distribution. Therefore, the choice of either median or mean to represent the mathematical average has a profound impact on the results.

The results indicate that the use of biomass reduces environmental impacts in the categories of nonrenewable energy consumption and global warming potential, regardless of whether biomass is used to produce energy, fuels, or materials. In contrast, biobased products show, on average, disadvantages compared with their conventional counterparts with respect to eutrophication. In the category acidification, biomaterials score better, whereas bioenergy and biofuels score worse than their fossil counterparts. The error bars in Figure 7.3 display the value ranges of the individual products. The ranges are, in general, large for all three options of biomass utilization. The median values in Figure 7.3 indicate, however, that bioenergy and biomaterials score on average better than biofuels regarding the consumption of nonrenewable energy resources and greenhouse gas emissions (see also Weiss et al. [7]).

7.3.3 Discussion

The findings presented in this chapter are based on the most important LCA studies on biobased energy, fuels, and materials published to date. The studies do, however, vary considerably regarding system boundaries, allocation schemes, waste scenarios, and the treatment of agricultural residues (see also Table 7.1). Errors and uncertainties of data collected for the inventory analysis might also have a significant effect on the results of individual life-cycle assessment studies. Process energy requirements and emissions for conventional products can vary substantially, depending on the data source used for the inventory analysis [2]. Dinkel et al. [11] quantify error ranges of inventory data at 40%, and different allocation schemes are reported to cause deviations up to 90%. When comparing results of the various LCA publications, it was, however, not possible to correct for all of these differences and uncertainties. The conclusions drawn in this chapter can be false if the LCA results differ only within their error margins from each other.

In our analysis, we assume identical functionality of biobased and conventional products. Possible differences regarding lifetime, product quality, and applications are neglected. However, starch-based polymers, for example, do not offer the same range of applications as conventional polymers (e.g., polystyrene) due to their susceptibility to moisture. Also, biofuels might not have exactly the same quality as petrochemical fuels. However, these differences in product properties can usually be assumed not to change the general interpretation of LCA results because, in LCA, it is ensured that the material or product choices fit the respective application (e.g., use of starch polymers for certain packaging applications but not for durable goods like car components).

The available body of LCA studies allows the identification of subsections within the system that have a particularly strong effect on the results. The studies point, in particular, to agricultural production and to the waste treatment. For these two areas, the findings and conclusions about further possibilities for improvement are discussed next.

The utilization of agricultural residues has an important effect on the environmental performance of biobased fuel and materials because, in most cases, only certain parts of the plants are used in production processes.[††] Leaves and stems of crops often remain on the fields and are either (a) taken into account as a substitute for mineral fertilizers or (b) simply excluded from the product system (see Table 7.1). However, if these residues are used for energy generation, the environmental performance of bio–product can improve significantly. Along the life cycle of bio-based polymers, for example, the consumption of nonrenewable energy resources can be reduced by up to 190 GJ/(Ha·a), and greenhouse gas emissions are decreased by up to 15 t CO_2/(Ha·a) [1].

The ecological disadvantages of biobased products in the impact categories of eutrophication and acidification mainly arise from biomass production using conventional agricultural practices. Würdinger et al. [17] have shown in their LCA study of loose fills from wheat starch the high potential of extensive farming practices for reducing environmental impacts compared with conventional farming practices in all four environmental categories. These results show that biobased products do not score *per se* worse than conventional products with respect to eutrophication. Future biomass production should, therefore, aim (a) to improve the management of fertilizer application and (b) to reduce on-site soil erosion. A reduction of eutrophication potential associated with biomass production remains doubtful, however, if farming practices do not change.

Additional environmental benefits can be realized by so-called agricultural intercrops, which are mainly used to cover fields inbetween vegetation periods. These crops can also be used for energy or material purposes [23]. Environmental impacts can be further reduced by choosing the most appropriate crop for each specific purpose and region and by processing it in the most efficient way. Related to environmental risks of biomass production, the introduction of genetically modified crops also has to be taken into account. This is, so far, not included in life-cycle assessment studies, mainly due to lack of scientifically proven insight. Further research on this subject is therefore highly recommended.

Another important factor affecting environmental impacts of both bio–based and conventional materials is the waste-treatment technology chosen. In their LCA study on loose fills, Würdinger et al. [17] state that the differences between the various waste-treatment options for conventional polystyrene loose fills are comparable with the differences between biobased and conventional loose-fill materials, based on a cradle-to-factory-gate comparison. This result is also confirmed by the findings of Dinkel and Waldeck [12] for disposable plates. Therefore, not only replacing fossil feedstocks by renewable ones, but also optimized waste management can effectively reduce environmental impacts. This finding calls for a detailed assessment of all major waste-management options, including landfilling, composting, waste-to-energy facilities, municipal waste incineration, digestion, and recycling [2].

While considerable attention is paid in LCA studies and ongoing research to issues related to agricultural production and waste management (see above), relatively little is still known about the total environmental impact and improvement potentials for the conversion of biobased feedstock to new fuels, materials, and chemicals. This holds, in particular, for biotechnological processes as, for example, applied for the production of PLA and ethanol (Note that all other products shown

---

[†]†† This example is less relevant for biobased energy (power and heat) production because usually the whole plant is harvested and incinerated.

in Figure 7.2 imply conventional chemical conversion, simple blending, or inciner-ation). In view of the great progress being made in biotechnology, major opportu-nities for reducing environmental impacts are expected. To exploit them fully, tech-nological as well as logistical shortcomings need to be overcome. At the same time, new environmental impacts may arise, e.g., from the use of genetically modified organisms in fermentation processes.

The completeness of LCA studies with regard to environmental impacts is not only an issue for future and novel technologies, but also for existing processes. In Section 7.3.1, we discussed four environmental impact categories, i.e., nonrenewable energy use, greenhouse gas emissions, eutrophication, and acidification. The results provide important insight into the environmental performance of bio–based versus conventional products. However, relevant environmental impacts remain excluded from our analysis, such as human and environmental toxicity, ozone depletion, as well as impacts on soil erosion and biodiversity. The reason most studies are limited to just a few indicators is the lack of data as well as the effort and costs related to closing these data gaps. This also holds for processes that were implemented on a large scale. Future research should, therefore, address these environmental impacts to come to a more complete and comprehensive environmental evaluation of bio-based energy, fuels, and materials.

Once available, a conclusion in favor of or against biobased energy, fuels, and materials needs to be drawn. This conclusion often depends on the weighing of the different environmental impacts with respect to a sustainability framework [6]. Weighing requires value judgment about the relative importance of the various impact categories (weighting) is, therefore, subjective to some extent. For this reason, it is excluded from LCA studies [24] and generally seen as the task of economic, political, and environmental decision makers.

The results presented in this chapter generally show good correspondence with the findings of Dornburg et al. [1] and Kaenzig et al. [6], even if these two publi-cations differ from our comparative analysis regarding assumptions and harmoniza-tion steps (see footnote on page 141). This finding is a strong indication of the robustness and reliability of (a) LCA results on biobased products in general and (b) our results presented in this chapter.

7.3.4 Conclusion

In this chapter we compare nonrenewable energy consumption, greenhouse potential, as well as eutrophication and acidification potential of biobased and conventional energy, fuels, and materials based on a functional unit of 1 Ha of agricultural land. Biobased products score generally better than their fossil-based counterparts with respect to nonrenewable energy consumption and global warming potential. The use of biomass is, therefore, suitable for contributing to climate protection, independent of its specific utilization. Biobased products perform worse than conventional ones with respect to environmental eutrophication. Regarding mixed results are obtained, with biomaterials scoring on average better than conventional materials and bioenergy and biofuels scoring worse than their conventional counterparts. Product-specific results can vary considerably from these averages. Decisions in favor of or against a particular option of biomass use can only be made based on individual cases. It is, hence, advisable (a) to assess the environmental performance of the desired bioproduct, (b) to compare the result with its conventional counterpart, and (c) to evaluate the relative advantages and disadvantages by comparison with the findings for other product systems.

The results in this chapter show, that biobased energy, fuels, and materials are neither without any environmental impacts nor are they *per se* more environmental friendly than their conventional fossil-based counterparts. This finding calls for a comprehensive evaluation of the environmental performance of products from renewable resources.

Major disadvantages of biobased products result from agricultural emissions and the leaching of fertilizers. Various studies have shown that the environmental performance of biomass utilization can be increased by (a) adopting extensive cultivation methods and farming practices, (b) using agricultural residues and "intercrops" for energy, fuel, or material production, (c) choosing the most suitable crops for the various biomass utilization options, and finally (d) optimizing end-of-life waste treatment. New production processes, e.g., those involving biotechnology, may offer further substantial improvement potentials, but their realization will require substantial technology development. Potential negative impacts of these technological developments need to be assessed in parallel with the evaluation of these new opportunities.

For nonrenewable energy use and greenhouse gas emissions, which currently represent the most important targets of environmental and industrial policy, the results presented in this chapter show the relevance of biobased energy, fuels, and materials. In Section 7.3.1 of this chapter, the results are presented at the level of products, while the overall contribution to energy savings and emissions reduction will depend, to a large extent, on how the markets for these biobased products develop. Despite expected large growth rates of several biobased products (energy, fuels, and materials), the very large scale of conventional fossil-based products implies that the time required for biobased products to reach significant market shares may be substantial [25]. The emergence of biobased products will depend on physical limitations (availability of land), boundary conditions (fossil fuel prices, environmental legislation, etc.), their technological performance, and on developments in other technological areas. The fact that energy and fuels can be also supplied by renewable resources other than biomass, while raw materials (e.g., feedstock for the chemical industry) cannot, may lead to a focus on biomaterials in view of the limited availability of land. For the moment, we conclude that the accessible LCA studies strongly support the further development of biobased energy, fuels, and materials, which are all still in an infant stage compared with their conventional counterparts.

ACKNOWLEDGMENTS

The authors would like to thank Veronika Dornburg from Utrecht University, The Netherlands; Stefan Bringezu from the Wuppertal Institute for Climate, Environment and Energy, Germany; and Hermann Heilmeier from the Technical University Bergakademie, Freiberg, Germany, for their valuable contributions to the comparison of LCA studies presented in this chapter.

APPENDIX: ABBREVIATIONS AND UNITS

a	year
ABS	acrylonitrile butadiene styrene
con.	conventional farming
dt	decitonne (10^{-1} t)
EPS	expanded polystyrene
ETBE	ethyl tertiary butyl ether
ext.	extensive farming
GJ	gigajoule
HIPS	high-impact polystyrene
kg	kilogram
LCA	life-cycle assessment
LDPE	low-density polyethylene
misc.	miscellaneous
MTBE	methyl tertiary butyl ether
PE	polyethylene
PLA	polylactic acid
PP	polypropylene
PS	polystyrene
RME	rapeseed methyl ester
SME	sunflower methyl ester
t	tonne
TPS	thermoplastic starch

REFERENCES

1. Dornburg, V., Lewandowski, I., and Patel, M., Comparing the land requirements, energy savings, and greenhouse gas emissions reduction of bio-based polymers and bioenergy, *J. Industrial Ecol.*, 7, 93–116, 2004.
2. Patel, M., Bastioli, C., Marini, L., and Würdinger, E., Life cycle assessment of bio-based polymers and natural fibers, in *Biopolymers*, Vol. 10, Wiley-VCH, Weinheim, Germany, 909–952, 2003.
3. Reinhardt, G.A. and Zemanek, G., Ökobilanz Bioenergieträger: Basisdaten, Ergebnisse, Bewertungen, Erich Schmidt Verlag, Berlin, 2000.
4. Murphy, R. and Bartle, I., *Biodegradable Polymers and Sustainability: Insights from Life Cycle Assessment*, The National Non-Food Crop Centre, York, U.K., 2004; available on-line at http://www.nnfcc.co.uk/, accessed 14 April 2005.
5. Müller-Sämann, K.M., Reinhardt, G., Vetter, R., and Gärtner, S., Nachwachsende Rohstoffe in Baden-Württemberg: Identifizierung vorteilhafter Produktlinien zur stofflichen Nutzung unter Berücksichtigung umweltgerechter Anbauverfahren, Forschungsbericht FZKA-BWPLUS, 2002; available on-line at http://www.inaro.de, accessed 12 August 2003.

6. Kaenzig, J., Houillon, G., Rocher, M., Bewa, H., Bodineau, L., Orphelin, M., Poitrat, E., and Jolliet, O., Comparison of the Environmental Impacts of Bio-Based Products, in *Proceedings of the 2nd World Conference and Technology Exhibition on Biomass for Energy, Industry and Climate Protection*, Rome, 10–14 May 2004.

7. Weiss, M., Bringezu, S., and Heilmeier, H., Energie, Kraftstoffe und Gebrauchsgüter aus Biomasse: Ein flächenbezogener Vergleich von Umweltbelastungen durch Produkte aus nachwachsenden und fossilen Rohstoffen, *Zeitschrift Angewandte Umweltforschung*, 15/16, 3–5, 2003/2004.

8. CEN (Comité Européen de Normalisation), ISO 14040, Umweltmanagement — Ökobilanz — Prinzipien und allgemeine Anforderungen, DIN (Deutsches Institut für Normung), Beuth Verlag GmbH, Berlin, 1997.

9. Corbière-Nicollier, T., Gfeller-Laban, B., Lundquist, L., Leterrier, Y., Månson, A.E., and Jolliet, O., Life cycle assessment of bio fibres replacing glass fibres as reinforcement in plastics, *Resources, Conserv. Recycling*, 33, 267–287, 2001.

10. Diener, J. and Siehler, U., Ökologischer Vergleich von NMT und GMT-Bauteilen, *Die Angewandte Makromolekulare Chemie*, 272, 4744, 1–4, 1999.

11. Dinkel, F., Pohl, C., Ros, M., and Waldeck, B., Ökobilanz stärkehaltiger Kunststoffe, Carbotech AG, for Bundesamt für Umwelt, Wald und Landschaft (BUWAL), Bundesamt für Landwirtschaft (BLW) and Fluntera AG, *Schriftenreihe Umwelt*, No. 271/I–II, Bern, Switzerland, 1996.

12. Dinkel, F. and Waldeck, B., Ökologische Beurteilung verschiedener Geschirrtypen mit Empfehlungen, Carbotech AG, Arbeitspapier 4/99, 1999; available on-line at www.kompost.ch, accessed 27 October 2003.

13. Gärtner, S.O., Müller-Sämann, K., Reinhardt, G.A., and Vetter, R., Corn to Plastics: a Comprehensive Environmental Assessment, in *Proceedings of the 12th European Conference on Biomass for Energy, Industry and Climate Protection*, Amsterdam, 17–22 June 2002, Vol. II, Pala, W. et al., Eds., pp.1324-1326, ETH, Florence, 2002.

14. Reinhardt, G.A., Calzoni, J., Caspersen, N., Dercas, N., Gaillard, G., Gosse, G., Hanegraaf, M., Heinzer, L., Jungk, N., Kool, A., Korsuize, G., Lechner, M., Leviel, B., Neumayr, R., Nielsen, A.M., Nielsen, P.H., Nikolaous, A., Panoutsou, C., Panvini, A., Patyk, A., Rathbauer, J., Riva, G., Smedile, E., Stettler. C., Pedersen-Weidema, B., Wörgetter, M., and van Zeijts, H., *Bioenergy for Europe: Which One Fits Best? A Comparative Analysis for the Community*, final report, 2000; Institut für Energie und Umweltforschung, Heidelberg, Germany, available on-line at http://www.ifeu.de, accessed 25 July 2003.

15. Reinhardt, G.A. and Gärtner, S.O., Biodiesel or Pure Rapeseed Oil for Transportation: Which One Is Best for the Environment? Study of IFEU (Institut für Energie und Umweltforschung) GmbH, Heidelberg, in *Proceedings of the 4th International Colloquium Fuels 2003*, Ostfildern, D. and Bartz, W.J., Eds., Technische Akademie Esslingen, Germany, 15–16 January 2003, pp. 111–114.

16. Wötzel, K., Wirth, R., and Flake, M., Life cycle studies on hemp fibre reinforced components and ABS for automotive parts, *Die Angewandte Molekulare Chemie*, 272, 4763, 121–127, 1999.

17. Würdinger, E., Roth, U., Wegener, A., Peche, R., Rommel, W., Kreibe, S., Nikolakis, A., Rüdenauer, I., Pürschel, C., Ballarin, P., Knebel, T., Borken, J., Dethel, A., Fehrenbach, H., Giegrich, J., Möhler, S., Patyk, A., Reinhardt, G.A., Vogt, R., Mühlenberger, D., and Wante, J., Kunststoffe aus nachwachsenden Rohstoffen: Vergleichende Ökobilanz für Loose-fill-Packmittel aus Stärke bzw: Polystrol — Projektgemeinschaft BIfA/IFEU/Flo-Pak, Endbericht 2002 (DBU-Az. 04763), IFEU (Institut für Energie und Umweltforschung) GmbH, Heidelberg; available on-line at http://www.ifeu.de, accessed 29 July 2003.

18. Vink, E.T.H., Rabago, K.R., Glasner, D.A., and Gruber, P.R., Application of life cycle assessment to NaturWorks™ polylactide (PLA) production, *Polymer Degradation Stability*, 80, 403–419, 2003.

19. Estermann, R., Schwarzwälder, B., and Gysin, B., *Life Cycle Assessment of Mater-Bi and EPS Loose Fills*, study prepared by COMPOSTO for Novamont, Olten, Switzerland, 2000.

20. Heyde, M., Ecological considerations on the use and production of biosynthetic and synthetic biodegradable polymers, *Polymer Degradation Stability*, 59, 3–6, 1998.

21. Association of Plastics Manufacturers Europe, *Eco-Profiles of Polymers and Related Intermediates* (about 55 Products), APME, Brussels, 1999; available on-line at http://www.apme.org, accessed 4 April 2005.

22. Katalyse, Institut für angewandte Umweltforschung: Leitfaden Nachwachsende Rohstoffe, C.F. Müller Verlag, Heidelberg, 1998.

23. Karpenstein-Machan, M., Sustainable cultivation concepts for domestic energy production from biomass, *Crit. Rev. Plant Sci.*, 20, 1–14, 2001.

24. CEN (Comité Européen de Normalisation), EN ISO 14042, Umweltmanagement, Ökobilanz: Wirkungsabschätzung, DIN (Deutsches Institut für Normung), Beuth Verlag GmbH, Berlin, 2000.

25. Crank, M., Patel, M., Marscheider-Weidemann, F., Schleich, J., Hüsing, B., and Angerer, G., Techno-Economic Feasibility of Large-Scale Production of Bio-Based Polymers in Europe (PRO-BIP), final report prepared for the European Commission's Institute for Prospective Technological Studies (IPTS), Sevilla, Spain, 2004.

Part III

Technologies for Renewable Energy

8 Hydrogen Production and Cleaning from Renewable Feedstock

Stanko Hočevar

CONTENTS

8.1 INTRODUCTION

A large amount of primary energy stored in hydrocarbon-based solid, liquid, and gaseous fuels is consumed for electricity and heat generation in residential and industrial sectors and for internal-combustion-engine-based transportation. World-wide, 7235 million t of oil equivalent (MTOE), or 75% of energy use, in 1997 was based on fossil fuels. Per capita CO_2 emissions in that year amounted to 1.13 t of carbon [1]. In the 20th century, the human population has quadrupled, while the

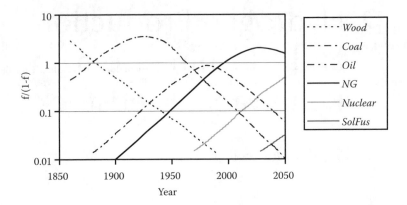

FIGURE 8.1 World primary energy substitution (f = market share). (From Marchetti, C., *Society of a Learning System: Discovery, Invention and Innovation Cycles Revisited*, International Institute for Applied System Analysis, Pub. RR-81-29, Laxenburg, Austria, 1981. With permission.)

primary power consumption has increased 16-fold. The concentration of atmospheric CO_2 has increased from about 275 to about 370 ppm. If nothing changes in our consumption pattern of energy, the CO_2 concentration will pass 550 ppm this century [2]. However, atmospheric CO_2 stabilization targets as low as 450 ppm could be needed to forestall coral reef bleaching, thermohaline circulation shutdown, and sea level rise from disintegration of the West Antarctic ice sheet [3]. These targets require immediate action on all levels of human life, especially in the field of research and development of new, environmentally benign, and highly efficient energy conversion technologies. There is but one extenuating circumstance in this seemingly suicidal human activity, which is, regrettably, not the result of reflection over the situation, but rather the result of thirst for fuels with ever-higher energy content. Figure 8.1 illustrates the data of primary energy substitution in the world since the time of the Industrial Revolution together with somewhat speculative prediction until the year 2050 [4]. A pattern in these fuel substitutions has been the replacement of fuels with low hydrogen-to-carbon (H/C) ratio with progressively higher H/C ratio fuels (see Table 8.1) [5]. Hidden behind this pattern is that technologies, not fuels, compete in the marketplace. Technologies win because they are better. "Better," a deceptively simple word, is a more encompassing concept than, for example, "cheaper." "Better," in this case, means technologies that can "squeeze" higher specific power and higher power density per unit mass and unit volume of primary energy source.

Any energy pathway can be represented by five links from sources to the services that require the input of energy. These five links are depicted on the upper part of Figure 8.2. To make the nomenclature of this five-link architecture clear, the lower part in Figure 8.2 gives some examples that are covered by the headings on the upper part. The real comparison between different energy pathways is not just simple comparison of production costs of energy vectors (e.g., hydrogen vs. gasoline), but a comparison between the life cycles of possible energy pathways needed to deliver certain services (e.g., transportation). A part of life-cycle analysis (LCA) for service

TABLE 8.1
Basic Characteristics of Fossil Fuels and Hydrogen

Fuel Type	Hydrogen (%)	Energy Content (Btu/lb)	Particulates (lb/Btu)	Carbon Dioxide (Rel. %)
Dry wood	5	6,900	5.22	100
Coal	50	10,000	5.00	31
Oil	67	19,000	0.18	21
Natural gas	80	22,500	<0.01	15
Hydrogen	100	61,000	0.0	0

Source: Cannon, J.S., *Harnessing Hydrogen: The Key to Sustainable Transportation*, New York: Inform, 1995; available on-line at http://www.ttcorp.com/nha. With permission.

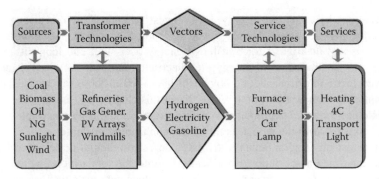

FIGURE 8.2 General presentation of energy pathways.

like transportation is called, very indicatively, "well-to-wheel" analysis, and it comprehends the amount of energy required by certain traction system, e.g., for hydrogen-powered electric vehicles, versus the amount of energy required by another traction system, e.g., diesel-powered internal-combustion-engine vehicles, and all the costs associated with production and distribution.

The time sequence (I to IV) of competing technologies for electricity generation based on chemical fuels is a good illustration for the concept of "better" technology, and it simultaneously gives an idea about different possible energy pathways for grid electricity production from chemical fuels (Figure 8.3).

The "natural" tendency for new technologies to use fuels with higher energy content thus coincides with the fuels having higher H/C ratio. One can eventually arrive at the following conclusion: the most potent fuel among the "chemical" fuels is hydrogen. Fortunately, it is the most abundant element on Earth and in the universe, and it is also the cleanest fuel: the product of hydrogen combustion is water.

Pure hydrogen as the strongest chemical fuel gives a possibility to suppress the CO_2 and particulate emissions almost completely (depending on the process of hydrogen production) and to lower the NO_x emissions (depending on the energy conversion system used).

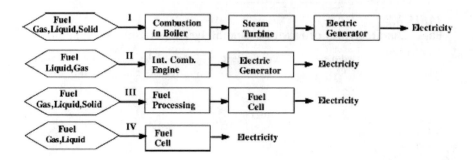

FIGURE 8.3 Competing technologies for electricity generation. Their development and market penetration through time (I-IV) dictated the type of fuel that these technologies preferably use. (From Song, C., *Catal. Today*, 77, 17, 2002. With permission.)

However, hydrogen is not only the strongest "chemical" fuel, it also serves as an *energy carrier* (vector). Namely, one can produce hydrogen in a "thousand and one" ways and then use it for transfer of energy over short and long distances or for on-site transformation of energy into electricity and heat. Among the methods for hydrogen production, we are obviously interested in those that are sustainable or, even better, renewable. In the remainder of this chapter, we shall describe some technologies for hydrogen production from sustainable and from renewable primary energy sources, with the emphasis on hydrogen cleaning processes.

Since our civilization depends to a large degree on the power delivered by heat engines and, in transportation, specifically by internal-combustion engines (ICE), the primary energy sources are predominantly "chemical" fuels that can burn. The efficiencies of the current thermal electric power generation plants are about 35 to 40% [6], and the well-to-wheels efficiency of the current four-passengers car is between 14 and 22% [7]. These numbers reveal that we are using our natural resources, fossil fuels, extremely inefficiently; most of the chemical energy stored in fossil fuels is used to produce waste heat and proportionally large amounts of greenhouse gases.

Among the energy converters, fuel cells are unique direct-energy-conversion devices capable of converting the energy of chemical reactions into electricity with the highest maximum feasible efficiency of 90% [8]. The value of deriving electric current directly from the chemical reactions of fuels was recognized well before electricity became a commodity sold by power utilities. The first investigations go back to 1839 and Sir William Grove. It was not until the 1960s, however, that fuel cells were employed in a practical capacity. NASA had used them first to provide electric power on board the Gemini space mission. Afterward, the fuel cells were used on board nearly every space mission, regardless of the country having space technology. Steady progress over the last 40 years has made it possible for fuel cells to start displacing combustion from its central technological role. Nowadays, we have experienced the use of fuel cell technology in demonstration projects encompassing electric cars and buses and mobile and stationary use in the span of power from a few tens of watts to a few megawatts. Fuel cell technology is now at the dawn of commercialization [6].

TABLE 8.2
Fuel Cells Ordered According to Operating Temperature and Type of Electrolyte

Type	Electrolyte	Operating Temperature (°C)
AFC	KOH solution	50–90
PEMFC	Solid polymer	50–95
PAFC	H_3PO_4	190–210
MCFC	Li_2CO3/K_2CO_3	630–700
SOFC	Y-stabilized ZrO_2	900–1000

Most fuel cells being developed consume either hydrogen or fuels that have been preprocessed into a suitable hydrogen-rich form. Some fuel cells can directly consume sufficiently reactive fuels such as methane, methanol, carbon monoxide, or ammonia, or they can process such fuels internally. Different types of fuel cells are most suitably characterized by the electrolyte that they use to transport the electric charge and by the temperature at which they operate. This broad classification is presented in Table 8.2.

Further, we shall concentrate on the low-temperature proton-exchange-membrane fuel cells (PEMFC) as a most representative H_2/O_2 or H_2/air fuel cell. We shall do this deliberately, since PEMFCs, working at low temperature, have high thermodynamic equilibrium potential, and therefore they can reach high open-circuit voltage and potentially high efficiency in energy conversion. Low working temperature also poses fewer restrictions on the construction materials.

A PEM fuel cell consists of a negatively charged electrode (cathode), a positively charged electrode (anode), and a thin proton-conducting polymer electrolyte membrane. Hydrogen is oxidized on the anode, and oxygen is reduced on the cathode. Protons are transported from the anode to the cathode through the electrolyte membrane, and electrons are carried to the cathode over an external circuit. On the cathode, oxygen reacts with protons and electrons, forming water and producing heat. Both the anode and the cathode contain a catalyst to speed up the electrochemical processes. The schematic construction and both half-cell reactions are depicted in Figure 8.4.

The electricity and heat are produced by the cathode reaction. Theoretically, the Gibbs energy of the reaction is available as electrical energy, and the rest of the reaction enthalpy is released as heat. In practice, a part of the Gibbs energy is also converted into heat via the loss mechanisms.

One unit cell produces, roughly, 2 kA/m² at a cell voltage of 0.8 V. This implies that we have to couple several cells in series to increase the voltage to usable levels. This is done in a fuel cell stack, where bipolar plates serve both as separators and current conductors between adjacent anodes and cathodes. These bipolar plates also serve as gas distributors to the electrodes through channels in their structure. In addition, the edges of the plates serve as manifolds for the fuel cell stack. Figure 8.5 shows the principle behind a bipolar stack.

Although PEMFCs have been in use for more than 40 years in diverse advanced applications (space, military), further accelerated development of this technology is

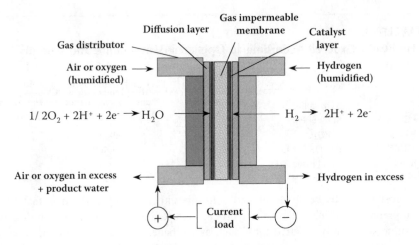

FIGURE 8.4 Schematic representation of the PEMFC cross-section.

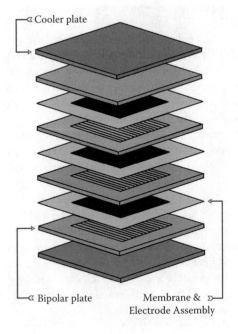

FIGURE 8.5 Bipolar configuration of the stack.

yet to come within different areas of its application. According to a recent report, the U.S. market for stationary use of fuel cells in the electric power generation (e.g., remote supply, grid support, cogeneration) will commercialize first, followed by mobile use of fuel cells in portable power (e.g., cellular phones, handheld computers, camcorders) and in automobiles (not before 2011) [9]. In all these areas of application, the PEMFC has an outstanding role. Obviously, the most demanding market

TABLE 8.3
Well-to-Wheel Analysis of Vehicle Traction Systems Efficiency

System Performance [a]	Gasoline ICE	Fuel Cell System Type			
		Gasoline ATR	Ethanol ATR	Methanol SR	Hydrogen cH_2
Weight, W/kg	560	320	290	290	420
Efficiency at 100% load, LHV	35%	39%	40%	46%	58%
Efficiency at 20% load, LHV	25%	36%	37%	43%	62%

Note: ICE = internal combustion engine; ATR = autothermal reforming; SR = steam reforming; cH_2 = compressed H_2; LHV = lower heating value.

[a] Includes engine/fuel cell stack, radiator, fuel processor, fuel, and fuel tank.

for fuel cells is their use in transportation. The technology of the internal-combustion engine (ICE) is very sophisticated, developed through more than 100 years. To penetrate the market with a new fuel-cell-based electric engine is a challenging but enormously complicated task. This is at the same time the application area where most of the future advanced research and development efforts will be concentrated. Table 8.3 illustrates the well-to-wheel analysis of efficiencies for differently fueled polymer-electrolyte-membrane fuel cell (PEMFC) vehicle traction systems in comparison with the gasoline-fueled internal-combustion engine [10].

Of the fuel cell systems, direct hydrogen has the highest power density and efficiency. This system has 1.5 to 2 times higher efficiency than the PEMFC systems fueled with reformate fuels.

PEMFC is, in fact, a superb example of a catalytic membrane reactor performing a variety of reactions and separations. To fuel it, we need hydrogen. Distributed combined heat and power (CHP) generation based on today's common (logistic) fuels demands that hydrogen be produced either on-site (for stationary applications) or onboard (for mobile applications). A general scheme for PEMFC-grade hydrogen production from sustainable energy sources (biomass, organic waste, ethanol) or from fossil fuels (coal, oil, natural gas) is represented in Figure 8.6. As can be seen, nearly all processes in the chain are catalytic.

On-site and onboard hydrogen production processes are viewed as a transitional solution until new H_2 production and H_2 storage technologies are developed. On the other hand, the catalytic processes for hydrogen production from hydrocarbons have been practiced in the chemical industry for many years. Therefore, every step forward in the formulation of the new catalysts and the design of new reactors and processes are very demanding because of pressure for time to enable fuel cell technology for market penetration and because of the relatively long and rich history of each and all of the catalytic processes involved in this technology. Some of the classical processes for hydrogen production and cleaning are discussed elsewhere in this book (Chapter 11 by N. Hickey, P. Fornasiero, and M. Graziani). In the present chapter,

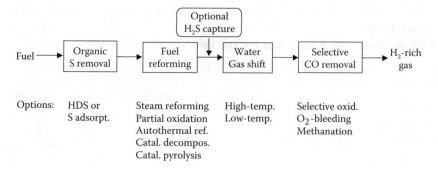

FIGURE 8.6 Steps and options for on-site and onboard processing of renewable and fossil fuels.

the discussion will therefore be limited on some possible routes for hydrogen production from sustainable and renewable feedstock and on some new catalysts for hydrogen cleaning in downstream processes.

8.2 GENERAL PATHWAYS FOR HYDROGEN PRODUCTION

A simple general scheme for hydrogen production is given in Figure 8.7, while Table 8.4 lists the major hydrogen production processes from various feedstocks, convenient energy sources to drive these processes, and the emissions that each process produces [11]. It can be quickly realized that the only feedstock from which hydrogen can be produced without emissions is water. However, one has quite a number of possible processes to produce hydrogen from water. Further on, we shall focus on the thermochemical processes, since these are of immediate interest due to our technological heritage based on exploitation of fossil fuels, which will continue for several decades from now.

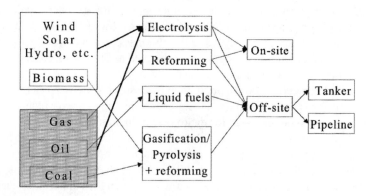

FIGURE 8.7 General scheme for hydrogen production from fossil primary energy sources (shadowed area) and from renewable and sustainable sources.

TABLE 8.4
Major Hydrogen Production Processes

Primary Method	Process	Feedstock	Energy	Emissions
Thermochemical	Reforming	Natural gas	HT steam	CO_2
	HT water splitting	Water	HT heat from gas cooled nuclear reactors	none
	Gasification	Coal, Biomass	HT&P steam and O_2	CO_2, SO_2, NO_x
	Pyrolysis	Biomass	Moderate HT steam	CO_2
Electrochemical	Electrolysis	Water	Electricity from wind, solar, hydro or nuclear	None
	Electrolysis	Water	Electricity from coal or gas	CO_2, SO_2, NO_x
	Photoelectrochemical	Water	Direct sunlight	None
Biological	Photobiological	Water in algae strains	Direct sunlight	None
	Anaerobic digestion	Biomass	Ht heat	CH_4, NH_3
	Fermentative microorganisms	Biomass	HT heat	CO_2

Today, hydrogen is produced as a chemical reactant in processes like ammonia synthesis, hydrogen peroxide synthesis, and various hydrogenation processes in the petrochemical industry, pharmaceutical industry, fine organics synthesis, etc. The exact quantities of hydrogen production are difficult to estimate, since hydrogen represents an intermediate reactant in the vast majority of cases. There are, however, some estimates regarding the resources from which hydrogen is produced. Table 8.5 gives one such estimate. Obviously, over 95% of hydrogen is currently derived from fossil fuels, and only about 5% is generated from sustainable or renewable resources. This situation will not change overnight, of course, since the processes (technology)

TABLE 8.5
Hydrogen Sources Today and Their Percentages

H_2 source	Percentage
Natural gas	48
Oil	30
Coal	18
Water splitting	4
Biomass	low

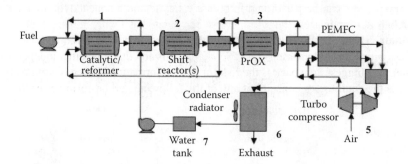

FIGURE 8.8 Fuel processing scheme for a PEMFC system.

for hydrogen production are based on fossil feedstock. Substantial research and development efforts in the near future must be engaged to develop new technologies adapted on sustainable and renewable feedstock.

8.2.1 THERMOCHEMICAL PATHWAYS FOR HYDROGEN PRODUCTION

We shall discuss in this section only conventional processes and catalysts for hydrogen generation and their limitations with respect to PEMFC technology to demonstrate how different are specific technical demands for hydrogen production and cleaning in this case.

Fuel processing system for a PEMFC system is illustrated in Figure 8.8. The major components of the system are:

Catalytic reformer
Shift reactor
PrOx reactor
Fuel cell stack
Turbocompressor to provide compressed air to the fuel cell stack, the reformer, and PrOx reactor
Condenser/radiator to recover water
Water-storage tank

The catalytic reformer in Figure 8.8 should be understood in a broader sense, including all the options: partial oxidation, steam reforming, and the combination of these — autothermal reforming.

Partial oxidation is based on extreme rich fuel combustion (low air/fuel ratios), where the following reactions may occur:

$$C_xH_yO_z + (x - z/2)O_2 \rightarrow y/2\,H_2 + xCO_2$$

$$C_xH_yO_z + (x/2 - z/2)O_2 \rightarrow y/2\,H_2 + xCO \qquad (8.1)$$

$$C_xH_yO_z + (x + y/4 - z/2)O_2 \rightarrow y/2\,H_2O + xCO_2$$

The process can be performed in both a catalytic and a noncatalytic manner. If a catalytic system is used, the reformer can be operated at a much lower temperature and the heat can be supplied directly into the catalyst bed. The advantage with this process is that it is rather insensitive to contaminants and that it is rather independent of fuel. The biggest drawback is the risk for carbon formation if the gas composition exceeds the equilibrium constant of any of the following carbon-forming reactions :

$$2CO(g) \rightleftharpoons CO_2(g) + C(s)$$

$$CH_4(g) \rightleftharpoons 2H_2(g) + C(s)$$

$$CO(g) + H_2(g) \rightleftharpoons H_2O(g) + C(s)$$

$$CO_2(g) + 2H_2(g) \rightleftharpoons 2H_2O(g) + C(s)$$

(8.2)

The first reaction (Boudouard equilibrium) will favor carbon formation at lower temperatures than encountered in partial oxidation. The hydrogen concentrations that can be theoretically attained when using different fuels in a partial oxidation process are represented in Table 8.6, together with values obtained experimentally.

Steam reforming can be represented with the following equation:

$$C_xH_yO_z + (2x - z)H_2O \rightarrow (y/2 + 2x - z)H_2 + xCO_2$$

(8.3)

The biggest advantage with steam reforming is the high concentrations of hydrogen that can be achieved. For instance, with methanol as feedstock, 75 vol% hydrogen can be obtained at stoichiometric conditions and total conversion. However, steam reforming is an endothermic process, which lowers otherwise high system efficiency.

TABLE 8.6
Concentration of Hydrogen in the Product Gas Obtained with Different Fuels

	H_2 Concentration, % (dry)			
Fuel	Theory	Experiment	H_2 selectivity [a] (%)	Temperature (°C)
Methanol	70	64	91	450
Ethanol	71	62	88	580
i-Octane	68	60	88	630
Cyclohexane	67	61	91	700
2-Pentene	67	58	88	670
Toluene	61	50	82	660

[a] Selectivity = (% H_2 in product gas) × 100/(% H_2 theoretically possible). For example, with C_8H_{18} and $x = 4$, the theoretical hydrogen percentage is 68%. Thus, 60% hydrogen in the product is equal to a selectivity of 88%.

Copper-based catalysts have been mostly used for ethanol or methanol steam reforming. For methanol, $CuO/ZnO/Al_2O_3$ catalyst is usually used at temperature between 180 and 250°C. When unreacted methanol is present, the rate of the water–gas-shift (WGS) reaction is negligible. As the conversion of methanol approaches 100%, the rate of the WGS reaction becomes significant, and the CO concentration in the reformer product gas approaches equilibrium. The reforming of ethanol proceeds better over CuO/ZnO catalyst at temperatures above 300°C.

Autothermal reforming is a combination of steam reforming and partial oxidation that, in theory, can be totally heat balanced. The reaction can be represented by the following equation:

$$C_xH_yO_z + m\left(O_2 + 3.76N_2\right) + \left(2x - 2m - z\right)H_2O \rightarrow$$
$$\left(2x - 2m - z + y/2\right)H_2 + xCO_2 + 3.76mN_2$$

(8.4)

where m is an oxygen/fuel ratio. The concentration of hydrogen in the product gas is

$$\left\{\left(2x - 2m - z + y/2\right)/\left(x + \left(2x - 2m - z + y/2\right) + 3.76m\right)\right\}\cdot 100$$

(8.5)

and the reaction enthalpy is calculated as

$$H_r = x \cdot H_{f,CO_2} - \left(2x - 2m - z\right)\cdot H_{f,H_2O} - H_{f,fuel}$$

(8.6)

Table 8.7 gives the ratio of oxygen/fuel at which the reaction is heat balanced (thermoneutral).

TABLE 8.7
Calculated Thermoneutral Ratios for Oxygen/Fuel (m_{O2}) and Theoretical Yield

$C_xH_yO_z$	x	y	z	ΔHr, fuel (kcal/mol)	y/2z	M_{O2} ΔHr = 0	Efficiency (%)
Methanol CH_3OH	1	4	1	-57,1	2	0,230	96,3
Methane CH_4	1	4	0	-17,9	2	0,443	93,9
Iso-Octane C_8H_{18}	8	18	0	-62,0	1,125	2,947	91,2
Gasoline $C_{7,3}H_{14,8}O_{0,1}$	7,3	14,8	0,1	-52,0	1,014	2,613	90,8

Power density, specific power, dynamic response, cost, and consumer safety are obviously critical considerations when developing a fuel processor [12]. New technologies have to be developed to meet these requirements, since the traditional ones have several limitations. Some of them are [13]:

- Current steam reforming catalysts based on Ni are extremely sulfur sensitive and deactivate considerably in the presence of traces of sulfur.
- The hydrodesulfurization (HDS) process operates at pressure highly exceeding the pressure of natural gas available in existing infrastructure.
- Ni-based reforming catalysts are pyrophoric; they will sinter if exposed to air, and they represent a fire hazard for consumers.
- Steam reforming is an endothermic process that requires complicated heat management of the system.
- High- and low-temperature water–gas-shift (WGS) catalysts based on Fe and Cu, respectively, require slow and carefully controlled activation procedure. After reduction, they are highly reactive toward air and can be a fire hazard to the consumer.
- Methanation of CO requires removal of CO_2 due to the highly exothermic competitive methanation.
- Ni-based methanation catalysts are also pyrophoric.
- Pressure-swing adsorption requires high pressure.
- Industrial H_2 production plants operate at steady state. They were not designed for numerous start-ups and shutdowns nor for the cycling in load. The catalyst and other reformer materials would be chemically and physically damaged.

The catalytic steam-reforming process operates at low space velocity (gas hourly space velocity [GHSV] between 3000 and 8000 h^{-1}) owing to slow kinetics. Although these conditions are not desirable for transient duty cycles, this process gives the highest yield of hydrogen compared with partial oxidation and autothermal reforming processes. Ni-based catalysts are cost effective and commercially available, but besides being pyrophoric, they are prone to coke formation at lower H_2O/C ratios. Commercial Cu-based methanol steam-reforming catalysts deactivate when exposed to liquid water during shutdown mode, and they are also pyrophoric. Recently, the researchers from Tsukuba Research Center in Japan have tested a new nanostructured cerium oxide-based Cu catalyst in steam reforming of methanol. They have found that this catalyst, containing 3.8 wt% Cu, gives higher methanol conversion (53.9%) than the Cu/ZnO (37.9%), Cu/Zn(Al)O (32.3%), and Cu/Al$_2$O$_3$ (11.2%) catalysts containing the same amount of Cu [14, 15].

For mobile applications, the most suitable reforming technology appears to be autothermal reforming (ATR) because of the adiabatic design permitting a compact smaller reactor, which has a small pressure drop. The design combines a highly exothermic partial-oxidation reaction, which provides the energy for the endothermic steam reforming. A new generation of natural-gas ATR reactor design is based on the overlapped reaction-zone concept: the bottom wash-coat layer with Pt/Rh steam-reforming (SR) catalyst is covered with the Pt/Pd partial oxidation (CPO) catalyst.

The heat released in the CPO layer is consumed by the SR reactions immediately without going through any heat-transfer barriers [13]. Another, efficient radial-flow ATR reactor, which uses Cu/SiO_2 and Pd/SiO_2 for CPO and SR reactions, was used for methanol reforming [16].

The reformate gas contains up to 12% CO for SR and 6 to 8% CO for ATR, which can be further converted to H_2 through the water–gas-shift reaction (WGSR). The shift reactions are thermodynamically favored at low temperatures. The equilibrium CO conversion is 100% below 200°C. However, the kinetics are very slow, requiring space velocities less than a few thousand per hour. The commercial Fe-Cr high-temperature-shift (HTS) and Cu-Zn low-temperature-shift (LTS) catalysts are pyrophoric and therefore impractical and dangerous for fuel cell applications. A Cu/CeO_2 catalyst was demonstrated to have better thermal stability than the commercial Cu-Zn LTS catalyst [17]. However, it had lower activity and had to be operated at higher temperature. New catalysts are needed that will have higher activity and tolerance to flooding and sulfur.

The gas at the outlet of the WGSR still contains from 0.1 to 1.0% of CO, depending on operating conditions. In the last step of hydrogen production for low-temperature fuel cells, the CO concentration has to be reduced to a minimum. The most effective mechanism for CO removal in PEMFC-grade H_2 production is selective oxidation. Because of the high ratio of H_2 to CO (>>100:1) at the outlet from the LTS reactor, the oxidation catalyst has to be highly selective. The process is therefore called selective oxidation or preferential oxidation (PrOx). The process runs in the temperature window between 80°C, the PEMFC working temperature, and about 200°C, the LTS reactor working temperature. The PrOx reactor must run at the varying flow. This poses additional demand on catalyst selectivity. Pt-based PrOx catalysts, for instance, will produce CO by the reverse WGSR at longer residence time, since the oxygen is consumed in the first part of the catalyst bed. Besides, the cost of Pt is high. The CO oxidation over this catalyst is a multistep process, commonly obeying Langmuir-Hinshelwood kinetics for a single-site-competitive mechanism between CO and O_2. An optimum range of O_2/CO ratio is required to obtain proper balance of adsorbed CO and adsorbed O_2 on adjacent sites. However, pure precious metals lack the selectivity that is required for PrOx. Recently, we have developed a nanostructured $Cu_xCe_{1-x}O_{2-y}$ catalyst that is highly selective, active, and stable at given reaction conditions [18]. We shall describe the performance of this catalyst in the last section of this chapter.

8.2.2 HYDROGEN PRODUCTION FROM SUSTAINABLE AND RENEWABLE FEEDSTOCK

Although hydrogen is currently derived from nonrenewable coal, oil, and natural gas, it could in principle be generated from sustainable and renewable sources such as biomass and water.

Biomass has the potential to accelerate the realization of hydrogen as a major fuel of the future. Since biomass is sustainable and consumes atmospheric CO_2 during growth, it can have a small net CO_2 impact compared with fossil fuels. However, hydrogen from biomass has major challenges. There are no completed

TABLE 8.8
Composition of Common Fuels

	Coal	Oil	Methane	Wood
LHV (GJ/t)	28	42	55	20
C (wt%)	85	85	75	49
H (wt%)	6	15	25	6
O (wt%)	9	0	0	45

technology demonstrations. The yield of hydrogen is low from biomass, since the hydrogen content in biomass is low to begin with (Table 8.8), and the energy content is low due to the 40% oxygen content of biomass.

In steam reforming of biomass, the energy content of the feedstock is an inherent limitation of the process, since over half of the hydrogen from biomass comes from spitting water in the steam reforming reaction. The yield of hydrogen in a steam reforming process as a function of oxygen content is shown in Figure 8.9. The low yield of hydrogen on a weight basis is misleading, since the energy conversion efficiency is high. For example, the steam reforming of bio-oil at 825°C with a five-fold excess of steam demonstrated in the laboratory has an energy efficiency of 56% [19].

However, the cost of growing, harvesting, and transporting biomass is high. Thus, even with reasonable energy efficiencies, it is not presently economically competitive with natural gas steam reforming for stand-alone hydrogen without the advantage of high-value coproducts. Additionally, as with all sources of hydrogen, production from biomass will require appropriate hydrogen storage and utilization systems to be developed and deployed.

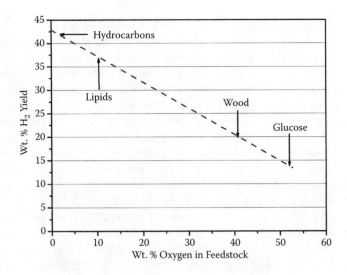

FIGURE 8.9 Theoretical yield of H_2 as a function of the oxygen content in the feed.

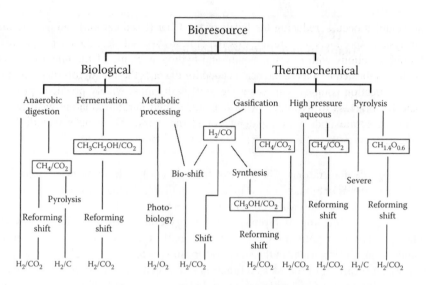

FIGURE 8.10 Pathways from biomass to hydrogen. Storable intermediates are shown in boxes.

Biomass conversion technologies can be divided into two categories: (a) direct production routes and (b) conversion of storable intermediates. Direct routes have the advantage of simplicity. Indirect routes have additional production steps, but have an advantage in that there can be distributed production of the intermediates, minimizing the transportation costs of the biomass. The intermediates can then be shipped to a central, larger-scale hydrogen production facility. Both classes have thermochemical and biological routes. Figure 8.10 shows a "tree" of possible technologies.

A third area of hydrogen from biomass is metabolic processing to split water via photosynthesis or to perform the shift reaction by photobiological organisms. The photobiological production of hydrogen is an area of long-term research and is not discussed here. The use of microorganisms to perform the shift reaction is of great relevance to hydrogen production because of the potential to reduce carbon monoxide levels in the product gas far below the level attained using water–gas-shift catalysts and, hence, eliminate final CO scrubbing for fuel-cell applications. However, this is also a long-term research goal and is not covered in this chapter.

Among the direct processes, we shall mention here only gasification of biomass. It is a two-step process in which a solid fuel (biomass or coal) is thermochemically converted to a low- or medium-energy-content gas. Natural gas contains 35 MJ/N·m^3. Air-blown biomass gasification results in approximately 5 MJ/m^3; oxygen-blown results in 15 MJ/m^3. In the first reaction, pyrolysis, the dissociated and volatile components of the fuel are vaporized at temperatures as low as 600°C. Included in the volatile vapors are hydrocarbon gases, hydrogen, carbon monoxide, carbon dioxide, tar, and water vapor. Because biomass fuels tend to have more volatile components (70 to 86% on a dry basis) than coal (30%), pyrolysis plays a larger role in biomass gasification than in coal gasification. Gas-phase thermal cracking of the volatiles occurs, reducing the levels of tar. Char (fixed carbon) and ash are the pyrolysis by-products that are not vaporized. In the second step, the char is gasified through reactions with oxygen, steam, and hydrogen. Some of the unburned char can be combusted to release the heat needed for the endothermic pyrolysis reactions.

Gasification coupled with water-gas shift is the most widely practiced process route for biomass to hydrogen. Thermal, steam, and partial-oxidation gasification technologies are under development around the world. Feedstocks include both dedicated crops and agricultural and forest-product residues of hardwood, softwood, and herbaceous species.

Thermal gasification is essentially high-severity pyrolysis, although steam is generally present:

$$\text{Biomass} + \text{Energy} \rightarrow CO + H_2 + CH_4 + \ldots \tag{8.7}$$

By including oxygen in the reaction gas, the separate supply of energy is not required, but the product gas is diluted with carbon dioxide and, if air is used to provide the oxygen, then nitrogen is also present:

$$\text{Biomass} + O_2 \rightarrow CO + H_2 + CO_2 + \text{Energy} \tag{8.8}$$

Other relevant gasifying processes are bubbling fluid beds and the high-pressure, high-temperature slurry-fed process.

All of these gasifier examples will need to include significant gas conditioning, including the removal of tars and inorganic impurities and the conversion of CO to H_2 by the water–gas-shift reaction:

$$CO + H_2O \leftrightarrow CO_2 + H_2 \tag{8.9}$$

Significant attention has been given to the conversion of wet feedstocks by high-pressure aqueous systems. This includes the super critical gasification-in-water approach as well as the super critical partial-oxidation approach.

Among the processes for conversion of biomass via storable intermediates, two recently developed processes deserve special attention. The first one is based on pyrolysis with a special temperature profile that results in a carbon char with an affinity for capturing CO_2 through gas-phase reaction with mixed nitrogen-carrying nutrient compounds within the pore structures of the carbon char. This provides a high-added-value by-product that significantly increases the economics of the whole process. The patent-pending process is particularly applicable to fossil-fuel power plants as it also removes SO_x and NO_x, does not require energy-intensive carbon dioxide separation, and operates at ambient temperature and pressure. The method of sequestration uses existing farm-fertilizer-distribution infrastructure to deliver a carbon that is highly resistant to microbiological decomposition. The physical structure of carbon material provides a framework for building an NPK fertilizer inside the pore structure and creating a physical slow-release mechanism of these nutrients.

The complete process produces three times as much hydrogen as it consumes, making it a net energy producer for the affiliated power plant [20].

The second process is autothermal reforming of ethanol to produce hydrogen. Ethanol is a well-known gasoline fuel additive. However, a significant fraction of its production cost comes from the need to remove all water. When converting ethanol to hydrogen by autothermal reforming, it is possible in principle to consume less than 20% of the energy content of the sugar, first forming ethanol by fermentation and then reacting it to H_2. The efficiency of these processes for a fuel cell suggests that it may be possible to capture over 50% of the energy from photosynthesis as electricity in an economical chemical process that can be operated at large or small scales [21].

8.3 HYDROGEN CLEANING PROCESSES

The product composition from the fuel reformer generally consists of 35 to 40% H_2 and 6 to 10% CO balanced with H_2O, CO_2, and N_2 [22]. The CO is further reduced to 2 to 3% by high-temperature water–gas-shift reaction (HT-WGS) and then down to less than 0.5% CO with low-temperature water–gas-shift reaction (LT-WGS). It is not possible to reduce the concentration of CO down to a few ppm with LT WGS because of equilibrium constraints. This has to be done by selectively removing CO in the last step. However, with the recent development of high-temperature PEMFC, capable of working at temperatures higher than 423 K, the need for the last step becomes obsolete, since the anode noble metal-based catalysts can tolerate much higher concentrations of CO (usually more than 2%) in the hydrogen fuel. This possibility, however, raises a need to develop a completely new concept of fuel reformer, which would comprise only two stages: a fuel reforming stage and a water–gas-shift stage. To accomplish the development of a compact fuel processing system, the WGS catalyst has to be improved. One possible solution in this direction is discussed in the next section.

Various technologies have been investigated to reduce the concentration of CO in the fuel gas exiting the shift reactor to 10 ppm or less to feed low-temperature PEMFC with reformed fuel. Among the candidates are membrane separation, methanation, and preferential CO oxidation (PrOx). For membrane separation, Pd alloy membranes can effectively remove CO from the fuel gas, but such membranes require a large pressure difference and a high temperature that reduces system efficiency. For methanation, CO reacts with H_2 to generate methane and water; however, the amount of H_2 required is three times the amount of CO removed. For PrOx, a small quantity of air is bled into the fuel gas, and CO is selectively oxidized to CO_2 over H_2 using supported and promoted noble metal catalysts such as Pt, Rh, or Ru. For onboard fuel processing, PrOx is the preferred method because of the lower parasitic system load and energy requirement compared with membrane separation and methanation [23]. The last section in this chapter gives an example of a new nonpyrophoric catalyst for PrOx process.

8.3.1 WATER–GAS-SHIFT REACTION

The main obstacles for the use of hydrogen as a fuel for proton-exchange-membrane fuel cells (PEMFC) in portable and mobile applications are in low hydrogen mass fraction and low volumetric energy density of the existing hydrogen storage technologies. One of the possibilities to overcome the problem of nonexistent large-scale hydrogen production, storage, and distribution infrastructure is to produce hydrogen from logistic fuels, for which infrastructure already exists, by onboard (or on-site) reforming. Development of a compact fuel reformer for onboard hydrogen production from fossil (methane, gasoline, etc.) or renewable fuels (methanol, ethanol, etc.) demands an integrated approach in which four classical industrial processes for hydrogen production, namely, fuel reforming (FR), high-temperature water–gas-shift reaction (HT-WGS), low-temperature water–gas-shift reaction (LT-WGS), and preferential oxidation of carbon monoxide (PrOx) would be designed in such a way as to increase the specific power (kW/kg), power density (kW/m^3), and dynamics (low power to peak power transient time) of the fuel processing system.

The water–gas-shift reaction,

$$CO + H_2O \underset{k_b}{\overset{k_f}{\rightleftharpoons}} CO_2 + H_2 \; \Delta H = -41.2 \text{ kJ/mol} \tag{8.10}$$

is thermodynamically favored at low temperature, and the kinetics over existing catalysts are so slow (space velocities below 2000 h^{-1}) that the shift reactor alone would occupy 50% of the entire fuel processor volume. On the other hand, the WGS reaction is a very important and established industrial reaction that increases the hydrogen content in synthesis gas [24]. The catalysts used in the LT-WGS process usually consist of different combinations of CuO, ZnO, and Al$_2$O$_3$ components. Unfortunately, these catalysts are extremely pyrophoric in activated (reduced) state and can cause explosion upon exposure to air, which makes them impossible to use in the onboard fuel processor and several other applications [25]. They are also very susceptible to shutdown–start-up cycling because the catalyst's active component is leached out by condensed water during quenching or deactivated by the formation of surface carbonates. The desire is to accomplish the WGS reaction at temperatures below 623 K and to shorten the overall residence time, i.e., to achieve space velocities above 30,000 h^{-1} while maintaining high CO conversion into hydrogen, so that the product gases can be fed directly on the anode of the high-temperature PEMFC. In short, a WGS catalyst for automobile applications has to be a cost effective, more active, nonpyrophoric, and stable catalyst, processable on a monolith support surface to achieve the required high flow rates.

Catalyst systems for the WGS reaction that recently have received a lot of attention are the cerium oxides, mostly loaded with noble metals, especially platinum [26–30]. Jacobs et al. [28] even claim that it is probable that promoted ceria catalysts with the right development should realize higher CO conversions than the commercial CuO-ZnO-Al$_2$O$_3$ catalysts. Ceria doped with transition metals like Ni, Cu, Fe, and Co are also very interesting catalysts [27–32], especially the copper-ceria catalysts that have been found to perform excellently in the WGS reaction, as reported by Li et al. [32]. They have found that the copper-ceria catalysts are more stable than other Cu-based WGS catalysts and at least as active as the precious metal–ceria catalysts.

8.3.1.1 Catalytic Activity

In our laboratory, we have recently performed a kinetic study of WGS over the nanostructured $Cu_xCe_{1-x}O_{2-y}$ catalysts [33]. Here, we summarize the interesting and relevant results of this ongoing study.

While the H_2O/CO ratio is crucial for the performance of LT-WGS, it was particularly interesting to study the activity of catalysts at stoichiometric ratio and at $H_2O/CO = 3/1$ ratio. Both are lower than those used in the commercial LT WGS processing of the gas exiting HT-WGS. This was done deliberately for three reasons. The first was that we did not use CO_2 in the feed. Hence, we could lower the H_2O/CO ratio because there was no need to compensate the CO_2 influence on equilibrium with higher H_2O concentration (due to reverse WGS reaction). The second reason was that we wanted to study the behavior of LT-WGS catalysts at stoichiometric H_2O/CO ratio to check the influence of H_2O concentration on the WGSR kinetics. The third reason was that we also wanted to study the behavior of LT-WGS catalysts at relatively low inlet CO concentration (0.5 vol%) with respect to the usual inlet CO concentrations used in the industrial process (1.5 to 3 vol%). The feed composition used in this study was similar to that reported in the literature [29, 30], except that the CO concentration and the H_2O/CO ratio were lower.

Figure 8.11 shows CO conversion over the G-66 A, $Cu_{0.2}Ce_{0.8}O_{2-y}$, and $Cu_{0.1}Ce_{0.9}O_{2-y}$ catalysts, with a feed mixture containing 1.8% CO and 1.8% H_2 ($CO/H_2O = 1/1$) diluted in 96.4% He at a space velocity of 5000 h^{-1}. The G-66 A catalyst is, as shown in the figure, very active, but, as mentioned previously, it is pyrophoric and has to be carefully activated by reduction with H_2. Nevertheless the G-66 A was tested over identical experimental conditions to compare its performance

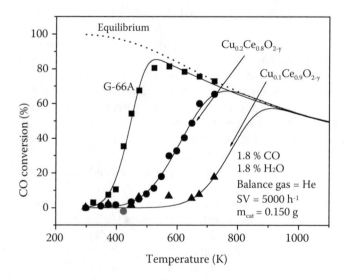

FIGURE 8.11 CO conversion for the WGS reaction over G-66 A, $Cu_{0.2}Ce_{0.8}O_{2-y}$, and $Cu_{0.1}Ce_{0.9}O_{2-y}$ catalysts. The solid lines are model fits assuming first-order reversible kinetics. The dotted line represents equilibrium conversion for the specific feed-gas composition.

with those of the copper-ceria catalysts. The G-66 A is truly a LT-WGS catalyst, as can be seen in Figure 8.11. It has T_{10} = 390 K (the temperature at which 10% conversion is achieved), and the conversion rapidly increases up to 85% close to the equilibrium curve (dotted) at about 500 K. The $Cu_{0.2}Ce_{0.8}O_{2-y}$ catalyst prepared by a coprecipitation method shows good activity. The fitted curve reaches the equilibrium curve at 760 K with a 77% CO conversion. The activity of the $Cu_{0.2}Ce_{0.8}O_{2-y}$ catalyst is in the midtemperature (MT) WGS range. The nanostructured $Cu_{0.1}Ce_{0.9}O_{2-y}$ catalyst prepared by a sol-gel method is less active: at 673 K it has only reached 17% CO conversion, which confirms the importance of the preparation of the catalyst to form active surface sites. Namely, the nanostructured $Cu_{0.1}Ce_{0.9}O_{2-y}$ catalyst has shown better conversion and selectivity than the $Cu_{0.2}Ce_{0.8}O_{2-y}$ catalyst [34–38] in the PrOx reaction, as we shall see in the next section, while in WGSR we found the reverse situation. The G-66 A contains 42 wt% copper, while the copper-ceria catalysts contain significantly less copper: the $Cu_{0.2}Ce_{0.8}O_{2-y}$ and the $Cu_{0.1}Ce_{0.9}O_{2-y}$ catalyst contain 7.5 wt% and 3.9 wt% copper, respectively. Li et al. [32] have shown that there is no significant difference in the light-off temperature over the WGS reaction over different $Cu-Ce(10\%La)O_x$ catalysts in which copper contents range between 5 and 40. All catalysts have probably enough active sites to sustain the WGS reaction. Our results are different, and they demonstrate that the catalysts prepared in different ways behave in a distinctly different manner. However, the scope of this study was not an optimization of the $Cu_xCe_{1-x}O_{2-y}$ catalyst. The dotted line is the equilibrium conversion for the feed-gas composition. The equilibrium CO conversions were calculated using the GASEQ software package [39] and calculated for this reactant composition using the expression:

$$K_{eq} = \frac{[CO_2][H_2]}{[CO][H_2O]} \tag{8.11}$$

The solid lines in Figure 8.11 are model fits of the experimental data. For fitting the experimental data, numerous research groups have proposed more-or-less complex models [29, 31, 40, 41]. In this work, we have used a simple rate expression derived by Wheeler et al. [29] that approximates the WGS process as a single reversible surface reaction, assuming an elementary reaction with first-order kinetics with respect to all species in the WGS reaction:

$$r^{"} = k_f^{"} \cdot P_{CO} \cdot P_{H_2O} - k_b^{"} \cdot P_{CO_2} \cdot P_{H_2} \tag{8.12}$$

In a recent study, Qi et al. [42] found that, at higher working temperatures between 573 and 873 K, H_2O has very little effect on the reaction rate and that both CO_2 and H_2 only weakly inhibit the reaction. The reaction order for CO was found to be close to 1, while the reaction order for H_2O was found to be close to 0, which is indicative of a surface saturated with hydroxyls.

Wheeler et al. [29], in their recent study of ceria-supported transition metal WGS catalysts, proposed that the enhanced activity of the catalysts is due to the fact that

H_2O can adsorb on ceria rather than on metal surfaces saturated with adsorbed CO. They proposed the following reaction mechanism, which might be involved over the ceria-supported transition metal (M/CeO_2) catalysts:

$$CO_2 - Ce \rightleftharpoons CO_2 + Ce \qquad (8.13)$$

$$H_2O + Ce \rightleftharpoons H_2O - Ce \qquad (8.14)$$

$$H_2O - Ce \rightleftharpoons O - Ce + H_2 \qquad (8.15)$$

$$CO - M + O - Ce \rightleftharpoons CO_2 - Ce + M \qquad (8.16)$$

$$CO_2 - Ce \rightleftharpoons CO_2 + Ce \qquad (8.17)$$

They have observed that the overall activation energy is the same for each metal with or without ceria, which suggests that the effect of ceria is to increase the rate of the reaction between adsorbed CO on the metal and O on ceria (Equation 8.16).

Koryabkina et al. [31] have found the apparent activation energy of 56 kJ/mol for WGSR over 8 wt% $CuO\text{-}CeO_2$ catalyst. This value is very close to the value of the apparent activation energy of 57.2 kJ/mol that we have found for the reaction of the selective CO oxidation in excess of H_2 over the $Cu_{0.2}Ce_{0.8}O_{2-y}$ catalyst [36]. The same low activation energy might be regarded as a strong support of the above WGSR mechanism proposed by Wheeler in which the reaction step (Equation 8.16) is rate limiting.

Starting with these arguments it is easy to see that the rate expression (Equation 8.12) can be simplified into the following rate expression:

$$r = \frac{area}{volume} \cdot r'' = k_f \cdot P_{CO} - k_b \cdot P_{CO_2} \qquad (8.18)$$

This can be done because the influences of H_2O and H_2 concentrations in the feed are neutralized by a different rate-limiting reaction step on the catalyst surface that does not involve H_2O and H_2. Here, the pseudohomogeneous rate r is related to the surface-reaction rate r'' through the area of active catalyst per unit volume of reactor. Assuming further a plug-flow regime, the integration of mass-balance equation for this simple rate expression gives an expression for CO conversion:

$$X_{CO}(t) = \frac{k_f}{k_f + k_b} \left[1 - e^{-(k_f + k_b)t} \right] \qquad (8.19)$$

where t is the residence time in the reactor, which varies with temperature as predicted by the ideal gas law. This expression was used to fit all experimental data, but instead of varying t, we measured the residence time, and it was then set constant at all temperatures, giving the expression:

$$X_{CO}(T) = 100 \times \frac{k_f e^{(-E_{af}/RT)}}{k_f e^{(-E_{af}/RT)} + k_b e^{(-E_{ab}/RT)}} \left[1 - e^{-\left(k_f e^{(-E_{af}/RT)} + k_b e^{(-E_{ab}/RT)}\right) \cdot t} \right] \quad (8.20)$$

where $X_{CO}(T)$ is the CO conversion and T the temperature. The activation energy, E_{af}, was calculated by plotting $ln(X)$ vs. $1/T$, giving a straight line. For fitting the experimental data, the activation energy was set constant at all temperatures. The fitted values for the pre-exponential coefficients — k_f, and k_b, and the backward activation energy, E_{ab} — were obtained by using the Levenberg-Marquardt algorithm. They are presented in Table 8.9 together with adequate literature values for copper-ceria catalysts.

Even though Wheeler et al. [29] used a monolithic reactor system with very high flow rates and short contact times, the model fits of the experimental data obtained in our study show that this simplified model is capable of fitting data also derived from catalyst powder in a packed-bed reactor with lower reaction rates. The curves fit all the experimental points within the accuracy limits of the data.

TABLE 8.9
WGS Reaction Kinetics, Apparent Activation Energies, E_{af} (Forward), and Modeled Values for the Backward Activation Energy E_{ab} and Preexponential Factors K_{0f}, K_{0b}, Assuming an Elementary Reaction with First-Order Kinetics of the WGS Reaction

Catalyst	E_{af} (kJ/mol)	Conditions	E_{ab} (kJ/mol)	K_{0f} (s⁻¹)	K_{0b} (s⁻¹)
[a]$Cu_{0.2}Ce_{0.8}O_{2-y}$	34	473–623 K CO/H$_2$O = 1/3	61	$1.8*10^3$	$1.1*10^4$
[a]$Cu_{0.1}Ce_{0.9}O_{2-y}$	51	573–673 K CO/H$_2$O = 1/3	78	$4*10^3$	$2.4*10^4$
8%CuCeO$_2$ [9]	56	513 K CO/H$_2$O = 1/3			
5%Cu-Ce(10% La)O$_x$[10]	30.4 19.2	448–573 K CO/H$_2$O = 1/7.5 CO/H$_2$O = 1/5.3			
[a]42% CuO-ZnO-Al$_2$O$_3$ (G-66 A)	47	396–448 K CO/H$_2$O = 1/3	71	$4.9*10^6$	$2.2*10^7$
40% CuO-ZnO-Al$_2$O$_3$ [9]	79	463 K CO/H$_2$O = 1/3			

8.3.1.2 Influence of the Oxygen Storage Capacity

The cerium oxide catalysts are known for their high oxygen storage capacity, and it is clear that the cerium oxide has a direct role in the catalytic activity. However, the function of ceria and how the metal component promotes the WGS reaction is not clear. Some research groups [26, 27, 32] have claimed that the redox mechanism and oxygen storage capacity have a direct role in the WGS reaction. The other mechanism was proposed [28, 43] to proceed through the formation of surface-formate intermediates. Nevertheless, the oxygen storage capacity of the copper-ceria catalyst can be observed in Figure 8.12. Copper-ceria catalysts have oxygen stored in the catalyst lattice, as described by the formula $Cu_{0.2}Ce_{0.8}O_{2-y}$, where the oxygen storage capacity is reported to be $y = 0.17$ [34].

The physisorbed oxygen reacts quickly with CO in the gas flow to form CO_2, as can be seen in Figure 8.12 (open squares). When this takes place, no water is consumed (open circles). As soon as all oxygen stored in the catalyst has reacted, the WGS reaction takes over. As can be observed in the figure, the WGS equilibrium line is crossed (open squares), which unfortunately makes it clear that this is CO oxidation and not WGS reaction. Hence the catalyst had to be pretreated in the actual WGS feed stream before starting to observe the WGS reaction. In the above experiment, it took more than 70 min to discharge all of the oxygen stored in the catalyst to get stable results for measuring the WGS reaction, which probably means that oxygen from the crystalline bulk material also was liberated at these temperatures (up to 673 K) due to reducing atmosphere [44]. This was observed for both copper-ceria catalysts and always appeared after the catalyst was left in oxygen-containing

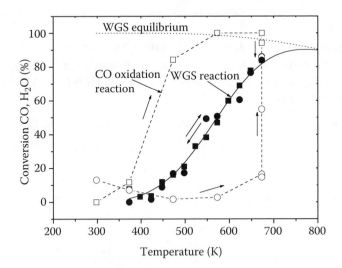

FIGURE 8.12 Oxygen storage capacity over the $Cu_{0.2}Ce_{0.8}O_{2-y}$ catalyst. Open squares represent conversion of CO and open circles the conversion of H_2O (which is none) in the CO oxidation reaction. Filled squares and filled circles are CO and H_2O conversions, respectively, in the WGS reaction. The dotted line represents the WGS equilibrium curve. The solid line is model fit assuming first-order reversible kinetics in WGS.

atmosphere. The catalyst reoxidation process at atmospheric conditions was nonpyrophoric. After the initial treatment with WGS feed (see Figure 8.2), the catalysts exhibit stable operation, with little or no deactivation (filled squares [CO] and filled circles [H_2O]).

8.3.1.3 Influence of Feed-Gas Composition, H_2O/CO Ratio, and Presence of Hydrogen

In Figure 8.13 we observe the importance of utilizing the right feed-gas composition over the $Cu_{0.2}Ce_{0.8}O_{2-y}$ catalyst prepared by coprecipitation method. The dotted lines represent different equilibrium curves calculated for three different feed mixtures: 0.5% CO and 1.5% H_2O; 1.8% CO and 1.8% H_2O; and 50% H_2, 0.5% CO, 1.5% H_2O, all diluted with He. The first mixture (0.5% CO and 1.5% H_2O) represents a suitable feed-gas composition for the WGS reaction, where high conversions (>99% CO conversion up to 550 K) can be accomplished. It is in this equilibrium conversion region that the WGS reaction has to be carried out.

The experiments with 1/1 CO/H_2O feed were carried out to show how fast the WGS equilibrium is approached due to the lower water content in the feed-gas composition. It can be seen that the lower (stoichiometric) CO/H_2O ratio in the feed does not influence the rate of approach to the equilibrium. This confirms the observations of Qi et al. [42] that water has little or no effect on the reaction rate in the case of ceria-supported transition metal catalysts.

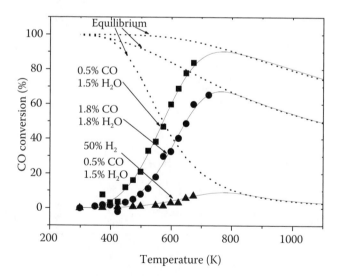

FIGURE 8.13 The influence of increasing the ratio between water and carbon monoxide, and addition of 50% H_2 to the feed-gas mixture. CO conversion in WGS reaction over $Cu_{0.2}Ce_{0.8}O_{2-y}$, catalyst at different feed compositions with a S.V. = 5000 h^{-1}. The solid lines are model fits assuming first-order reversible kinetics. The dotted lines represent the equilibrium conversions for the specific feed compositions.

The feed mixture containing a large amount of hydrogen (50% H_2, 0.5% CO, and 1.5% H_2O) was also used. This feed composition represents conditions that are close to those used in the industry with regard to H_2 content: 473 K, 30 bar, and a steam-to-dry-gas ratio of 0.4 with a dry gas composition of 2% CO, 20% CO_2, and 78% H_2 [24]. The intent was to examine the behavior of the catalysts in the presence of H_2 in the gas feed, which could cause methanation reactions. Yet the typical industrial gas composition has an equilibrium conversion even higher than the gas mixture with 0.5% CO and 1.5% H_2O, since the high amount of H_2O shifts the WGSR equilibrium to the right side of Equation 8.10.

8.3.1.4 Methanation Reactions

When carrying out the WGS reaction, methane can be formed in the reactor through the methanation reaction, which is the reverse methane steam reforming reaction and is highly exothermal:

$$CO + 3H_2 \longleftrightarrow CH_4 + H_2O \quad (\Delta H = -205.8 \text{ kJ/mol}) \tag{8.21}$$

Methane is an undesired product in the WGS reaction, and for every CH_4 molecule formed, two H_2 molecules are taken away from the product stream. Figure 8.14 shows two WGS equilibrium curves for the hydrogen-containing conditions. The upper dashed line is the WGS equilibrium curve including the methanation reaction. The lower dotted line is the WGS equilibrium curve without methanation reactions.

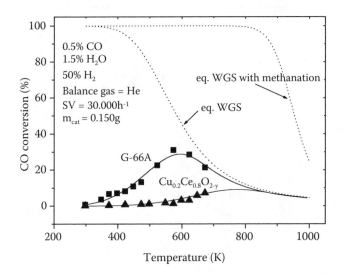

FIGURE 8.14 Equilibrium curves (dotted lines) for the WGS reaction with/without methanation reaction for feed-gas composition containing 50% H_2. The filled squares and circles are CO conversions over G-66 A and $Cu_{0.2}Ce_{0.8}O_{2-y}$ catalysts, respectively. The solid lines are model fits assuming first-order reversible kinetics.

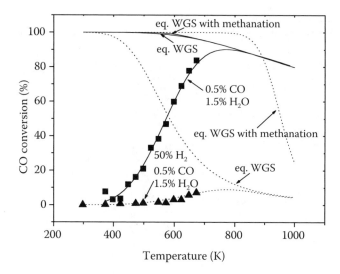

FIGURE 8.15 Equilibrium curves for the WGS reaction with/without methanation over the $Cu_{0.2}Ce_{0.8}O_{2-y}$ catalyst. Filled squares represent feed-gas composition of 0.5% CO and 1.5% H_2O, and the solid line is the model fit assuming first-order reversible kinetics. The other two solid lines are the respective equilibrium curves. Filled triangles represent the feed composition of 50% H_2, 0.5% CO, and 1.5% H_2O, and the dotted line is the model fit assuming first-order reversible kinetics. The two other dotted lines are the respective equilibrium curves.

The figure shows that even though the methanation reaction is highly exothermal, these reactions do not occur on the catalyst, even at temperatures lower than 600 K. This confirms the selectivity of the copper-containing ceria-supported catalysts for the WGS reaction; otherwise, the experimental data would cross the WGS equilibrium line (lower dotted) calculated without taking into account the methanation reaction. If the methanation reaction had also proceeded, the experimental points would reach the (upper dotted) equilibrium line calculated for complex equilibrium including the methanation reaction. This can also be seen in Figure 8.15, which shows equilibrium curves for the WGS reaction with and without methanation over the $Cu_{0.2}Ce_{0.8}O_{2-y}$ catalyst including the feed-gas composition of 0.5% CO and 1.5% H_2O.

For this specific gas composition, the difference between the WGS equilibrium with and without methanation reactions is very small, and in this case it would not be possible to draw the above conclusion about the selectivity of the catalysts. Anyhow, the reactor outlet gases were analyzed by a gas chromatograph, and no methane formation was detected in any experiment.

8.3.1.5 Influence of Contact Time

As revealed earlier in this chapter, one important aspect of the WGS catalyst is that it should be able to operate at high flow rates to reduce the total reactor size for the production of hydrogen. To examine the impact of the contact time, we increased the space velocity over the catalysts for the feed mixture of 0.5% CO and 1.5% H_2O in He from 5000 h^{-1} to 30,000 h^{-1}, as shown in Figure 8.16. All three of the different

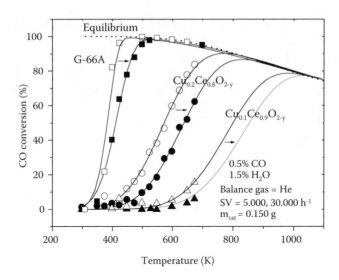

FIGURE 8.16 The effect of increasing the flow rate and decreasing the contact time for WGS reaction over the G-66 A, $Cu_{0.2}Ce_{0.8}O_{2-y}$, and $Cu_{0.1}Ce_{0.9}O_{2-y}$ catalysts. Empty symbols illustrate low flow rate, S.V. = 5000 h^{-1}, and filled symbols high flow rate, S.V. = 30,000 h^{-1}, and. The dotted line represents the equilibrium curve for a feed-gas composition of 0.5% CO and 1.5% H_2O in He. The solid lines are model fits assuming first-order reversible kinetics.

catalysts respond very similarly, with the CO conversion decreasing upon increasing the space velocity, i.e., decreasing the contact time. To achieve the same CO conversion at the higher space velocity, temperatures approximately 50 to 100 K higher are required.

In conclusion to this section, we can state that the copper-ceria catalysts are nonpyrophoric and stable, showing little or no deactivation during the experiments and shutdown–start-up cycling. The $Cu_{0.2}Ce_{0.8}O_{2-y}$ catalyst prepared by a coprecipitation method showed good catalytic activity for the WGS reaction. The $Cu_{0.1}Ce_{0.9}O_{2-y}$ catalyst prepared by sol-gel method was found to be less active, which could be due to a lower amount of active copper sites or to different CuO crystallite size and structure. The copper-ceria catalysts were shown to be selective for the WGS reaction, and no methanation reaction were observed over any catalyst under the experimental conditions used.

Model fits of the experimental data show that the simplified first-order elementary reaction kinetics for these catalysts can be used to approximate the WGS reaction as a single reversible surface reaction. This model reaction-rate expression could be used in all concentration ranges of H_2O present in the feed due to a possibly different rate-limiting step not including H_2O.

8.3.2 SELECTIVE CO OXIDATION IN EXCESS OF H_2 (PrOx) over the Nanostructured $Cu_xCe_{1-x}O_{2-y}$ Catalyst

To lower the cost and improve the selectivity of the catalyst, a novel nonstoichiometric nanostructured $Cu_xCe_{1-x}O_{2-y}$ catalyst for the selective low-temperature oxidation of CO in excess of H_2 was synthesized by coprecipitation and sol-gel methods; these methods have been patented [18, 34]. The sol-gel method of catalyst preparation is particularly convenient for deposition on diverse geometries of support (i.e., honeycomb supports) and on reactors that can be used in PrOx processes. This type of catalyst is also capable of converting methanol directly into hydrogen and CO_2 by steam reforming through a water–gas-shift reaction [45, 46]. By use of this catalyst, the four previously mentioned reactors (reformer, two-stage WGS reactors [LT and HT], and PrOx reactor) could be incorporated into a single unit.

Here we present a summary of recent results obtained in a study on the kinetics of the selective CO oxidation in excess of hydrogen over the nanostructured $Cu_xCe_{1-x}O_{2-y}$ catalyst as obtained in the fixed-bed reactor operating in a differential mode. The inlet gaseous mixture composition simulates the real composition at the outlet of the low-temperature water–gas-shift reactor concerning the concentrations of CO, H_2, and O_2, except that no CO_2, H_2O, and unconverted CH_3OH were present.

The experimental kinetic results obtained are presented in Figure 8.17 for three samples of $Cu_xCe_{1-x}O_{2-y}$ catalysts with different content of copper prepared by coprecipitation. It was found that the reaction kinetics could be best represented by the redox mechanism [36]. Such a redox reaction can be described by the following two-step reaction:

$$Cat–O + Red \_ Cat + Red–O \tag{8.22}$$

$$Cat + Ox–O \_ Cat–O + Ox \tag{8.23}$$

The first step in this reaction mechanism is the catalyst reduction. Cat–O represents oxidized catalyst, which is attacked by a reductant (Red). The catalyst itself undergoes reduction, while the reductant is oxidized. The second step represents reoxidation of the catalyst by the oxidant (Ox–O), which donates an oxygen atom to the catalyst while it reduces itself.

The kinetics of selective CO oxidation over the $Cu_xCe_{1-x}O_{2-y}$ nanostructured catalysts can be well described by employing a Mars and van Krevelen type of kinetic equation derived from a redox mechanism:

$$r_{CO} = \frac{k_{CO}k_{O2}P_{CO}P_{O2}^n}{0.5 \cdot k_{CO}P_{CO} + k_{O2}P_{O2}^n} \tag{8.24}$$

$$k_{CO} = A_{CO} \cdot \exp\left(-E_{a,CO}/RT\right) \tag{8.25}$$

$$k_{O2} = A_{O2} \cdot \exp\left(-E_{a,O2}/RT\right) \tag{8.26}$$

The parameters k_{CO} and k_{O2} are taken to be the reaction rate constants for the reduction of surface by CO and reoxidation of it by O_2. The parameters k_{CO}, k_{O2}, and n at one temperature were obtained by fitting experimental values of P_{CO}, P_{O2},

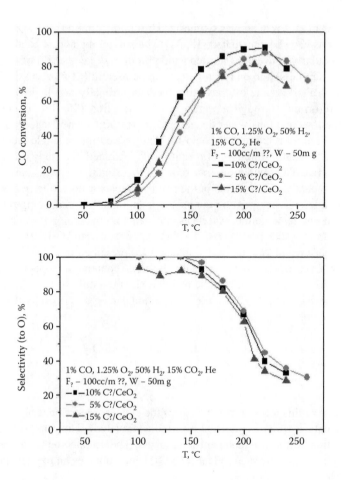

FIGURE 8.17 CO conversion and selectivity over nanostructured $Cu_xCe_{1-x}O_{2-y}$ catalysts with x = 0.05, 0.10, and 0.15, respectively.

and reaction rate with the rate equation. The parity plot for calculated vs. experimental values of reaction rate is presented in Figure 8.18. The agreement between experimental and calculated values is very good over three orders of magnitude of reaction rate.

We have performed the dynamic oxidation of carbon monoxide over $Cu_{0.1}Ce_{0.9}O_{2-y}$ nanostructured catalyst using a step change in CO concentration over the preoxidized catalyst. Figure 8.19 represents the CO and CO_2 responses after a step change from He to 1 vol% CO/He over the fully oxidized $Cu_{0.1}Ce_{0.9}O_{2-y}$ nanostructured catalyst. At low temperatures, CO breakthrough is delayed for a few seconds, as can be seen from Figure 8.19a. At the temperature of 250°C, however, 20 sec is needed for the first traces of CO to exit the reactor. On the other hand, the evolution of CO_2 in the reactor effluent stream has no delay, as seen in Figure 8.19b. However, the nature of the CO_2 peak as a function of temperature changes significantly. At temperatures lower than 100°C, only one peak in CO_2 response is visible.

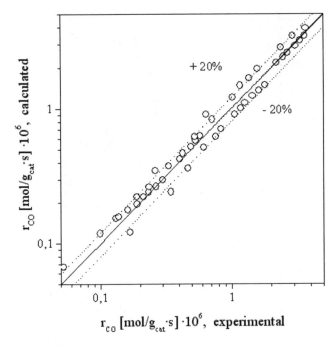

FIGURE 8.18 Calculated vs. experimental values of reaction rates for the selective CO oxidation in excess hydrogen.

At 100°C, the CO_2 peak broadens, and at 125°C, two separate peaks are clearly visible. The first peak is narrow, followed by a second broader peak. When the temperature is increased further, the first peak in the CO_2 response becomes invisible because it is covered by the second peak. Only the origin of first peak is signified by fast evolution of CO_2 in the reactor effluent stream. It is also important to notice how the maximum of the second peak shifts to the right when the temperature is increased. At 250°C, the catalyst surface responds almost instantly to a CO step change by producing CO_2. The concentration of CO_2 in the reactor effluent gas after 3 sec is 0.65 vol%, as shown in Figure 8.19b. However, the CO_2 concentration in the reactor effluent gas rises further and reaches 0.80 vol% after 25 sec. This is followed by a sharp decrease in the CO_2 concentration, which stabilizes after 100 sec at 0.2 vol%. Afterward, the concentration in CO_2 decreases very slowly and falls to zero after 13 min.

A detailed elementary-step model of the CO oxidation over $Cu_{0.1}Ce_{0.9}O_{2-y}$ nanostructured catalyst under dynamic conditions was developed. The model discriminates between adsorption of carbon monoxide on catalyst-inert sites as well as on oxidized and reduced catalyst-active sites. Apart from that, the model also considered the diffusion of subsurface species in the catalyst and the reoxidation of reduced catalyst sites by subsurface lattice oxygen species. The model allows us to calculate activation energies of all elementary steps considered as well as the bulk diffusion coefficient of oxygen species in the $Cu_{0.1}Ce_{0.9}O_{2-y}$ nanostructured catalyst. The diffusion coefficient obtained by the mathematical modeling of step experiments is

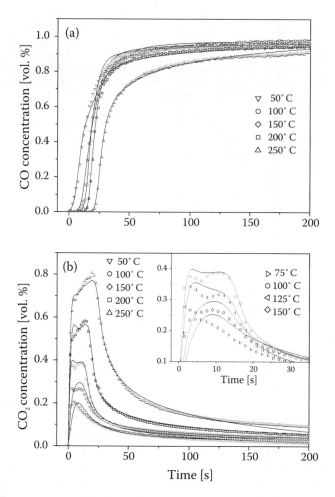

FIGURE 8.19 a) CO and (b) CO_2 concentrations in the reactor effluent stream as a function of temperature. The markers represent experimental points. The solid lines represent the model predictions obtained by the integration of the rate equations given in Table 8.11 with initial and boundary conditions given in Table 8.12. The kinetic parameter values are given in Table 8.13. Conditions: $m_{cat} = 200$ mg, $\Phi_v = 200$ ml min^{-1}.

shown to be within the range of bulk diffusion coefficients measured over other oxide catalysts. The elementary reaction steps, the mass balance equations, the initial and boundary conditions, and the estimated kinetic parameters are given in Table 8.10, Table 8.11, Table 8.12, and Table 8.13.

In our studies, we have demonstrated that the redox mechanism that was used to model dynamic behavior of CO oxidation is consistent with a kinetic model of the selective CO oxidation obtained under a steady-state mode of operation [37]. We propose the following tentative scheme (Figure 8.20) for the selective CO oxidation over the $Cu_{0.1}Ce_{0.9}O_{2-y}$ nanostructured catalyst: CO and H_2 adsorb on the copper/ceria interfacial region of the catalyst, the most reactive places for both CO

TABLE 8.10
Elementary Reaction Steps Considered in the Kinetic Modeling of the CO Concentration-Step-Change Experiments for the Oxidation of CO over Completely Oxidized $Cu_{0.1}Ce_{0.9}O_{2-Y}$ Nanostructured Catalyst without the Presence of Oxygen in the Reactor Feed

Step No.	Elementary Reaction Step
1	$CO + Cu^{2+}O_{s,s} \xrightarrow{k_1} CO\cdots Cu^{2+}O_{s,s}$
2	$CO \cdots Cu^{2+}O_{s,s} \xrightarrow{k_2} CO\cdots Cu^+\square_s$
3	$CO + Cu^+\square_s \xrightarrow{k_3} CO\cdots Cu^+\square_s$
4	$Ce^{4+}O_{b,b} \xrightarrow{D} Ce^{3+}\square_b + O_{s,b}$
5	$CO\cdots Cu^+\square_s + O_{s,b} \xrightarrow{k_4} CO\cdots Cu^{2+}O_{s,b}$
6	$CO\cdots Cu^+\square_s \xrightarrow{k_2} CO_2 + Cu^+\square_s$
7	$\dfrac{CO + * \xrightarrow{k_5} CO*}{2\,CO + Cu^{2+}O_{s,s} + Ce^{4+}O_{b,b} \longrightarrow 2CO_2 + Cu^+\square_s + Ce^{3+}\square_b}$

Note: Oxygen vacancy is represented by $\square$. The meaning of subscripts accompanying oxygen species and oxygen vacancies is explained in text.

and H_2 oxidation reactions. It is further proposed that CO (and H_2) use mostly copper cations as the adsorption sites, while cerium oxide must also be present in the close vicinity. Copper oxide might also form a solid solution with cerium oxide, at least in the form of small intergrowths at the interface, which are x-ray diffraction (XRD)-invisible. In this concerted mechanism of copper and cerium oxide, copper cation has the following role: it is the adsorption site for the CO (and H_2). When either of the two reactants is adsorbed on the copper cation, it extracts oxygen from the surface, and copper is reduced from Cu^{2+} to Cu^+. Cerium cation, which lies next to copper cation, can supply an additional oxygen atom from the catalyst lattice while it reduces itself simultaneously from the Ce^{4+} into the Ce^{3+} form. Cerium oxide acts as an oxygen supplier (buffer) when it is needed at the place of reaction. Only one single copper ion is enough to convert one molecule of CO (or H_2) into CO_2 (or H_2O), respectively. When the product molecule is desorbed, the site becomes available for the next reactant molecule, either CO or H_2. Upon extraction of surface oxygen from the catalyst lattice, the oxygen vacancy can be refilled directly from the gas phase or by oxygen diffusion through the bulk of the catalyst. The latter mechanism is observed at higher temperatures.

The $Cu_xCe_{1-x}O_{2-y}$ nanostructured catalyst prepared by the sol-gel method is a very efficient selective CO oxidation catalyst even under the highly reducing conditions that are present in a PrOx reactor. The catalyst is energy efficient toward the PEM fuel cell technology because it oxidizes CO with 100% selectivity close to the PEM fuel cell working temperature. These performances are obtained with a catalyst that contains cheap copper and cerium oxides rather than costly noble metals.

TABLE 8.11

Mass-Balance Equations for Gas-Phase, Surface, and Subsurface Species Corresponding to Elementary Reaction Steps Given in Table 8.10

$$\frac{\partial y_{CO}}{\partial t} + \frac{1}{\tau}\frac{\partial y_{CO}}{\partial z} = \frac{\rho_B}{\varepsilon_B}\left(-k_1 L_{TOT}\, y_{CO}\theta_{Cu^{2+}O_{s,s}} - k_3 L_{TOT}\, y_{CO}\theta_{Cu^+\square_s} - k_5 F_{TOT}\, y_{CO}\delta_*\right)$$

$$\frac{\partial y_{CO_2}}{\partial t} + \frac{1}{\tau}\frac{\partial y_{CO_2}}{\partial z} = \frac{\rho_B}{\varepsilon_B}k_2\left(L_{TOT}/C_{TOT}\right)\left(\theta_{CO\cdots Cu^{2+}O_{s,s}} + \theta_{CO\cdots Cu^{2+}O_{s,b}}\right)$$

$$\frac{\partial\theta_{Cu^{2+}O_{s,s}}}{\partial t} = -k_1 C_{TOT}\, y_{CO}\theta_{Cu^{2+}O_{s,s}}$$

$$\frac{\partial\theta_{CO\cdots Cu^{2+}O_{s,s}}}{\partial t} = k_1 C_{TOT}\, y_{CO}\theta_{Cu^{2+}O_{s,s}} - k_2\theta_{CO\cdots Cu^{2+}O_{s,s}}$$

$$\frac{\partial\theta_{Cu^+\square_s}}{\partial t} = k_2\left(\theta_{CO\cdots Cu^{2+}O_{s,s}} + \theta_{CO\cdots Cu^{2+}O_{s,b}}\right) - k_3 C_{TOT}\, y_{CO}\theta_{Cu^+\square_s}$$

$$\frac{\partial\theta_{CO\cdots Cu^+\square_s}}{\partial t} = k_3 C_{TOT}\, y_{CO}\theta_{Cu^+\square_s} - k_4 H_{TOT}\theta_{CO\cdots Cu^+\square_s}\xi_{O_{s,b}}$$

$$\frac{\partial\theta_{CO\cdots Cu^{2+}O_{s,b}}}{\partial t} = k_4 H_{TOT}\theta_{CO\cdots Cu^+\square_s}\xi_{O_{s,b}} - k_2\theta_{CO\cdots Cu^{2+}O_{s,b}}$$

$$\frac{\partial\delta_*}{\partial t} = -k_5 C_{TOT}\, y_{CO}\delta_*$$

$$\frac{\partial\xi_{O_{b,b}}}{\partial t} = \frac{1}{t_d}\left(\frac{\partial^2\xi_{O_{b,b}}}{\partial x^2} + \frac{1}{x}\frac{\partial\xi_{O_{b,b}}}{\partial x}\right)$$

TABLE 8.12
Initial and Boundary Conditions Corresponding to Mass Balance
Equations Given in Table 8.11

$$y_{CO}(z,0) = 0, \quad y_{CO}(0,t) = 0.01$$

$$y_{CO_2}(z,0) = 0, \quad y_{CO_2}(0,t) = 0$$

$$\theta_{Cu^{2+}O_{s,s}}(z,0) = 1$$

$$\theta_{CO=Cu^{2+}O_{s,s}}(z,0) = 0$$

$$\theta_{Cu^+{}_{\square_s}}(z,0) = 0$$

$$\theta_{CO\cdots Cu^+{}_{\square_s}}(z,0) = 0$$

$$\theta_{CO=Cu^{2+}O_{s,b}}(z,0) = 0$$

$$\delta_*(z,0) = 1$$

$$\xi_{O_{b,b}}(z,x,0) = 1$$

$$\frac{\partial \xi_{O_{b,b}}}{\partial x}(z,0,t) = 0$$

$$a \cdot \frac{1}{t_d} \frac{\partial \xi_{O_{b,b}}}{\partial x}(z,1,t) = -k_3 L_{TOT} \theta_{CO\cdots Cu^+{}_{\square_s}} \xi_{O_{b,b}}(z,1,t)$$

TABLE 8.13
Estimates of the Kinetic Parameters Obtained by Regression of He → 1 vol% CO/He Concentration-Step Experiments for the CO Oxidation over Fully Oxidized Catalyst without the Presence of Oxygen in the Reactor Feed Gas

Ln (A_I) ($m^3 \cdot mol^{-1} \cdot sec^{-1}$)	14.6 ± 1.2
E_1 ($kJ \cdot mol^{-1}$)	39.6 ± 4.6
Ln (A_2) (sec^{-1})	1.85 ± 2.1
E_2 ($kJ \cdot mol^{-1}$)	9.7 ± 8.1
Ln (A_3) ($m^3 \cdot mol^{-1} \cdot sec^{-1}$)	10.0 ± 1.6
E_3 ($kJ \cdot mol^{-1}$)	25.8 ± 5.9
Ln (A_4) ($kg_{cat} \cdot mol^{-1} \cdot sec^{-1}$)	19.9 ± 5.2
E_4 ($kJ \cdot mol^{-1}$)	72.9 ± 19.8
Ln (A_5) ($m^3 \cdot mol^{-1} \cdot sec^{-1}$)	4.07 ± 0.63
E_5 ($kJ \cdot mol^{-1}$)	13.9 ± 2.3
Ln (A_{td}) (sec^{-1})	2.29 ± 2.74
E_{td} ($kJ \cdot mol^{-1}$)	40.0 ± 10.3

Note: Parameters are obtained between 125 and 250°C. Mass-balance equations are given in Table 8.11; corresponding initial and boundary conditions are given in Table 8.12

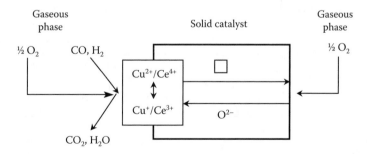

FIGURE 8.20 Scheme of the Mars and van Krevelen mechanism for the selective CO oxidation in excess of hydrogen over the $Cu_{0.1}Ce_{0.9}O_{2-y}$ nanostructured catalyst. The enhanced reactivity at the interface of CuO and CeO_2 is presented. The reoxidation of the catalyst can proceed by two parallel paths: directly from the gaseous phase or by the diffusion of oxygen trough the bulk of the catalyst crystal lattice. The latter path is more pronounced at higher temperatures.

8.4 CONCLUSIONS

In this chapter we have tried to present the complexity of problems that are encountered on the way to resolving the everlasting problem: how to develop a reliable, clean, cheap, and long-lasting source of energy. We have presented a view of possible solutions based on the "hydrogen economy," in which the transition period of the way toward this goal will include all possible and feasible processes of hydrogen "extraction" and cleaning from sustainable and renewable feedstock. The recent efforts of one among many of the research groups in the world to contribute in resolving these problems has been presented here.

8.5 ANNOTATIONS

Primary energy sources are found or stored in nature. The prevalently used sources are biomass, coal, oil, natural gas, sunlight, wind, water, nuclear power, geothermal power and potential energy from the Earth's gravity.

Energy carriers (vectors) are produced from primary energy sources using different technologies. Most common among them are electricity (produced from coal with many technologies or via photovoltaics), ethanol (produced from corn, sugar cane, etc.), hydrogen, etc.

Feedstocks are the resources from which the energy carriers are extracted.

Carbon sequestration — technologies to capture and sequester (store) carbon or carbon dioxide from feedstocks in the processes of hydrogen production.

Hydrogen is an energy carrier and a fuel that can be produced from a wide variety of feedstocks.

Distributed production — located at the point of use; scalable and produces smaller quantities of hydrogen.

Centralized production — large quantities of hydrogen are produced, which need to be stored and/or distributed through distribution infrastructure (pipelines, railway, trucks.)

a	specific surface area of a catalyst particle, m2external surface m-3particle
A_i	preexponential factor of the rate constant i, various units
A_{td}	preexponential factor of the reverse oxygen diffusion coefficient, cm2 s-1
C_b	concentration of O2- ions in bulk catalyst, mol m-3
C_s	concentration of O2- ions on the surface of the catalyst, mol m-3
C_{TOT}	total molar concentration of gas phase, mol mgas-3
D	diffusion coefficient of oxygen species in the solid catalyst lattice, m2 s-1
E_i	activation energy of the rate constant i, kJ/mol
E_{td}	activation energy of the reverse oxygen diffusion coefficient, kJ/mol
F_{TOT}	surface capacity of catalyst inert phase, mol kgcat-1
H_{TOT}	volumetric capacity of catalyst, mol kgcat-1
J_{O2-}	molar flux of O2- ions from the bulk to the catalyst surface, mol m-2 s-1
k_1	rate constant of elementary reaction step 1, m3 mol-1 s-1
k_2	rate constant of elementary reaction steps 2 and 6, s-1
k_3	rate constant of elementary reaction step 3, m3 mol-1 s-1
k_4	rate constant of elementary reaction step 5, kgcat mol-1 s-1

k_5	rate constant of elementary reaction step 7, m3 mol-1 s-1
L_{TOT}	surface capacity of catalyst active phase, mol kgcat-1
m_{cat}	mass of catalyst, kg
R	crystallite radius, m
$R_{k,i}$	production rate of component i regarding to catalyst phase k, mol kgcat-1 s-1
t	time, s
t_d	characteristic diffusion time, s
x	dimensionless radial position in catalyst crystallite particle
$<\bar{x}>$	average displacement of O^{2-} species in the bulk catalyst lattice, m
yi	molar fraction of gas component i
z	dimensionless distance of catalyst bed

Greek letters

δn	fractional surface coverage of species n adsorbed on catalyst inert phase
εB	void fraction of catalyst bed = 0.35
Φv	total volumetric flow rate at reaction conditions, mgas3 s-1
ρB	catalyst bulk density, kgcat mr-3
θj	fractional surface coverage of species j adsorbed on catalyst active phase
τ	space time, s
ξm	fractional volumetric "coverage" of species m that diffuses through the catalyst particle

Subscripts

b	oxygen vacancy located in the catalyst bulk lattice
s	oxygen vacancy located on the catalyst surface
b,b	oxygen species located in the catalyst bulk lattice, that originated in the bulk
s,b	oxygen species located on the catalyst surface, that originated from the bulk catalyst lattice
s,s	oxygen species located on the catalyst surface, that originated on the catalyst surface

ACKNOWLEDGMENTS

My sincere thanks go to Professor Janez Levec, head of the Laboratory for Catalysis and Chemical Reaction Engineering at the National Institute of Chemistry, and to colleagues in this laboratory, including Dr. Jurka Batista, Dr. Gorazd Berčič, and Dr. Albin Pintar, to postdoctoral research fellow Dr. Henrik Kušar, and to Dr. Gregor Sedmak, former Ph.D. student, who all contributed to the research results and provided fruitful discussions. The work has been performed in part with financial support of the Slovenian Research Agency under Grant No. P2-0152, which I gratefully acknowledge.

REFERENCES

1. Song, C.S., *Catal. Today*, 77, 17, 2002.
2. Hoffert, M.I., Caldeira, K., Benford, G., Criswell, D.R., Green, Ch., Herzog, H., Jain, A.K., Kheshgi, H.S., Lackner, K.S., Lewis, J.S., Lightfoot, H.D., Manheimer, W., Mankins, J.C., Mauel, M.E., Perkins, L.J., Schlesinger, M.E., Volk, T., and Wigley, T.M.L., *Science*, 298, 981, 2002.
3. O'Neill, B.C. and Oppenheimer, M., *Science*, 296, 1971, 2002.

4. Marchetti, C., *Society of a Learning System: Discovery, Invention and Innovation Cycles Revisited*, International Institute for Applied System Analysis, Pub. RR-81-29, Laxenburg, Austria, 1981.

5. J.S. Cannon, *Harnessing Hydrogen: the Key to Sustainable Transportation*, New York, Inform, 1995; available on-line at http://www.ttcorp.com/nha.

6. Kartha, S. and Grimes, P., *Physics Today*, 47, 54, 1994.

7. Future Wheels, NAVC report for DARPA, Boston, MA, November 1, 2000.

8. Bockris, J.-O'M. and Srinivasan, S., *Fuel Cells: Their Electrochemistry*, McGraw-Hill, New York, 1969.

9. *Fuel Cells to 2006*, The Fredonia Group Report: 1514, Jan. 2002; available on-line at http://www.fuelcell-info.com.

10. Lasher, S., Teagan, P., and Unnasch, S., Near-term and long-term perspectives of fuel choices for fuel cell vehicles, paper presented at *2003 Fuel Cell Seminar*, Miami Beach, FL, Nov. 7, 2003.

11. Hydrogen Facts Sheet, Hydrogen Production Overview, The National Hydrogen Association, Aug. 2004; available on-line at http://www.hydrogenUS.org.

12. Edlund, D., Characteristics and performance of a commercial fuel processor, paper presented at *2002 Fuel Cell Seminar*; available on-line at http://www.idatech.com.

13. Farrauto, R., Hwang, S., Shore, L., Ruettinger, W., Lampert, J., Giroux, T., Liu, Y., and Ilinich, O., *Annu. Rev. Mater. Res.*, 33, 1, 2003.

14. Liu, Y.Y., Hayakawa, T., Suzuki, K., Hamakawa, S., *Appl. Catal. A: General*, 223, 137, 2002.

15. Liu, Y.Y., Hayakawa, T., Tsunoda, T., Suzuki, K., Hamakawa, S., Murata, K., Shiozaki, R., Ishii, T., and Kumagai, M., *Topics in Catal.*, 22, 205, 2003.

16. Golunski, S., *Platinum Metals Rev.*, 42, 2, 1998.

17. Li, K., Fu, Q., Flytzani-Stephanopoulos, M., *Appl. Catal. B: Environmental*, 27, 179, 2000.

18. Hočevar, S., Batista, J., Matralis, H., Ioannides, T., and Avgouropoulos, G., PCT application No. PCT/SI01/00005, 2001.

19. Milne, T.A. et al., *Hydrogen from Biomass: State of the Art and Research Challenges*, Report No. IEA/H2/TR-02/001, NREL, Golden, CO, USA, 2002.

20. Day, D., Evans, R.J., Lee, J.W., and Reicosky, D., *Energy*, 30, 2558, 2005.

21. Deluga, G.A., Salge, J.R., Schmidt, L.D., and Verykios, X.E., *Science*, 303, 993, 2004.

22. Heck, R.M., Farrauto, R.J., and Gulati, S.T., *Catalytic Air Pollution Control: Commercial Technology*, 2nd ed., John Wiley & Sons, Inc., New York, 2002.

23. Shore, L. and Farrauto, R.J., in *Handbook of Fuel Cells: Fundamentals, Technology and Applications*, Vielstich, W., Lamm, A., and Gasteiger, H.A., Eds., John Wiley & Sons, Ltd., Chichester, U.K., 2003, pp. 211–218.

24. Ovesen, C.V., Clausen, B.S., Hammershøi, B.S., Steffensen, G., Askgaard, T., Chorkendorff, I., Nørskov, J.K., Rasmussen, P.B., Stoltze, P., and Taylor, P., *J. Catal.*, 158, 170, 1996.

25. Ruettinger, W., Ilinich, O., and Farrauto, R.J., *J. Power Sources*, 118, 61, 2003.

26. Bunluesin, T., Gorte, R.J., and Graham, G.W., *Appl. Catal. B*: Environmental, 15, 107, 1998.

27. Hilaire, S., Wang, X., Luo, T., Gorte, R.J., and Wagner, J., *Appl. Catal. A*: General, 215, 271, 2004.

28. Jacobs, G., Chenu, E., Patterson, P.M., Williams, L., Sparks, D., Thomas, G., and Davis, B.H., *Appl. Catal. A*: General, 258, 203, 2004.

29. Wheeler, C., Jhalani, A., Klein, E.J., Tummala, S., and Schmidt, L.D., *J. Catal.*, 223, 191, 2004.

30. Fu, Q., Kudriavtseva, S., Saltsburg, H., and Flytzani-Stephanopoulos, M., *Chem. Eng. J.*, 93, 41, 2003.

31. Koryabkina, N.A., Phatak, A.A., Ruettinger, W.F., Farrauto, R.J., and Ribeiro, F.H., *J. Catal.*, 217, 233, 2003.

32. Li, Y., Fu, Q., and Flytzani-Stephanopoulos, M., *Appl. Catal. B*: Environmental, 27, 179, 2000.

33. Kušar, H., Hočevar, S., and Levec, J., *Appl. Catal. B: Environmental*, 63, 194, 2006.

34. Avgouropoulos, G., Ioannides, T., Matralis, H.K., Batista, J., and Hočevar, S., *Catal. Lett.*, 73, 33, 2001.

35. Avgouropoulos, G., Ioannides, T., Papadopoulou, C., Batista, J., Hočevar, S., and Matralis, H.K., *Catal. Today*, 75, 157, 2002.

36. Sedmak, G., Hočevar, S., and Levec, J., *J. Catal.*, 213, 135, 2003.

37. Sedmak, G., Hočevar, S., and Levec, J., *J. Catal.*, 222, 87, 2004.

38. Sedmak, G., Hočevar, S., and Levec, J., *Top. Catal.*, 30, 31, 445, 2004.

39. GASEQ Ver. 0.78; available on-line at http://www.gaseq.co.uk.

40. Callaghan, C., Fishtik, I., Datta, R., Carpenter, M., Chmielewski, M., and Lugo, A., *Surf. Sci.*, 541, 21, 2003.

41. Mhadeshwar, A.B. and Vlachos, D.G., *J. Phys. Chem. B*, 108, 15246, 2004.

42. Qi, X. and Flytzani-Stephanopoulos, M., *Ind. Eng. Chem. Res.*, 43, 3055, 2004.

43. Shido, T. and Iwasawa, Y., *J. Catal.*, 136, 493, 1992.

44. Pintar, A., Batista, J., and Hočevar, S., *J. Colloid Interface Sci.*, 285, 218, 2005.

45. Sekizawa, K., Yano, S., Eguchi, K., and Arai, H., *Appl. Catal. A: General*, 169, 291, 1998.

46. Liu, Y., Hayakawa, T., Suzuki, K., and Hamakawa, S., *Catal. Commun.*, 2, 195, 2001.

9 Gasification of Biomass to Produce Hydrogen

Simone Albertazzi, Francesco Basile, and Ferruccio Trifirò

CONTENTS

9.1 INTRODUCTION

Fossil fuels such as coal, oil, and natural gas are used to generate a very large proportion of the electricity in the world. Firing with these fuels gives rise to carbon dioxide, which is a greenhouse gas discharged to the atmosphere. So there is keen interest in reducing the CO_2 emissions from sources such as fossil fuels. On the other hand, the carbon dioxide generated in the combustion of biofuels is not considered to contribute to the net CO_2 content of the atmosphere, since it is absorbed by photosynthesis when new biomass is growing [1, 2].

The use of biofuels for power generation is limited, although a number of smaller combined heat and power (CHP) plants have been built in recent years [3], all of them based on conventional technology, i.e., a boiler plant and a steam turbine cycle. The electrical efficiency of these plants is around 30% [4], and the ratio of electrical energy to thermal energy generated (called alpha value) is around 0.5 or below. Although there is potential development for these plants, the electrical efficiency and the alpha value cannot be expected to increase to any significant extent.

A technique that offers opportunities for achieving higher electrical efficiencies is based on the gasification of solid fuels and combustion of the gas thus produced in a gas turbine and in a steam cycle that follows. These integrated gasification combined cycle (IGCC) plants were originally developed for fossil fuels, but the principle can also be applied to biofuels. Several studies [5] have shown that well-optimized generation plants rated at 30 to 60 MW_e, based on pressurized gasification of wood fuel and integrated into a combined cycle, can achieve net electrical efficiencies of 40 to 50% and an overall efficiency of 85 to 90%, with competitive generation costs and low emission levels.

An interesting alternative to the production of power from biomass gasification is that of upgrading the gas to obtain a high-added-value mixture of CO and H_2. Hydrogen is an important raw material for the chemical industry and is a clean fuel that can be used in fuel cells and internal combustion engines [6, 7]. The main process for hydrogen production is currently the catalytic steam reforming of methane, light hydrocarbons, and naphtha. Partial oxidation of heavy oil residues and coal gasification are also alternative processes to produce hydrogen [8]. Renewable lignocellulosic biomass can be used as an alternative feedstock for hydrogen production. Two possible technologies that have been explored in recent years are steam gasification [9–13] and catalytic steam reforming of pyrolysis oils [14, 15]. The latter route begins with fast pyrolysis of biomass to produce bio-oil, which can be converted to hydrogen via catalytic steam reforming followed by, if necessary, a shift-conversion step. Moreover, the economy of scale can make possible a more efficient but more complex utilization of the pyrolysis slurry. Because the costs per kilogram of transporting low-density biomass (such as straw, wood chips, bagasse) are much higher than those of transporting high-density liquor, the slurries can be transported from a wide number of small and local pyrolysis plants to a large central gasification facility to produce the most valuable products [16].

The demand for hydrogen as a fuel derived from renewable sources will be increasing due to major advances in the field of hydrogen-based fuel-cell research [17]. While the technology for efficiently converting hydrogen to electricity at the required power levels is approaching commercialization, attempts have been made to directly power fuel cells with biogas (internal reforming), but these have been mostly unsuccessful [18, 19]. One major drawback in the reforming process is the formation of carbon monoxide (CO) as a gaseous by-product. In the 50-ppm range, CO acts as a poison to the fuel-cell catalysts by disproportionation of CO (carbon deposition). Variability of the biogas composition as well as the presence of trace quantities of sulfur are further drawbacks. Therefore, maximization of the hydrogen yield from biogas through controlled steam reforming followed by CO shift reactions (high temperature [HT] and low temperature [LT]) appears to be a less elegant but more economical and feasible solution, particularly with respect to CO minimization. A scheme of the different possibilities derived from biomass gasification is shown in Figure 9.1.

9.2 BIOMASS FEEDSTOCK

Biomass is mainly composed of the following constituents [20]:

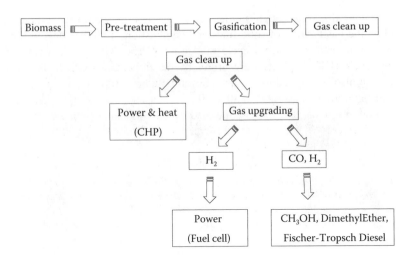

FIGURE 9.1 Scheme of the possibilities involved in biomass gasification.

Cellulose, a linear polymer of anhydroglucose $C6$ units with a degree of polymerization of up to 10,000. The long strains form fibers, which give biomass its mechanical strength but also a strongly anisotropic character.

Hemicellulose, a linear polymer of $C6$ and $C5$ units with a degree of polymerization of less than 200.

Lignin, which is a random three-dimensional structure of phenolic compounds.

Traditionally, biomass (mainly in the form of wood) has been utilized by humans through direct combustion, and this process is still widely used in many parts of the world. Biomass is a dispersed, labor-intensive, and land-intensive source of energy. Therefore, as industrial activity has increased, more concentrated and convenient sources of energy have been substituted for biomass. Biomass currently represents only 3% of primary energy consumption in industrialized countries [21]. However, much of the rural population in developing countries, which represents about 50% of the world's population, is reliant on biomass for fuel. Biomass accounts for 35% of primary energy consumption in developing countries, raising the world total to 14% of primary energy consumption. The Earth's natural biomass replacement represents an energy supply of around 3000 EJ (3×10^{21} J) per year, of which just under 2% is currently used as fuel. However, it is not possible to use all of the annual production of biomass in a sustainable manner. The main potential benefit in growing biomass especially for fuel is that, provided the right crops are chosen, it is possible to use poor-quality land that is unsuitable for growing food. Burning biomass produces some pollutants, including dust and the acid rain gases sulfur dioxide (SO_2) and nitrogen oxides (NO_x), but it produces less sulfur than burning coal. Carbon dioxide, the greenhouse gas, is also released. However, as this originates from harvested or processed plants, which have absorbed it from the atmosphere in the first place, no additional amounts are involved. A wide variety of biomass resources can be used as feedstock (Figure 9.2, [22]), divided into three general categories [23]:

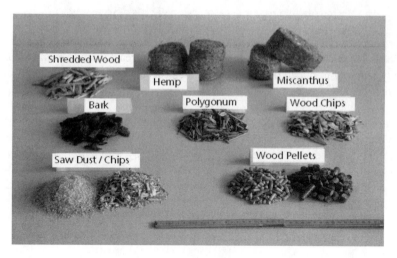

FIGURE 9.2 Biomass types from energy crops and forest products.

Wastes. Large quantities of agricultural plant residues are produced annually worldwide and are vastly underutilized. The most common agricultural residue is the rice husk, which makes up 25% of rice by mass. Other plant residues include sugar cane fiber (bagasse), coconut husks and shells, groundnut (peanut) shell, and straw. Included in agricultural residue is waste, such as animal manure (e.g., from cattle, chicken, and pigs). Refuse-derived fuel (RDF) is the combustible material in domestic or industrial refuse. It consists mainly of plant material but may also include some plastics. RDF may be used raw and unprocessed, partially processed, or highly processed in the form of pellets . These latter burn more efficiently and with lower emissions.

Standing forests. Wood fuels are derived from natural forests, natural woodlands, and forestry plantations (fuelwood and charcoal). These fuels include sawdust and other residues from forestry and wood-processing activities. Fuelwood is the principal source for small-scale industrial energy in rural areas of developing countries. However, large reforestation programs will be required to meet future energy demands as the world population grows. In industrialized countries, the predominant wood fuels used in the industrial sector are from wood-processing industries. The utilization of this residue for energy production at or near its source has the advantage of avoiding expensive transport costs. Domestic wood fuels are sourced principally from land clearing and logging residues.

Energy crops. Energy crops are those grown specially for the purpose of producing energy. These include short-rotation plantations (or energy plantations), such as eucalyptus, willows, and poplars; herbaceous crops, such as sorghum, sugar cane, and artichokes; and vegetable-oil-bearing plants, such as soya beans, sunflowers, cotton, and rapeseed. Plant oils are important, as they have a high energy density. The oil-extraction technology and the agricultural techniques are simple, and the crops are very hardy.

9.3 PRETREATMENT OF THE FEEDSTOCK

The characteristics of feedstocks as they are collected are often very different from the feed characteristics demanded by the conversion reactor, and several steps are usually required to match the feedstock to the process. The key requirements of the feed pretreatment system are [24]:

Reception and storage of incoming biomass until it is required by the conversion step. The logistics of ensuring a constant feed supply is very important, because most of the biomass is only available on a seasonal basis. In such cases, continuous operation of the conversion facility will require either extensive long-term storage of the feedstock or a feed reactor and pretreatment system that is flexible enough to accommodate multiple feedstocks.

Screening of the feedstock to keep particle sizes within appropriate limits and prevent contamination of the feedstock by metal or rocks.

Drying of the feedstock to a moisture content suitable for the conversion technology. Drying is generally the most important pretreatment operation, necessary for high cold-gas efficiency at gasification. Drying reduces the moisture content to 10 to 15%. Drying can either be done with flue gas or with steam. Steam drying results in very low emissions, and it is safer with respect to risks for dust explosion. However, using flue gas is the cheapest way to dry the feedstock.

Comminution of the feedstock to an appropriate particle size. To avoid pressure drop, sawdust and other small particles must be pelletized. Smaller particle can be used in fluidized-bed gasifiers.

Buffer storage of prepared feed immediately prior to the reactor.

The pyrolysis of biomass to produce the slurry to be pumped into a gasifier to produce hydrogen or synthesis gas (syngas) instead of burning can be classified as a pretreatment step. Pyrolysis of biomass can be described as the direct thermal decomposition of the organic matrix in the absence of oxygen to obtain an array of solid (char), liquid (oil), and gas products, depending on the pyrolysis conditions. The solid char can be used as a fuel in the form of briquettes or as a char–oil/water slurry, or it can be upgraded to activated carbon and used in purification processes. The gases generated have a low-to-medium heating value, but may contain sufficient energy to supply the energy requirements of a pyrolysis plant. The pyrolysis liquid is a homogenous mixture of organic compounds and water in a single phase, and it is commonly burned in a diesel stationary engine, but extraction can be carried out to obtain chemicals and other valuable products (food additives, perfumes).

Pyrolysis processes can be classified as slow, fast, and flash, depending on the applied residence time and heating rate. Slow pyrolysis has traditionally been used for the production of charcoal. Fast and flash pyrolysis of biomass at moderate temperatures have generally been used to obtain high yield of liquid products (up to 75 wt% on a dry biomass feed basis), and they are characterized by high heating rates and rapid quenching of the liquid products to terminate secondary conversion of the products [25]. High heating rates of up to 10^4 K·sec^1, at temperatures $<650°C$ and with rapid quenching, favor the formation of liquid products and minimize char and gas formation; these process conditions are often referred to as "flash pyrolysis." High heating rates to temperatures $>650°C$ tend to favor the formation of gaseous products at the expense of liquids. Slow heating rates coupled with low maximum temperatures maximize the yield of char. In slow pyrolysis, the reactions taking place are always in equilibrium because the heating period is sufficiently slow to allow equilibration during the thermochemical process. In this case, the ultimate yield and product distribution are limited by the heating rate. In fast and flash pyrolysis, there are a negligible number of reactions during the heat-up period.

Volatile residence time is a very important factor affecting yields of gaseous and liquid products in a biomass sample [26]. A low volatile residence time, obtained by rapid quenching, increases the liquid fraction. The particle size is also known to influence pyrolysis yield. This effect can be related to heating rate, in that larger particles will heat up more slowly, so the average particle temperatures will be lower, and hence volatile yields can be expected to be less. If the particle size is sufficiently small, it will be heated uniformly. Finally, the use of a sweeping gas allows a faster removal of the compounds from the hot zone, minimizing the unwished secondary reactions of cracking and polycondensation, thus achieving higher yields in the oil.

9.4 GASIFICATION

Gasification is the conversion by partial oxidation at elevated temperature of a carbonaceous feedstock into a gaseous energy carrier consisting of permanent, noncondensable gases. Development of gasification technology dates back to the end of the 18th century, when hot gases from coal and coke furnaces were used in boiler and lighting applications [27]. Gasification of coal is now well established, and biomass gasification has benefited from that sector [28]. However, the two technologies are not directly comparable due to differences between the feedstocks (e.g., char reactivity, proximate composition, ash composition, moisture content, density). Gasifiers have been designed in various configurations (Figure 9.3), but only the fluid-bed configurations are being considered in applications that generate over 1 MW$_e$ [4, 5]. Fluid-bed gasifiers are available from a number of manufacturers in thermal capacities ranging from 2.5 to 150 MW$_{th}$ for operations at atmospheric or elevated pressures, using air or oxygen as a gasifying agent. Ideally, the process produces only a noncondensable gas and an ash residue. However, incomplete gasification of char and the pyrolysis tars produce a gas containing several contaminants such as particulate, tars, alkali metals, fuel-bound nitrogen compounds, and an ash residue containing some char. The composition of the gas and the level of contamination vary with the feedstock, reactor type, and operating parameters (Table 9.1).

Since the mid-1980s, interest has grown on the subject of catalysis for biomass gasification [1]. The advances in this area have been driven by the need to produce a tar-free product gas from the gasification of biomass, since the removal of tars and the reduction of the methane content increase the economic viability of the

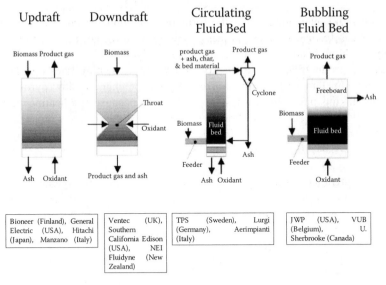

FIGURE 9.3 The main gasifier configurations. (Adapted from Bridgewater, A.V., Toft, A.J., and Brammer, J.G., *Renewable and Sustainable Energy Reviews*, 6, 181, 2002.)

TABLE 9.1
Gasifier Product Gas Characteristics

| Gasifier Configuration | Gas Composition (% v/v) | | | | | HHV |
	H_2	CO	CO_2	CH_4	N_2	($MJ/N \cdot m^3$)
Fluid bed (air-blown)	9	14	20	7	50	5.4
Updraft (air-blown)	11	24	9	3	53	5.5
Downdraft (air-blown)	17	21	13	1	48	5.7
Downdraft (oxygen-blown)	32	48	15	2	3	10.4
Multisolid fluid bed	15	47	15	23	0	16.1
Twin fluid bed	31	48	0	24	0	17.4

biomass gasification process. The criteria for the catalysts are fundamentally the same and can be summarized as follows:

1. The catalysts must be effective in the removal of tars.
2. If the desired product is syngas, the catalysts must be capable of reforming methane, also providing a suitable syngas ratio for the intended process.
3. The catalysts should be resistant to deactivation as a result of carbon fouling and sintering.
4. The catalysts should be easily regenerated.
5. The catalysts should be resistant to abrasion and attrition.
6. The catalysts should be cheap.

TABLE 9.2
Main Chemical Reactions of Biomass Gasification

Main Reactions Occurring in a Gasifier	ΔH^0_{298} (kJ/mol)
$VOC = CH_4 + C$	<0
$C + 1/2\ O_2 = CO$	−111
$CO + 1/2\ O_2 = CO_2$	−254
$H_2 + 1/2\ O_2 = H_2O$	−242
$C + H_2O = CO + H_2$	131
$C + CO_2 = 2\ CO$	172
$C + 2\ H_2 = CH_4$	−75
$CO + 3\ H_2 = CH_4 + H_2O$	−206
$CO + H_2O = CO_2 + H_2$	−41
$CO_2 + 4\ H_2 = CH_4 + 2\ H_2O$	−165

Adapted from Sutton, D., Kelleher, B., and Ross J.H.R., *Fuel Processing Technology*, 73, 155, 2001.

Catalysts for use in biomass conversion can be divided into two distinct groups, depending on the position of the catalytic reactor relative to that of the gasifier in the gasification process. The first group of catalysts (primary catalysts), directly added to the biomass prior to gasification, catalyzes the reactions listed in Table 9.2. The addition is performed either by wet impregnation of the biomass material or by dry mixing of the catalyst with it. These catalysts have the purpose of reducing the tar content and have little effect on the conversion of methane and C_{2-3} hydrocarbons in the product gas. They operate under the same conditions of the gasifier and are usually nonrenewable and consist of cheap disposable material. The second group of catalysts is placed in a reformer reactor downstream from the gasifier and it will be discussed later.

Dolomite, an ore with the general formula $MgCO_3 \cdot CaCO_3$, is a suitable catalyst for the removal of hydrocarbons evolved in the gasification of biomass. Dolomites increase gas yields at the expense of liquid products. With suitable ratios of biomass feed to oxidant, almost 100% elimination of tars can be achieved. The dolomite catalyst deactivates due to carbon deposition and attrition, but it is cheap and easily replaced. The catalyst is most active if calcined and placed downstream of the gasifier in a fluidized bed at temperatures above 800°C. The reforming reaction of tars over dolomite occurs at a higher rate with carbon dioxide instead of steam. Dolomite activity can be directly related to the pore size and distribution. A higher activity is also observed when iron oxide is present in significant amounts. Dolomite is also basic and thus does not react with alkali from the fuel. However, this material is not active in reforming the methane present in the product gas, and hence it is not a suitable catalyst if syngas is required. The main function of dolomite is to act as a guard bed for the removal of heavy hydrocarbons prior to the reforming of the lighter hydrocarbons to produce a product gas of syngas quality.

An alternative of dolomite is olivine, a mineral containing magnesium oxide, iron oxide, and silica. Olivine is advantageous because its attrition resistance is higher than that of dolomite. Moreover, pretreated olivine has a good performance in tar reduction, its activity being comparable with that of calcined dolomite [29]. Natural olivine presents good characteristics for use as a biomass gasification catalyst in a fluidized-bed reactor, but it is also useful as a nickel-support agent [30]. Iron presence helps in stabilizing nickel in reducing conditions. One part of nickel oxide seems to be included within the olivine structure and maintains the reducible nickel oxide on the olivine surface. On the other hand, nickel integration in the olivine structure leads to an increase of free iron oxide, which favors reverse water–gas shift reactions. In conclusion, this catalytic system seems to meet all the requirements of activity, stability, and attrition resistance for use in a fluidized bed for biomass steam gasification.

Alkali catalysts directly added to the biomass by wet impregnation or dry mixing reduce tar content significantly and also reduce the methane content of the product gas. However, the recovery of the catalyst is difficult and costly. Variable concentrations of alkali metals are included in the ash of several biomass types. Ash is an effective catalyst for the removal of tar when mixed with the biomass. Alkali catalysts directly added to the biomass in a fluidized-bed gasifier are subject to particle agglomeration. Alkali metal catalysts are also active as secondary catalysts. Potassium carbonate supported on alumina is more resistant to carbon deposition, but it is not as active as nickel, having a much lower hydrocarbon conversion.

9.5 GAS CLEANUP

Tar formation is one of the major problems to deal with during biomass gasification [2]. Tar condenses at reduced temperature, thus blocking and fouling process equipments such as engines and turbines. Tar-removal technologies can be divided into gas cleaning after the gasifier (secondary methods) and treatments inside the gasifier (primary methods). Although secondary methods are proven to be effective, treatments inside the gasifier are gaining much attention, as these can reduce the downstream cleanup.

In primary treatment, the gasifier is optimized to produce a fuel gas with minimum tar concentration. The different approaches of primary treatment are (a) proper selection of operating parameters, (b) use of bed additive/catalyst, and (c) gasifier modifications. The operating parameters such as temperature, gasifying agent, equivalence ratio, residence time, etc., play an important role in formation and decomposition of tar. There is a potential of using some active bed additives such as dolomite, olivine, char, etc., inside the gasifier. Ni-based catalysts are very effective not only for tar reduction, but also for decreasing the amount of nitrogenous compounds such as ammonia. Also, reactor modification can improve the quality of the product gas. The concepts of two-stage gasification and secondary air injection in the gasifier are of prime importance. However, primary measures cannot solve the purpose of tar reduction without affecting the useful gas composition and heating value. A combination of proper primary measures with downstream methods is observed to be very effective in all respects (Figure 9.4).

Secondary methods are conventionally used as treatments to the hot product gas from the gasifier [2]. These methods can be classified into two distinct routes: "wet" low-temperature cleaning and "dry" high-temperature cleaning. Conventional wet

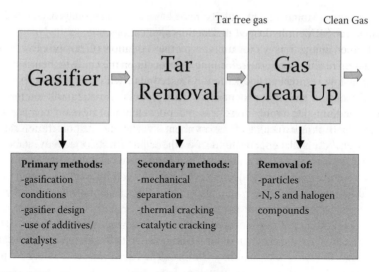

FIGURE 9.4 Tar removal approach in biomass gasification.

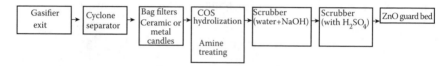

FIGURE 9.5 Schematic view of "wet" low-temperature cleaning.

low-temperature syngas cleaning (Figure 9.5) is the preferred technology in the short term. This technology has some efficiency penalties and requires additional waste-water treatment, but it is well established. Hot gas cleaning consists of several filters and separation units in which the high temperature of the syngas can partly be maintained, achieving efficiency benefits and lower operational costs. Hot gas cleaning is specifically advantageous when preceding a reformer or shift reactor, because these process steps have high inlet temperatures. Hot gas cleaning after atmospheric gasification does not improve efficiency because the subsequent essential compression requires syngas cooling anyway. Hot gas cleaning is not a commercial process yet, since some unit operations are still in the experimental phase. However, within some years, hot gas cleaning will become established and commercially available.

9.6 GAS UPGRADING BY REFORMING

The exit gas from the gasifier needs to be improved to a rather clean syngas to produce fuels or other products. The predominant commercial technology for syngas generation is steam reforming (SRM), in which methane and steam are catalytically and endothermically converted to hydrogen and carbon monoxide [31]. An alternative approach is partial oxidation (PO), the exothermic, noncatalytic reaction of methane and oxygen to produce a syngas mixture. SRM and PO produce syngas mixtures having appreciably different compositions. In particular, SRM produces a syngas having a much higher H_2/CO ratio. This, of course, represents a distinct advantage for SRM in hydrogen production applications.

A further innovation is the catalytic partial oxidation (CPO) process, in which the oxidation reactions and the reforming ones occur on the catalytic bed. This allows working at low residence time (a few milliseconds), with the advantages of small reactor dimension and high productivity. This process is particularly interesting for small-to-medium-size applications; it is under development by many companies (Air Liquide, Shell, Amoco), and at the moment it is at the demonstration stage by Snamprogetti. Catalysts containing Ni, Ni/Rh, or just Rh dispersed on high-temperature-resistant (above 1000°C) supports are suitable for this reaction.

A different approach is autothermal reforming (ATR), which combines PO with SRM in one reactor. The process is "autothermal" because the endothermic reforming reactions proceed with the assistance of the internal combustion (or oxidation) of a portion of the feed hydrocarbons, in contrast to the external combustion of fuel characteristic of conventional tubular reforming. ATR properly refers to a stand-alone, single-step process for feedstock conversion to syngas. However, the same basic idea can be applied to reactors fed by partially reformed gases from a primary reformer. Such reactors constitute a subcategory of ATR that is commonly called secondary reforming. Due to feed composition differences, in particular, the lower concentration of combustibles in secondary reformer feeds, ATR reactors and secondary reformers have different thermal and soot formation characteristics that require different burner and reactor designs. Nonetheless, the distinction between ATR and secondary reforming is not consistently drawn by technology users and vendors, with the result that secondary reformers often are referred to as ATRs. The prereformed feedstock of the secondary ATR may be assimilated to the stream coming from a gasifier (Table 9.3). In practice, oxygen-blown ATR has yet to see application in a large-scale methanol plant, although oxygen-blown secondary reformers have seen operation in a limited number of plants, such as the 2400 MTPD Conoco/Statoil methanol plant, of Haldor Topsøe design, that started in Norway in 1997. This plant, which also contains a prereformer upstream of the SMR, is said to be operating well.

It appears that considerable confidence is being placed in advances in the engineering tools now available for designing autothermal reforming burners and reactors [32, 33]. Both Lurgi and Haldor Topsøe claim to now have rigorous computer models to facilitate the scale-up and design of oxygen-blown ATRs. ICI claimed a similar capability with respect to their oxygen-blown secondary reformers. Accordingly, further dramatic cost reductions may require the application of still newer reforming technologies. One such development to watch is Exxon's oxygen-blown, fluidized bed ATR, which could offer increased potential for economies of scale.

Finally, it was found [34] that high-reforming conversion could be attained by using a fluidized-bed reactor in ATR of methane under pressurized condition. Fluidization of the catalysts could reduce and activate the catalyst, which was oxidized by oxygen near the inlet of the catalyst bed with the produced syngas. Catalysts with higher Ni content had higher reducibility and thus exhibited high conversion in ATR using a fluidized-bed reactor. Moreover, it was found [35] that carbon

TABLE 9.3
Typical Operating Conditions and Dry-Gas Compositions (% v/v) of Secondary Reformers

SECONDARY ATR (air-blown)	Reformer A [a]		Reformer B [b]	
	Inlet %	Outlet %	Inlet %	Outlet %
Pressure (bar)	30.0	29.0	30.4	29.7
Temperature (°C)	790	971	780	965
CH_4	9.4	0.2	7.8	0.2
CO_2	11.6	8.8	14.2	10.6
CO	8.3	11.5	8.5	11.7
H_2	70.2	57.1	68.6	55.5
N_2	0.5	22.1	0.9	21.8
Ar	0.0	0.3	0.0	0.2

[a] Adapted from *Ullmann's Encyclopedia of Industrial Chemistry*, 5th ed., A12, 1989, John Wiley & Sons, Inc., VCH, p. 238.
[b] Data adapted from *Catalyst Handbook, 2nd ed.*, M. Twigg (Ed.), 1989. Wolfe Publ., London.

deposition could be inhibited in a fluidized-bed reactor through the gasification of carbon with oxygen.

The ATR technology has considerable potential for further optimization, especially in combination with gas-heated reforming (GHR). During GHR, part of the heat in the ATR effluent is used for steam reforming and feed preheats in a heat-exchange-type reactor. There are two principally different layouts for incorporating GHR in combination with ATR [36]: a parallel and a series arrangement. In the parallel arrangement, the two reformers are fed independently, giving freedom to optimize the S/C ratio individually. In the series arrangement, all gas passes through the GHR unit and then the ATR. The commercial plants commonly use supported nickel catalysts [35, 36]. The catalyst contains 15 to 25 wt% nickel oxide on a mineral carrier (α-Al_2O_3, aluminosilicates, cement, magnesia). Before start-up, nickel oxide must be reduced to metallic nickel. This is preferably done with hydrogen but can also be done with natural gas or even with the feed gas itself at a high temperature (above 600°C, depending of the reducing stream). Required properties of the catalyst carriers include relatively high specific surface area, low pressure drop, and high mechanical resistance at temperatures up to 1000°C. The main catalyst poison in SMR plants is sulfur. Concentrations as low as 50 ppm give rise to a deactivating layer on the catalyst surface [33]. To some extent, activity loss can be offset by raising the reaction temperature. This helps to reconvert the inactive nickel sulfide to active nickel sites. In the specific case of biomass gasification, a number of alkaline salts and heavy metals and metal oxides particles can also act as additional poisons. Another cause of activity loss is carbon deposition, which can be avoided if a high steam(S)/C ratio is employed. However, economic evaluations indicate that the optimum S/C ratio tends to be low (2.5 to 3 v/v).

9.7 DOWNSTREAM OF THE REFORMER

According to the aims to convert the produced gas from biomass to synthesis gas for applications requiring different H_2/CO ratios, the reformed gas can be ducted to a conventional water–gas-shift (WGS) unit (see Section 9.7.1) to obtain the H_2 purity required for fuel cells, or directly to a hydrogenation/hydrogenolysis unit (see Section 9.7.2) to convert the residual hydrocarbons, in particular aromatic compounds and olefins, for applications requiring a H_2/CO ratio close to 2 (dimethyl ether, methanol, Fischer-Tropsch synthesis).

9.7.1 WGS UNIT

The WGS reaction is a critical step in fuel processors for preliminary CO cleanup and additional hydrogen generation prior to the CO preferential oxidation (PROX) or methanation step [37]. WGS units are placed downstream of the reformer to further lower the CO content and improve the H_2 yield. Ideally, the WGS stage(s) should reduce the CO level to less than 5000 ppm. The WGS catalysts have to be active at low temperatures, 200 to 280°C, depending on the inlet concentrations in the reformate. The reaction is moderately exothermic, with low CO levels resulting at low temperatures, but with favorable kinetics at higher temperatures. Under adiabatic conditions, conversion in a single bed is thermodynamically limited (as the reaction proceeds, the heat of reaction increases the operating temperature), but improvements in conversion are achieved by using subsequent stages with cooling and perhaps CO_2 removal between the stages. Because the flow contains CO, CO_2, H_2O, and H_2, additional reactions can occur, depending on the H_2O/CO ratio, and at high temperatures, the following reactions are favored: methanation, CO disproportionation, or decomposition. In industrial applications, the classical catalyst formulations employed are Fe–Cr oxide for the first stage (high-temperature shift [HTS]), typically in the range 360 to 400°C), and $Cu–ZnO–Al_2O_3$ for subsequent stages (low-temperature shift [LTS]), operating just above the dew point, the lowest possible inlet temperature, which is about 200°C, for good performance under steady-state conditions (CO exit concentration in the range of 0.1 to 0.3%).

9.7.2 HYDROGENATION/HYDROGENOLYSIS UNIT

Cobalt and Ni molybdates are the most widely used catalysts for the hydrogenolysis of sulfur compounds. They also promote most hydrogenation reactions [33]. The Ni-Mo catalysts behave in a similar way to Co-Mo catalysts, but these catalysts provide slightly more powerful hydrogenation, which can cause more hydrocracking. Co-Mo and Ni-Mo have low activity in the oxide form, and to obtain maximum activity they must be sulfided. Hydrogenolysis/hydrogenation reactions are usually carried out over these catalysts at temperatures between 280 and 400°C and space velocities of about 3000 h^{-1}. These catalysts are not normally active enough at lower temperatures, and there is the risk of hydrocarbon cracking at higher temperatures. The reaction pressure is normally set up by the inlet pressure to the reformer (30 to 50 bars).

9.8 FUEL CELLS

The principle of the fuel cell was first discovered in 1839 by Sir William R. Grove, who used hydrogen and oxygen as fuels catalyzed on platinum electrodes [18]. A fuel cell is defined as an electrochemical device in which the chemical energy stored in a fuel is converted directly into electricity. A fuel cell consists of an anode (negatively charged electrode) to which a fuel, commonly hydrogen, is supplied, and a cathode (positively charged electrode) to which an oxidant, commonly oxygen, is supplied. The oxygen needed by a fuel cell is generally supplied by feeding air. The two electrodes of a fuel cell are separated by an ion-conducting electrolyte. The input fuel is catalytically reacted (electrons removed from the fuel elements) in the fuel cell to create an electric current. The input fuel passes over the anode, where it catalytically splits into electrons and ions, and oxygen passes over the cathode. The electrons go through an external circuit to serve an electric load while the ions move through the electrolyte toward the oppositely charged electrode. At the electrode, ions combine to create by-products, primarily water and CO_2. Depending on the input fuel and electrolyte, different chemical reactions will occur. The main product of fuel-cell operation is the DC electricity produced from the flow of electrons from the anode to the cathode. The amount of current available to the external circuit depends on the chemical activity and amount of the substances supplied as fuels and the loss of power inside the fuel-cell stack. The current-producing process continues for as long as there is a supply of reactants, because the electrodes and electrolyte of a fuel cell are designed to remain unchanged by the chemical reactions. The stack is the main component of the power section in a fuel-cell power plant. The by-products of fuel cell operation are heat, water in the form of steam or liquid water, and CO_2 in the case of hydrocarbon fuel.

There are five types of fuel cells, on the basis of the electrolyte employed: alkaline fuel cells (AFC), phosphoric acid fuel cells (PAFC), proton-exchange membrane fuel cells (PEMFC), molten carbonate fuel cells (MCFC), and solid oxide fuel cells (SOFC). In all types, there are separate reactions at the anode and the cathode, and charged ions move through the electrolyte, while electrons move around an external circuit. Another common feature is that the electrodes must be porous, because the gasses must be in contact with the electrode and the electrolyte at the same time.

The advantages of the fuel cells are that they operate without combusting fuel and with few moving parts, but mainly because a fuel cell can be two to three times more efficient than an engine in converting fuel to electricity. A fuel cell resembles an electric battery in that they both produce a direct current by using an electrochemical process. A battery contains only a limited amount of fuel material and oxidant, which are depleted with use. Unlike a battery, a fuel cell does not run down or require recharging; it operates as long as the fuel and the oxidizer are supplied continuously from outside the cell. The main advantages of fuel cells are reflected by the following characteristics:

High-energy conversion efficiency
Extremely low emissions of pollutants

Extremely low noise or acoustical pollution

Effective reduction of greenhouse gas (CO_2) formation at the source compared with low-efficiency devices

Process simplicity for conversion of chemical energy to electrical energy

Depending on the specific types of fuel cells, other advantages may include fuel flexibility and existing infrastructure of hydrocarbon fuel supplies; cogeneration capability; modular design for mass production; and relatively rapid load response. Therefore, fuel cells have great potential to penetrate into markets for stationary power plants (for industrial, commercial, and residential home applications) but also for mobile power plants for transportation by cars, buses, trucks, trains, and ships, as well as human portable microgenerators. Because of these features, fuel cells have emerged in the last decade as one of the most promising new technologies for meeting the energy needs for power generation and transportation well into the 21st century [38]. Unlike power plants that use combustion technologies, fuel-cell plants that generate electricity and usable heat can be built in a wide range of sizes, from 200 kW, suitable for powering commercial buildings, to 100 MW, suitable to utility power plants. The disadvantages to be overcome include the costs of fuel cells, which are still considerably higher than conventional power plants per kilowatt. Moreover, the fuel hydrogen is not readily available, and thus on-site or onboard H_2 production via reforming is necessary. Finally, there are no readily available and affordable ways for onboard or on-site desulfurization of hydrocarbon fuels; therefore, the efficiency of fuel processing affects the overall system efficiency.

9.9 PERSPECTIVE

Biomass gasification is a central enabling technology for the clean, sustainable use of biomass in serving both CHP and fuel generation as well as on the much smaller scale of operation needed in distributed generation and in village and small-industry power applications in developing countries. Concerning CHP, the successful demonstration of the IGCC plant in Värnamo [39] and the widespread demonstration in India, China, and Europe have eliminated concerns about technical feasibilities. However, for the production of hydrogen gas or syngas from biomass gasification, there are still many improvements yet to be made, mainly due to the fact this route requires the rebuilding of existing plants.

REFERENCES

1. Sutton, D., Kelleher, B., and Ross J.H.R., *Fuel Processing Technology*, 73, 155, 2001.
2. Tijmensen, M.J.A., Faaij, A.P.C., Hamelinck, C.N., and van Hardeveld, M.R.M., *Biomass and Bioenergy*, 23, 129, 2002.
3. Devi, L., Ptasinski, K.J., and Janssen, F.J.J.G., *Biomass and Bioenergy*, 24, 125, 2003.
4. Beenackers, A.A.C.M., *Renewable Energy*, 16, 1180, 1999.
5. Bridgewater, A.V., *Fuel*, 74, 631, 1995.

6. Cox, J.L., Tonkovich, A.Y., Elliott, D.C., Baker, E.G., and Hoffman, E.J., in *Proceeding of Second Biomass Conference of the Americas*, Portland, OR, 21–24 Aug. 1995, p. 657.

7. Larson, E.D. and Katofsky, R.E., in *Advances in Thermochemical Biomass Conversion*, Bridgwater, A.V., Ed., Blackie, London, 1994, p. 495.

8. Garcia, L., French, R., Czernik, S., and Chornet, E., *Applied Catalysis A: General*, 201, 225, 2000.

9. Aznar, M.P., Corella, J., Delgado, J., and Lahoz, J., *Industrial Engineering Chemistry Resource*, 32, 1, 1993.

10. Aznar, M.P., Caballero, M.A., Gil, J., Olivares, A., and Corella, J., in *Making a Business from Biomass*, Overend, R.P. and Chornet, E., Eds., Pergamon Press, New York, 1997, p. 859.

11. Rapagna, S., Jand, N., and Foscolo, P.U., *International Journal of Hydrogen Energy*, 23, 551, 1998.

12. Turn, S., Kinoshita, C., Zhang, Z., Ishimura, D., and Zhou, J., *International Journal of Hydrogen Energy*, 23, 641, 1998.

13. Garcia, L., Salvador, M.L., Arauzo, J., and Bilbao, R., *Energy and Fuels*, 13, 851, 1999.

14. Wang, D., Czernik, S., Montané, D., Mann, M., and Chornet, E., *Industrial Engineering Chemistry Resource*, 36, 1507, 1997.

15. Wang, D., Czernik, S., and Chornet, E., *Energy and Fuels*, 12, 19, 1998.

16. Henrich, E. and Weirich, F., *Environmental Engineering Science*, 21, 53, 2004.

17. Effendi, A., Hellgardt, K., Zhang, Z.G., and Yoshida, T., *Fuel*, 84, 869, 2005.

18. Song, C., *Catalysis Today*, 77, 17, 2002.

19. Kirby, K.W., Chu, A.C., and Fuller, K.C., *Sens. Actuators*, B95, 224, 2003.

20. Janse, A.M.C., Westerhout, R.W.J., and Prins, W., *Chemical Engineering and Processing*, 39, 239, 2000.

21. Ramage, J. and Scurlock, J., in Biomass: renewable energy-power for a sustainable future, Boyle, G., Ed., Oxford University Press, 1996.

22. Tijmensen, M.J.A., Faaij, A.P.C., Hamelinck, C.N., and van Hardeveld, M.R.M., *Biomass and Bioenergy*, 23, 129, 2002.

23. Demirbas, A., *Energy Conversion and Management*, 42, 1357, 2001.

24. Bridgewater, A.V., Toft, A.J., and Brammer, J.G., *Renewable and Sustainable Energy Reviews*, 6, 181, 2002.

25. Yaman, S., *Energy Conversion and Management*, 45, 651, 2004.

26. Onay, O. and Kockar, O.M., *Renewable Energy*, 28, 2417, 2003.

27. Buekens, A.G., Maniatis, K., and Bridgwater, A.V., Introduction, in *Commercial and Marketing Aspects of Gasifiers*, Buekens, A.G., Bridgwater, A.V., Ferrero, G.L., and Maniatis, K., Eds., Commission of the European Communities, Brussels, 1990, p. 8.

28. Maniatis, K., Progress in biomass gasification: an overview, in *Progress in Thermochemical Biomass Conversion*, Bridgwater, A.V., Ed., Blackwell, Oxford, 2001.

29. Devi, L., Ptasinski, K.J., and Janssen, F.J.J.G., *Fuel Processing Technology*, 86, 707, 2005.

30. Courson, C., Udron, L., Wierczyski, D., Petit, C., and Kiennemann, A., *Catalysis Today*, 76, 75, 2002.

31. Wilhelm, D.J., Simbeck, D.R., Karp, A.D., and Dickenson, R.L., *Fuel Processing Technology*, 71, 139, 2001.

32. *Ullmann's Encyclopedia of Industrial Chemistry*, 5th ed., A12, Wolfe Publ., London, 1989, p. 238.

33. *Catalyst Handbook*, 2nd ed., Twigg, M., Ed., 1989.

34. Matsuo, Y., Yoshinaga, Y., Sekine, Y., Tomishige, K., and Fujimoto, K., *Catalysis Today*, 63, 439, 2000.
35. Tomishige, K., Matsuo, Y., Sekine, Y., and Fujimoto, K., *Catalysis Communications*, 2, 11, 2001.
36. Bakkerud, P.K., Gol, J.N., and Aasberg-Petersen, K., in H.S. Arpe (Ed.), Wiley Publ., VCH, 1989, *Natural Gas Conversion VII*, Elsevier, Amsterdam, 2004.
37. Ghenciu, A.F., *Current Opinion in Solid State and Materials Science*, 6, 389, 2002.
38. Katikaneni, S., Gaffney, A.M., Ahmed, S., and Song, C., *Catalysis Today*, 99, 255, 2005.
39. Albertazzi, S., Basile, F., Brandin, J., Einvall, J., Hulteberg, C., Fornasari, G., Rosetti, V., Sanati, M., Trifirò, F., and Vaccari, A., *Catalysis Today*, 106, 297, 2005.

10 Sustainable Biological Hydrogen Production

A. Inci Işli and T. Nejat Veziroglu

CONTENTS

10.1 INTRODUCTION

Hydrogen is accepted as the clean and new energy source of the future. The truth about the detrimental effects of greenhouse gases has directed the world to decarbonize the fossil fuels, and the technological developments in electricity generation are toward electrochemical energy conversion with fuel cells. The trend in fuel use is toward hydrogen. Hydrogen has a flexible energy carrier function as a remedy for the discontinuity in renewable resources such as sun and wind. It is also advantageous when hydroenergy is converted to hydrogen and transported very long distances.

Burning hydrogen does not produce any carbon dioxide; therefore it is a sustainable energy source. Consequently, there is worldwide research on the possibility of replacing conventional fossil fuels with hydrogen, a shift that seems to be both economically and environmentally beneficial.

Bockris [1], in 1969, stated that "due to the low viscosity of hydrogen, it would be cheaper to transport energy in the form of hydrogen in pipes rather than electricity in wires" and published an article pointing out that hydrogen should be a general medium of energy to eliminate pollution. From that time on, he continued his discussions on the "hydrogen economy," while Veziroglu started a series of conferences on hydrogen energy worldwide. Production of hydrogen from solar energy in Saudi Arabia, to be carried through pipelines throughout Europe, from wind in Argentina, and from hydropower in Canada are some of the expected future applications of this technology.

Winter [2] believes that hydrogen is the key to facilitating participation in the customary global energy trade system: it decarbonizes fossil fuels and makes them Kyoto conformable, and hydrogen and fuel cells decentralize the energy scheme. He advises that a hydrogen energy system should be put onto the agenda of the political class; two or more human generations will be engaged in the installation of the hydrogen economy. Energy sustainability without hydrogen is irrational.

Hydrogen by itself cannot solve the ever-increasing energy-requirement problem of the world population, but it may bring an immediate solution to the pollution sourced from the transportation sector.

Hydrogen is by far the most plentiful element in the universe, making up to 75% of the mass of all visible matter in stars and galaxies [3]. It is the simplest of all elements. The hydrogen atom, with a dense central nucleus and a single orbiting electron, is a large empty space, a chemically highly reactive arrangement, naturally combining into molecular pairs. Hydrogen has the second-lowest (to helium) boiling and melting points of all substances. It is a liquid below −253°C and a solid below −259°C at atmospheric pressure. Fuels that are gases at atmospheric conditions (like hydrogen and natural gas) are less convenient, as they must be stored as pressurized gas or as a cryogenic liquid.

Pure hydrogen is nontoxic, odorless, colorless, tasteless, and invisible in daylight. It diffuses to fill volumes quickly, and care should be taken to prevent leaks in confined areas. Hydrogen has the lowest atomic weight and therefore has very low density both as a gas and as a liquid. Gaseous hydrogen is approximately 7% of the density of air, and liquid hydrogen is approximately 7% of the density of water. Hydrogen's expansion ratio is 1:848, which means that hydrogen in its gaseous state at atmospheric conditions occupies 848 times more volume than it does in its liquid state. The molecules of hydrogen gas can diffuse through many materials considered airtight or impermeable to other gases; it is difficult to contain. Leaks of liquid hydrogen evaporate very quickly, with the hydrogen rising and becoming diluted very rapidly. When a small amount of activation energy in the form of a spark is provided to a mixture of hydrogen and oxygen, the molecules react vigorously and release a substantial amount of heat, with water (in the form of superheated vapor) as the final product. This reaction is reversible via electrolysis of water. Hydrogen has the highest energy-to-weight ratio of any fuel, which explains why it is used extensively in the space program. Hydrogen is flammable over a wide range of concentrations in air (4 to 75%) and it is explosive over a wide range of concentrations (15 to 59%) at standard atmospheric temperature. As a result, even small leaks of hydrogen have the potential to burn or explode. Hydrogen flames are very pale blue and are almost invisible in daylight due to the absence of soot. Constant exposure to hydrogen causes hydrogen embrittlement and leads to leakage or catastrophic failure in metal and nonmetal components.

This chapter presents the results of a literature survey carried out on the production of hydrogen by biological methods. After briefly summarizing the other methods of hydrogen production, biological production of hydrogen is evaluated from economic and environmental points of view, with an eye toward sustainable production. The latest progress on research and development is presented as a final conclusion on the vision for the future.

10.2 PRODUCTION OF HYDROGEN

During the present learning period, production of hydrogen from either conventional sources or new, sustainable, renewable sources should be studied in terms of the economics of production. It is a fact that the more renewable the source of production, the more sustainable will be the use of hydrogen energy.

Most of the hydrogen produced comes from:

Chemical synthesis, such as the production of chlorine, acetylene, styrene, or cyanide

Petrochemical cracking processes, such as catalytic reforming and cracking of crude oil during upgrading

Ethylene production at purities between 50.0 to 99.5% [4]

In general, this gas is mixed with the fuel in the system and burnt locally. An important production method of hydrogen is the steam reforming of natural gas and some light oil fractions, and this hydrogen is used within the same plant for desulfurization or ammonia production or methanol synthesis. Hydrogen production from biomass gasification and pyrolysis are greener methods than steam reforming of natural gas. Another well-known method is the electrolysis of water. These industrial methods of hydrogen production consume mainly fossil fuels and seldom hydroelectricity as an energy source.

On the other hand, "Biological hydrogen production methods are mostly operated at ambient temperatures and pressures, thus less energy intensive," say Das and Veziroglu [5]. Biological methods of hydrogen production are the research projects studied all over the world. These methods of direct production of hydrogen by algae and bacteria seem to be the promising methods of hydrogen production, since these are natural processes and the energy source of the bacteria is an organic waste that is expected to be removed from the environment. Biological hydrogen production combines the advantages of ecological production with clean combustion, i.e., reduction of waste treatment and carbon dioxide, even though it is difficult to expect that hydrogen produced biologically can compete with chemically synthesized hydrogen in the next few decades [6].

10.3 PRODUCTION OF BIOHYDROGEN

Biological hydrogen production is the production of hydrogen by microalgae and bacteria. Hydrogen produced from renewable resources such as water, organic wastes, or biomass, either biologically or photobiologically, is termed "biohydrogen." There has been active research in this field since the 1920s on bacterial hydrogen production and since the 1940s on microalgal hydrogen production [7]. Applied research and development on the photosynthetic processes were begun in the early 1970s, while dark fermentations were neglected for some time. Still, there are many publications in this field, but advances toward practical applications have been minimal [7]. In the "IEA Agreement on the Production and Utilization of Hydrogen, 2002 Annual Report" it is estimated that the development of practical processes will require significant scientific and technological advances and relatively long-term — more than 10 years — basic and applied research and development [8].

In natural environments, hydrogen is produced biologically during the anaerobic conversion of organic matter to volatile acids, but this hydrogen is used up by methane bacteria and is not available. There are three enzymes known to function in the biological hydrogen production, namely nitrogenase, Fe-hydrogenase and Ni–Fe hydrogenase, and biohydrogen production is basically dependent on the presence of one of these enzymes.

Biological hydrogen can be produced from:

1. *Biophotolysis–photoautotrophic production of hydrogen*, which involves the use of microalgae — cyanobacteria and eukaryotic microalgae — to generate hydrogen from sunlight and water under anaerobic conditions, called "direct biophotolysis." The catalyst is a hydrogenase enzyme that is extremely sensitive to oxygen, a by-product of photosynthesis.
   Reaction: $4 H_2O + \text{light energy} \Rightarrow 2O_2 + 4H_2$

   Alternative processes include hydrogen-evolving reactions separated in time or space; these are called "indirect biophotolysis" by nitrogen-fixing cyanobacteria. The indirect biophotolysis process begins with microalgal fixing of carbon dioxide into stored substrates like starch, which are then converted to hydrogen by algae themselves or by photosynthetic or fermentative bacteria [9].

2. *Photoheterotrophic production of hydrogen*, which involves nitrogen fixation by photoheterotrophic bacteria or cyanobacteria, a process catalyzed by the nitrogenase enzyme, particularly in the absence of nitrogen. The nitrogenase enzyme catalyzes the evolution of hydrogen in the absence of nitrogen [6]. That nitrogenase enzyme is known to be sensitive to oxygen and inhibited by ammonium ions.
   Reaction: $C_2H_4O_2 + 2H_2O + \text{light energy} \Rightarrow 2CO_2 + 4H_2$

3. *Photofermentation production of hydrogen* is the photo-decomposition of organic compounds by photosynthetic bacteria. Here the photosystem itself is not powerful enough to split water [6]. Under anaerobic conditions, these bacteria can use simple organic acids, food-processing and agricultural wastes, or even hydrogen disulfide as electron donors. The overall biochemical pathways for the photo-fermentation process can be expressed as [5]:
   Reaction: $C_2H_4O_2 + ATP \Rightarrow$ ferrodoxin; ferrodoxin $+ ATP \Rightarrow$ nitrogenase $\Rightarrow H_2$

   Carbon monoxide can also be used for the production of hydrogen using a microbial-shift reaction by the photosynthetic bacteria, $(CO + H_2O \Rightarrow CO_2 + H_2)$ [5].

4. *Dark fermentation production of hydrogen*, which involves anaerobic metabolism of pyruvate, formed during the catabolism of various substances. Thus, biomass feedstock is converted to various fermentation products like hydrogen, acetate, glycerol. Fermentative processes carry the advantage of using organic wastes as the feedstock.

   The breakdown of pyruvate is catalyzed either by a formate enzyme system or a reduced ferredoxin system [7]. In these biological systems, the pyruvate generated by glycolysis is used under anaerobic conditions to produce acetyl CoA (coenzyme A), from which ATP (adenosine triphosphate) can be derived, and one of the catalyst enzymes from which hydrogen can be derived. Hydrogen gas can be produced from the microbial fermentation of organic substances, like organic wastes, only if the methane bacteria can be prevented from using up the hydrogen produced. The heat treatment and low pH during culture growth are the methods accepted for remedy.

The hard collective work of IEA (International Energy Agency) Hydrogen Implementing Agreement (HIA) study groups have led to significant increases in the fundamental understanding of the genetics, biochemistry, and physiology of hydrogenase functions. A valuable databank of microalgae experimental studies has been maintained since 1995. In 2003, the IEA-HIA group published an annual report on the photobiological production of hydrogen, "Task 15" [10]. Canada, Japan, Norway, Sweden, the Netherlands, the United Kingdon, and the United States participate in the 5-year hydrogen research and development "Task 15 Group."

The initial activity of the "Task 15 Group" was the "BioHydrogen 2002" Congress held in Ede, the Netherlands, where 150 scientists discussed the state of the art of biological hydrogen production. A new Nordic collaborative group (Sweden, Finland, Iceland, Denmark, Estonia, and Latvia) has decided to develop a strong "Nordic research network" with significant training of graduate students.

The NREL (National Renewable Energy Laboratory) Hydrogen Group has initiated a "fuel gas to hydrogen" study. The aim is to utilize photosynthetic bacteria to shift crude-oil synthesis gas from biomass into hydrogen in darkness. The Norwegian Research Council has focused on biophotolytic hydrogen production by marine microalgae. Research on the molecular characterization of cyanobacterial hydrogenases continues in Sweden. In Japan, an international joint research grant program for the development of a molecular device of photo-hydrogen is in progress. Seven research groups from Japan, France, and Germany are studying a biomolecular device to produce hydrogen gas from water by light energy. This research, which is being carried out mainly in the United States at UC Berkeley, focuses on maximizing the solar conversion efficiency and hydrogen production of photosynthetic organisms by minimizing the chlorophyll antenna size of photosynthesis.

Plans are under way for the study of hydrogen fermentation in the Netherlands, Japan, and Canada. Some small farms generate raw hydrogen gas from the fermentation of sweet sorghum juice in the Netherlands. In addition, photoheterotrophic fermentation, whereby hydrogen is produced from organic acids (predominantly acetate), is studied at Wageningen University, The Netherlands. In Japan, a high-efficiency hydrogen-methane fermentation process using organic wastes is being studied as a highly efficient anaerobic digestion system. In Canada, a project to study hydrogen production through fermentation has started.

Suitable photobioreactor systems aiming to scale up hydrogen production from the laboratory scale to the commercial sector are being developed worldwide.

In 2001, Das and Veziroglu [5] presented a survey on biological hydrogen production processes and discussed developments in this field. They listed the names of microorganisms used for hydrogen generation and corresponding references in the form of tables. Comparison of different biological hydrogen production processes are tabulated in detail by giving the name of the working organisms, raw materials used, hydrogen production rates, and the major products of the process. The identified organisms — *Anabaena* spp.; *Rhodopseudomonas* spp.; *Rhodobacter* spp.; *Rhodospirillum* spp.; mixed microorganisms containing *Phormidium valderianum*, *Halobacterium halobium*, and *Escherichia coli*; *Clostridium butyricum*; and *Enterobacter* spp. — use biomass, organic wastes from processing plants, lactate glucose, or cellulose to produce hydrogen (60 to 90% by volume), carbon dioxide, oxygen, and fatty acids biomass. Hydrogen production rates vary between 0.30 and 29.63 mmol hydrogen per gram per dry-cell hour. Das and Veziroglu [5] have pointed out that the rate of fermentative hydrogen production is always faster than that of the photosynthetic hydrogen production. Most of the biological processes are operated at ambient temperature (30 to 40°C) and normal pressure. The organic acids produced as by-products should not be discharged without treatment, and the gases produced should be separated. Also, the moisture in the gas mixture must be reduced if a reasonable heating value of the fuel is to be obtained [5].

In 2002, Hallenbeck and Benemann [7] reviewed several approaches in biological hydrogen production, taking into consideration the limiting processes. They concluded that the limiting factor for direct biophotolysis is the low-energy density in the light-driven processes for biological conversion of solar energy to hydrogen. Low-energy densities result in increased bioreactor surface areas and consequently higher costs for the production of hydrogen fuel. A similar study by researchers in Japan showed that, compared with wild-type strains, microalgal mutants with reduced antenna sizes exhibited a 50% increase in productivity in continuous laboratory cultures operating at high light intensities.

The oxygen sensitivity of the hydrogenase enzyme in direct biophotolysis has been the other key problem for the past three decades, a problem that can be overcome using oxygen absorbers. Hallenbeck and Benemann [7] concluded that direct biophotolysis seems to suffer from the barriers of oxygen sensitivity, limitations in light-conversion efficiencies, and problems with gas capture and separation. They believe that indirect biophotolysis processes are still at the conceptual step. In the case of photofermentation, the limitations are observed to be the high energy demand of the nitrogenase enzyme, low solar conversion efficiencies, and the requirement for anaerobic photobioreactors with large areas. They suggest the use of dark fermentations instead of photofermentations in the production of hydrogen from organic wastes. Hydrogen-producing fermentative microorganisms are known by their biochemistry and physiology, but very little is presently known about the attainable yields. These topics are under active study.

In 2002, Akkerman et al. [6] investigated photobiological hydrogen production, photochemical efficiency, and bioreactor design, and then organized a table giving the details of hydrogen production rates, yields, and efficiencies of photoheterotrophic bacteria. These strains are *Rhodobacter*, *Rhodopseudomonas*, and *Rhodospirillum* type bacteria, and the light efficiencies are in the range of 0.1 to 35%.

Hawkes et al. [11] have collected information on continuous fermentative hydrogen production in the laboratory and pointed out that continuous processes are best operated at 30°C, at a pH around 5.5, and hydraulic retention time approximately 8 to 12 h for simple substrates. They propose on-line control, e.g., gas flow, gas composition, and liquid-redox-potential monitoring, to prevent deviations from acetate/butyrate/hydrogen-producing metabolism and washout of spore formers following process disturbances [11].

Related to the fermentative hydrogen production, future research is required on the mobilization of fermentable feedstock from cheap lignocellulosic biomass through the development of extrusion techniques and a cost-effective photofermentation [12]. Lay et al. [13] studied the feasibility of biological hydrogen production from an organic fraction of municipal solid waste and developed a simple method for estimating the hydrogen production potential and rate. Van Ginkel, Sung, and Lay [14] converted the chemical energy in a sucrose-rich synthetic wastewater to hydrogen energy, and noted the reaction conditions giving the highest conversion efficiency, which was 46.6 ml hydrogen per gram of chemical oxygen demand per liter. Oh et al. [15] investigated the relative effectiveness of low pH and heat treatment for enhancing biohydrogen gas production from microbial fermentation of organic substrates. They suggested that low pH was, without heat treatment, sufficient to control hydrogen losses to methanogens in mixed batch cultures. They reached high hydrogen gas concentrations (57 to 72%) in all tests. Oh et al. [16] designed a membrane bioreactor system for the production of hydrogen gas and used complex wastewaters with high concentrations of hydrogen-producing bacteria and short retention times. They observed membrane resistance due to internal fouling and the reversible cake resistance.

The Tenth Anaerobic Digestion World Congress was held in Montreal in August 2004, and fermentative hydrogen production was a topic that drew much attention. Noike et al. [17] presented the results of continuous hydrogen production experiments conducted by using bean curd manufacturing waste as an organic substrate at pH 5.5 and 35°C. The authors showed that increasing the substrate concentration and adding nitrogen enabled continuous hydrogen production. Hussy et al. [18] produced fermentative hydrogen from energy crops and food-processing coproducts in both batch and continuous operation and, using a commercial on-line hydrogen sensor, measured the evolved hydrogen gas as 50 to 70% hydrogen in carbon dioxide. Lay et al. [19] performed a full-factorial design to determine the effects of pH and hydraulic retention time on hydrogen production in a chemostat reactor using 3% waste yeast. They also applied a reverse-transcriptase polymerase chain reaction to identify the bacterial diversity. The results showed a maximum hydrogen production rate of 460 ml per gram volatile suspended solids per day at pH 5.8 and hydraulic retention time of 32 h.

Teplyakov et al. [20] studied the creation of continuous and sequential membrane bioreactors for fuel gas production. The first one was a solar bioreactor where algae consumed carbon dioxide and produced oxygen; the second one was an anaerobic bioreactor where methanogenic organisms produced biogas; the third was a bioreactor where purple bacteria consumed residual low-molecular organics to produce carbon dioxide and hydrogen. Three membrane systems separated the gases evolved from each reactor as oxygen, methane, and hydrogen, respectively.

Eroglu et al. [21] used a photobioreactor and converted the chemical energy in the olive mill wastewater, a sole nutrient source, to hydrogen gas by *Rhodobacter sphaeroides* and achieved remarkable reductions in the chemical and biological oxygen demand and phenol concentrations. Besides the main pure hydrogen gas product, they produced valuable by-products like carotenoid and polyhydroxybutyrate. Previously, three different biosystems were designed for hydrogen production at the Middle East Technical University in Turkey: one by *Rhodobacter sphaeroides*, another by *Halobacterium halobium* and *Escherichia coli*, and the third one involving photoelectrochemical hydrogen production by *H. halobium* [22].

10.4 SUSTAINABILITY OF BIOHYDROGEN PRODUCTION

The future sustainability of hydrogen production depends mainly on the sources used in production. Therefore, hydrogen production from nonrenewable resources should only be considered if it makes economic sense. Hydrogen production using biological methods is a promising pathway for the sustainability of this clean energy carrier and source. When it comes to the sustainability of biohydrogen production, both photosynthetic and fermentative production methods compete. A stable, continuous, and long-term generation of hydrogen on high-strength complex biomass substrates is cost competitive with other renewables. Beneman [9] has estimated a cost of $10/GJ of hydrogen energy produced by a two-stage indirect biophotolysis plant, a cost that is significantly higher than current competitive fossil fuel prices. There are other cost estimates in the literature that are close to this estimate, like $10 to 15/GJ hydrogen produced photobiologically and $19.2/GJ hydrogen produced by dark-fermentation processes [12].

Photobiological hydrogen production from water using solar energy is IEA-HIA's expected novel biological source of sustainable and renewable energy with no greenhouse gas problem. The tentative goal of "Task 15" is:

To obtain a light conversion of 3% into hydrogen gas in biophotolysis, i.e., the biological production of hydrogen from water and sunlight using microalgal photosynthesis

To increase the typical 10% hydrogen yields from storage carbohydrates in the algae in hydrogen fermentations (based on a stoichiometry of 12 moles hydrogen per mole glucose) to 30% or higher [10]

No direct biophotolysis process has been advanced beyond laboratory experimentation [6]. Until a means of overcoming the oxygen inhibition is found, the photoautotrophic process will remain impractical for application [6]. In the future, genetic engineering might produce mutants with increased oxygen tolerance. Photosynthetic efficiencies decline at full sunlight. A great number of the photons captured during photosynthesis are being wasted as heat or fluorescence. This light-saturation effect is a major reason why algal production is low at high light conditions. Some remedial measures are rapid mixing, dilution of light incident on the surface of the algal cultures, and algal mutants with reduced chlorophyll contents that would absorb and waste fewer photons [7]. The use of light-attenuation devices that transfer sunlight into the depths of a dense algal culture is also a potential remedial measure. The productivity of photobioreactors is dependent on the light regime inside the reactor, so optimization of the reactor surface area is crucial for an efficient design [6]. A central research and development need in photobiological hydrogen production is the reduction of chlorophyll antenna sizes to increase the photosynthetic efficiencies.

On the other hand, side products are important in the sustainability of every process. Photosynthetic bacteria produce low concentrations of fatty acids. However, fermentative hydrogen production processes produce high concentrations of such fatty acids as lactic acid, acetic acid, butyric acid, all of which are high-organic-content by-products that require treatment. During the fermentation process, the oxygen in the product gas mixture should be separated by alkaline pyrogallol solution, the carbon dioxide adsorbed on KOH solution, and the moisture removed by a dryer or a chiller unit [5].

Hydrogen production from organic substrates is expected to be bioenergetically more favorable than from water. Waste utilization during the photosynthetic or fermentative hydrogen production is an important factor in the future sustainability of hydrogen energy.

Genetic studies on hydrogenase negative gene are attempting to maximize the amount of hydrogen production by manipulation of the metabolic scheme. The immobilized whole-cell systems have several advantages over suspended-cell systems. This design simultaneously causes hydrogen production [5].

Finally, it can be concluded that, for a sustainable biohydrogen production, future research must address the low-energy-density limitation on the light-driven processes, the oxygen barrier in the hydrogenase's operations, and the gas-separation problem at the end of hydrogen production processes. Photobioreactor areas and costs will be studied in the future. Research has shown that the rates of fermentative processes are faster than those of photosynthetic processes in the production of biological hydrogen [5]. The high energy requirement of nitrogenase and difficulties in the design of anaerobic photofermentators make the dark-fermentation processes preferable to photofermentation processes.

REFERENCES

1. Bockris, J.O'M., The origin of ideas on a hydrogen economy and its solution to the decay of the environment, *Int. J. Hydrogen Energy*, 27, 731, 2002.
2. Winter, C.-J., Into the hydrogen energy economy-milestones, *Int. J. Hydrogen Energy*, 30, 681, 2005.
3. *Hydrogen Fuel Cell Engines and Related Technologies, Module 1: Hydrogen Properties*, U.S. DOE 2001. College of the Desert, Palm Desert, CA, USA.
4. Zittel, W. and Wurster, R., The Prospects for a Hydrogen Economy Based on Renewable Energy: Ireland's Transition to Renewable Energy, HyWeb, Thurles, Tipperary; available on-line at http://www.hydrogen.org/index-e.html, 1 Nov. 2002.
5. Das, D. and Veziroglu, T.N., Hydrogen production by biological processes: a survey of literature, *Int. J. Hydrogen Energy*, 26, 13, 2001.
6. Akkerman, I. et al., Photobiological hydrogen production: photochemical efficiency and bioreactor design, *Int. J. Hydrogen Energy*, 27, 1195, 2002.

7. Hallenbeck, P.C. and Benemann, J.R., Biological hydrogen production: fundamentals and limiting processes, *Int. J. Hydrogen Energy*, 27, 1185, 2002.

8. Elam, C.C., IEA *Agreement on the Production and Utilization of Hydrogen*, 2002 Annual Report, IEA/H2/AR02, National Renewable Energy Laboratory, USA .

9. Beneman, J.R., *Process Analysis and Economics of Biophotolysis of Water, IEA Agreement on the Production and Utilization of Hydrogen*, Report No. IEA/H2/TR2-98, Walnut Creek, CA, USA, March 1998.

10. Lindblad, P., *IEA Agreement on the Production and Utilization of Hydrogen*, 2003 Annual Report, Uppsala University, Sweden; available on-line at www.ieahia.org, 2003.

11. Hawkes, F.R. et al., Sustainable fermentative hydrogen production: challenges for process optimization, *Int. J. Hydrogen Energy*, 27, 1339, 2002.

12. Reith, J.H., Wijffels, R.H., and Barten, H., *Bio-Methane and Bio-Hydrogen: Status and Perspectives of Biological Methane and Hydrogen Production*; available on-line at www.ecn.nl/biomassa/biohydrogen.en.html, p. 150, 2005.

13. Lay, J.J., Lee, Y.J., and Noike, T., Feasibility of biological hydrogen production from organic fraction of municipal solid waste, *Water Res.*, 33, 2579, 1999.

14. Van Ginkel, S., Sung, S., and Lay, J.J., Biohydrogen production as a function of pH and substrate concentration, *Environ. Sci. Technol.*, 35, 4726, 2001.

15. Oh, S.E., Van Ginkel, S., and Logan, B.E., The relative effectiveness of pH control and heat treatment for enhancing biohydrogen gas production, *Environ. Sci. Technol.*, 37, 5186, 2003.

16. Oh, S.E. et al., Biological hydrogen production using a membrane bioreactor, *Biotechnol. Bioeng.*, 87, 119, 2004.

17. Noike, T. et al., Continuous hydrogen production from organic waste, in *Proc. Anaerobic Digestion, 2004, 10th World Congress*, Montreal, 2004, p. 856.

18. Hussy, I. et al., Fermentative hydrogen production from energy crops and food processing co-products, in *Proc. Anaerobic Digestion, 2004, 10th World Congress*, Montreal, 2004, p. 862.

19. Lay, J.J. et al., Influences of pH and hydraulic retention time on anaerobes converting beer processing wastes into hydrogen, in *Proc. Anaerobic Digestion, 2004, 10th World Congress*, Montreal, 2004, p. 869.

20. Teplyakov, V.V. et al., Lab-scale bioreactor integrated with active membrane system for hydrogen production: experience and prospects, *Int. J. Hydrogen Energy*, 27, 1149, 2002.

21. Eroglu, E. et al., Photobiological hydrogen production by using olive mill wastewater as a sole substrate source, *Int. J. Hydrogen Energy*, 29, 163, 2004.

22. Eroglu, I. et al., Substrate consumption rates for hydrogen production by *Rhodobacter sphaeroides* in a column photobioreactor, *J. Biotechnol.*, 70, 103, 1999.

11 Hydrogen-Based Technologies for Mobile Applications

Neal Hickey, Paolo Fornasiero, and Mauro Graziani

CONTENTS

11.1 INTRODUCTION

11.1.1 DRIVING TOWARD THE HYDROGEN ECONOMY

The idea of a world in which energy needs are met by hydrogen instead of carbon-based fuels is certainly appealing and has captured both the public and scientific imagination. This idea is summed up in the phrase, "the hydrogen economy." In the ideal scenario, hydrogen would be produced from water by electrolysis, powered by renewable sources such as solar or wind energy, and used in fuel cells with air, producing only water as a product. The far-reaching implications of this environmentally neutral water-hydrogen-water cycle perhaps best represent the attractiveness of the hydrogen economy. It would lay to rest the specter of diminishing fossil fuel reserves and the energy crisis that will inevitably come if current energy use is not modified. It would also address another great specter of our times that has captured the public imagination, if not completely the scientific community: global warming or climate change. The sociopolitical consequences would be enormous. Indeed, the establishment of the hydrogen economy has been described as potentially having a bigger impact than the Industrial Revolution.

By now it seems to be generally believed that the hydrogen economy will become a reality. Extensive research and development programs, especially in Europe, North America, and Japan, testify to this belief. However, international cooperation, through agreements such as the International Partnership for the Hydrogen Economy (IPHE), mean that the challenge is being affronted at a global level [1]. Against this, it is also known that the task is not straightforward, and that many breakthroughs must be made to achieve the final goal.

In this chapter we discuss the technologies that are being investigated to achieve the transition to a hydrogen economy in the case of mobile sources. To set the tone, a number of general points should be made. First, the discussion will focus primarily on environmental aspects of hydrogen use. Second, the technologies discussed here are not necessarily specific to mobile sources, and indeed many are common to all sectors of the hydrogen economy (power generation, home heating, etc). Finally, given the enormity of the task and the particular needs, it is commonly accepted that transport applications based on renewable sources are a long-term objective and that a transition period, in which carbon-based fuels are used to produce hydrogen, is necessary. The end game would be to unite the technologies for hydrogen production from renewable sources (developed in the meantime) with those developed for a transport infrastructure. Only then it can be said that the hydrogen economy has truly been achieved.

11.1.2 Air Pollution from Mobile Sources: Prevention and End-of-Pipe Technologies

11.1.2.1 Air Pollution from Mobile Sources

Air pollution can generally be defined as "the presence in the earth's atmosphere of one or more contaminants in sufficient quantity to cause short- or long-term deleterious effects to human, animal or plant life, or to the environment" [2]. Such contaminants are divided into two main types, viz., (a) primary pollutants, which are emitted directly into the atmosphere; and (b) secondary pollutants, which are formed in the atmosphere by subsequent reaction of primary pollutants. Air pollution generated by combustion processes is currently a major cause for concern. The hazardous effects of human-made air pollution, mainly as a result of the burning of coal, have been recognized for centuries, especially since the Industrial Revolution. However, the emergence of petroleum products as a primary source of fuel and the large acceleration of their use during the 20th century has added a whole new dimension to the problem. Essentially, the massive increase in the use of petroleum products in the internal combustion (IC) engines used for various forms of transport has led to greatly increased problems of *community* pollution. Gasoline-fueled automobile emissions have been the major offender. These problems have led to the introduction of catalytic control of exhaust gas emissions (catalytic converters) as a standard feature of modern automobiles.

Table 11.1 lists the typical compositions of untreated, engine-out exhaust emissions from the most common types of diesel- and gasoline-fueled IC engines [3–7]. The primary pollutants — CO, HCs, NO_x, and PM (particulate matter) — are regulated by environmental legislation, while CO_2 and SO_x are considered undesirable emissions, and efforts are being made to reduce their occurrence, as discussed in the following subsections.

11.1.2.1.1 Carbon Monoxide

Carbon monoxide (CO) is produced as a result of insufficient oxygen (O_2) for the complete oxidation of hydrocarbons to carbon dioxide (CO_2) and water (H_2O), either on an overall or a local basis. Thus, CO emissions are high when the engine is operated with an excess of fuel with respect to O_2 (so called rich conditions), but for local mixing reasons, they do not reduce to zero under lean conditions when there is an excess of O_2 [3]. The primary concern surrounding this pollutant is its high toxicity, as its affinity for hemoglobin (carboxyhemoglobin), which carries O_2 from the lungs to the tissues in the form of oxyhemoglobin, is 200 times greater than that of O_2.

TABLE 11.1
Typical Composition of Untreated Engine-Out Exhaust Emissions

Exhaust Component [a]	Four-Stroke, Spark-Ignited Gasoline Engine	Four-Stroke, Spark-Ignited, Lean-Burn Gasoline Engine	Diesel
NO_x	100–400 ppm	≈1200 ppm	350–1000 ppm
HC	500–5000 ppm C	≈1300 ppm	50–330 ppm C
CO	0.1–6.0%	≈1300 ppm	300–1200 ppm
O_2	0.2–2.0%	4–12%	10–15%
H_2O	10–12%	12%	1.4–7%
CO_2	10–13.5%	11.5%	7%
SO_x	15–60 ppm	20 ppm	10–1000 ppm
PM	Low	Low	65 mg m^3

[a] Dinitrogen makes up the balance of the exhaust.

Source: Reprinted from Kaspar, J., Fornasiero, P., and Hickey, N., *Catal. Today*, 77, 419, 2003. With permission from Elsevier.

11.1.2.1.2   Hydrocarbons

The presence of hydrocarbons (HCs) in exhaust emissions is usually referred to as unburned fuel. However, evaporation losses, which occur at all stages (from refining, to refueling, to use), also contribute to the presence of airborne hydrocarbons. These depend on the characteristics of the fuel, and the problem is more acute for gasoline, which is more volatile than diesel. As HC combustion is thermodynamically a very favorable process, the appearance of the unburned HCs is due to a lack of temperature homogeneity during combustion, which in turn is related to ignition inhibition; or it may be the result of physical phenomena in the engine that allow them to escape combustion (adsorption, trapping). They may also result from various reaction processes that take place during combustion (cracking, hydrogenation, dehydrogenation, etc.).

In reality, the collective designation of HCs covers a multitude of specific compounds. As they are considered together, their concentrations are often reported as ppm of carbon equivalent (ppm C, as in Table 11.1). Initially, HC emissions were thought to present a relatively low toxic threat. As will be seen in the next section, the main concern was their role as precursors in the build-up of the secondary pollutant ozone (O_3) through their interaction with NO_x in the presence of sunlight to produce photochemical smog. In fact, both NO_x and HCs are often referred to as ozone precursors. However, the effects of long-term exposure to many of the suspected carcinogens contained in HC emissions have not been well established and may yet prove to be injurious to health. In particular, in the United States, attempts to monitor levels of about 190 individual compounds, collectively known as toxic air pollutants (TAPs), have been under way since the early 1990s. In the case of automotive emissions, benzene, buta-1,3-diene, formaldehyde, acetaldehyde, and polynuclear aromatic hydrocarbons (PAHs, taken together) are of particular concern and closely monitored, although not regulated.

11.1.2.1.3   Nitrogen Oxides

There are three routes to formation of nitrogen oxides (NO_x) during fuel combustion [3]:

Thermal NO_x species are formed by the radical-chain Zeldovich mechanism. The concentrations of thermal NO_x formed is exponentially dependent on temperature.

Prompt NO_x species are formed in hydrocarbon-rich flames via the interactions of hydrocarbon radicals with nitrogen and can be formed at temperatures (1600 K) well below those typical of fuel combustion (>1800 K).

Fuel NO_x is formed by the oxidation of fuel-bound nitrogen, i.e., nitrogen-containing species in the fuel itself.

In relation to the formation of NO_x in gasoline-fueled automobile IC engines, there are two salient features: (a) thermal NO_x is the dominant source under most engine conditions, and therefore combustion-temperature regulation is an important control factor, and (b) NO is effectively the only NO_x species emanating from the engine and therefore relevant to post-combustion control. The reason for the latter point is that at the temperature of fuel combustion, the equilibrium between NO and NO_2 lies far to the left:

$$NO + SO_2 \leftrightharpoons NO_2 \qquad\qquad (11.1)$$

NO presents a very low toxicological threat. On the other hand, NO_2, is a highly toxic substance, with an affinity for hemoglobin that is 300,000 times that of O_2. It can irritate the mucous membrane and has the ability to form nitric acid in the lungs. As mentioned above, a major environmental problem associated with NO_x species related to the simultaneous presence of NO and HC_s and their further reaction to form secondary pollutants during photochemcial smog episodes. Once in the atmosphere, NO can be slowly converted into NO_2 via Equation 11.1, for which the equilibrium then lies to the right. The NO_2 formed can undergo a further photochemical dissociation to NO and atomic oxygen. The atomic oxygen produced reacts with O_2 to form ozone. A circular reaction process is completed when the NO reacts with O_3 to form NO_2 and O_2. The levels of O_3 are regulated by this cycle. In the presence of hydrocarbons, however, this cycle is broken, resulting mainly in increased NO oxidation to NO_2, tropospheric ozone buildup, and the formation of other hazardous secondary pollutants such as peroxyl acyl nitrates (PANs) and formaldehyde. In this context, the European Environment Agency monitors levels of NO_x, CO, CH_4, and non-methane volatile organic compounds (NMVOCs) as ozone precursors [8]. The potential hazard of automobile exhaust pollution was pointed out as early as 1915, but it was not until the mid-1940s that the first acute community problem attributed to this source was first identified. The Los Angeles type of photochemical smog problem, was the prototype for what has become a worldwide phenomenon. The detrimental health effects in such episodes include increased respiratory ailments and eye irritation.

11.1.2.1.4   Particulate Matter

Particulate matter (PM) is a collective term for solid and liquid particles, or their mixtures, suspended in the air [3]. They are subdivided according to size, which can range across four orders of magnitude, from visible soot particles (coarse PM) to much finer particles with diameters of up to 10 or 2.5 μm (PM_{10} and $PM_{2.5}$). The issue has come to the forefront quite recently, with the observation that their concentrations in urban areas are unacceptably high. As with all forms of pollution, there are multiple sources, which in the case of PM include natural sources (pollen, sand, etc.). In the case of transport, diesel applications are the worst offenders, with PM emissions from gasoline engines being rather low. In fact, they are not regulated for such engines.

The three main types of diesel PM are the soluble organic fraction (SOF), soot, and sulfates. Sulfates arise from the presence of sulfur in the fuel, which is converted into SO_2 during combustion and then into SO_3 (SO_x). These gases dissolve easily in water from the exhaust gases to form hydrated sulfuric acid ($H_2SO_4 \cdot 7H_2O$), which can in turn interact with other gases and particles in the air to form sulfates and other products. Thus, primary and secondary pollutants are formed. The collective name for all of these is sulfates. Other sources of solid PM in transport include mechanical formation from brake attrition, tire attrition, and road surface degradation. They are of health concern because PM easily reaches the deepest recesses of the lungs, and many studies have linked PM with a series of respiratory-related aliments such as aggravated asthma, aggravated coughing and difficult or painful breathing, chronic bronchitis, and decreased lung function.

11.1.2.1.5   Sulfur Oxides

As outlined above, the issue of sulfur oxides (SO_x) cannot be completely separated from particulates. One of the main concerns surrounding airborne SO_x is their acid nature, which can be damaging to people and the environment. NO_x also has this acid-forming potential. Transport makes a relatively small contribution to SO_x levels, with industry being the main source.

11.1.2.1.6   Carbon Dioxide

Combustion of fossil-based fuels necessarily leads to formation of carbon dioxide (CO_2). Although not considered a pollutant, in recent years CO_2 has come under scrutiny because it is a direct greenhouse gas. Although the so-called global-warming potential (GWP) of CO_2 is lower than other greenhouse gases (it is, in fact, used as the base to calculate this relative scale) [9], there is much debate centered around CO_2 because its atmospheric concentration is increasing, with automotive emissions making a significant contribution. It has been estimated that the level of atmospheric CO_2 has risen from 280 ppm to 370 ppm since the Industrial Revolution [10]. This has led to sometimes-catastrophic predictions on the effects of global warming and climate change and, ultimately, to the Kyoto Protocol, in which many countries committed to reduce CO_2 levels [11].

11.1.2.2   Factors Affecting Pollutant Formation

The typical exhaust-gas compositions given in Table 11.1 are average values, which are of course interdependent. The exact composition at any given time largely

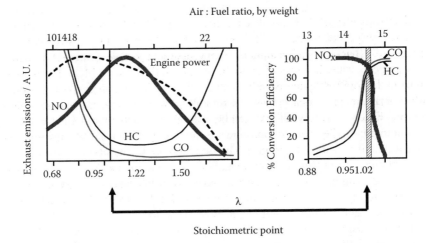

FIGURE 11.1 Effect of air/fuel ratio on automotive gasoline engine emissions and dependence of the simultaneous conversions of NO_x, CO, and HC on the air–fuel ratio in a three-way catalyst. (Adapted from Church, M.L., Cooper, B.J., and Willson, P.J., *Catalyst Formulations 1960 to Present*, SAE 890815, Society of Automotive Engineers, Warrendale, PA, 1989; McCabe, R.W. and Kisenyi, J.M., *Chem. Ind. (London)*, 15, 605, 1995. With permission.

depends on two closely related factors which predominate in the formation of the three primary pollutants (CO, NO_x, and HCs) — air-to-fuel ratio (A/F, denoted by λ) and temperature — these, in turn, are determined by driving conditions [12, 13]. The exact composition at any give time thus largely depends on these two factors. Figure 11.1 shows the variation in the production of NO_x, CO, and HCs as the A/F ratio varies. A value of $\lambda = 1$ indicates the point where the ratio is stoichiometrically balanced, i.e., when the amount of oxygen (air) present is exactly that require to fully combust all of the fuel present (to CO_2 and H_2O). This corresponds to an A/F ratio of 14.7. Combustion mixtures with low A/F ratios ($\lambda < 1$) are said to be fuel rich (net reducing), and thus form the highest amounts of CO and HC. Their levels of production decrease through the stoichiometrically balanced ratio ($\lambda = 1$) and into the fuel-lean situation ($\lambda > 1$, net oxidizing), where most of the fuel is consumed by the available O_2, resulting in production of small amounts of CO and HCs. At high values of λ, the output of HCs begins to increase because combustion becomes inefficient.

For NO_x, the situation is slightly more complicated. In fuel-rich conditions, little NO_x is formed due to a lack of availability of O_2, which is largely consumed in combustion. The amount of NO_x formed continues to increase as the A/F ratio increases through stoichiometry ($\lambda = 1$) into fuel-rich stoichiometries ($\lambda > 1$). The amount formed then goes through a maximum and starts to decrease. This is because the temperature of combustion falls, as the heat produced from the fuel combustion is required to heat up an increasing amount of air. As outlined above, the NO_x produced in an internal-combustion engine is mainly thermal in origin, and the drop in temperature results in a decrease in its production. Another temperature-related feature is the situation immediately after start-up, when the engine temperature has not reached its normal operating level. In this period, the lowest concentrations of NO are formed, while CO and HCs levels are high.

11.1.2.3   Existing Technologies for Automotive Pollution Control

11.1.2.3.1   Three-Way Catalysts

Two general options can be considered for automotive pollution control: prevention and cure. End-of-pipe technologies can be viewed as the cure option. Environmental legislation has largely been responsible for the introduction and subsequent development of end-of-pipe technologies. The first legislative controls of auto emissions were introduced in California as early as 1947. The chronic local pollution episodes experienced there have led to the most severe emission standards in the world and, as such, have shaped legislative policy elsewhere. The first national (federal) exhaust emission standards in the United States were introduced in 1968. Since then, environmental legislation has been adopted in many countries worldwide [14]. Most countries base their automotive environmental legislation on legislation and test procedures introduced in the United States, Europe, or Japan.

Initial strategies to reduce emissions were based on prevention: engine modification (crankcase controls) allowed CO and HC_s emissions to be reduced without tackling exhaust gases themselves. After the introduction of mandatory exhaust controls in the United States, Canada, and Japan and prior to the 1975 model year, emission limits were also met by engine modification without the use of catalytic control. However, increasingly stringent limits meant that end-of-pipe catalytic control eventually had to be introduced. The significant attenuation of CO and HCs required in that year led to the introduction of the so-called conventional oxidation catalysts (COCs — Pd, Pt/Al_2O_3) together with the use of unleaded gasoline. Engines were operated lean to provide the O_2 necessary for the oxidation catalyst to function. Some degree of NO_x control was achieved by exhaust-gas recirculation. As NO_x limits became stricter, COCs were superseded by a combination of COC and three-way catalysts (TWCs) and, finally, by TWCs only.

TWCs represent the current state-of-the-art technology for end-of-pipe emission control for gasoline automobiles. Their function is the simultaneous attenuation of the emission levels of three primary classes of pollutants — CO, NO_x, and HCs, and their name derives from the ability to simultaneously convert the three:

$$CO + SO_2 = CO_2 \tag{11.2}$$

$$\text{Hydrocarbons} + O_2 = H_2O + CO_2 \tag{11.3}$$

$$NO_x + CO \text{ and } H_2 = N_2 + CO_2 + H_2O \tag{11.4}$$

TWCs represent one the technological success stories of the past 30 years. Pushed by ever more stringent legislation, their formulations have evolved significantly since they were first introduced. However, their improvement is a result of a combination of factors, not just better formulations: materials advances (better mechanical stability), improved engine characteristics (onboard diagnostics, improved carburation), improved fuel characteristics (reformulated fuels), etc., have all contributed to their success.

The physical arrangement of a modern automotive catalyst consists of a thin layer of the porous catalytic material (wash-coat) coated on the channel walls of a ceramic (cordierite, $2MgO \cdot 2Al_2O_3 \cdot 5SiO_2$) or sometimes metal monolith. The channels of the monolith are axially orientated in the direction of the exhaust gas flow to ensure efficient flow-through and prevent a pressure buildup in the exhaust system. As would be expected, the exact composition and manufacturing processes used vary with the manufacturer and are subject to confidentiality, but a number of general observations can be made (see [14, 15] and references therein). They all contain highly dispersed noble metals (NMs) particles supported on doped and stabilized (multi-component) high surface area alumina support. Noble metals represent the key component of the TWCs, as the catalytic activity occurs at the metal center. Specifically, Rh is added to promote NO dissociation, while Pt and Pd are the metals of choice to promote the oxidation reactions. Interaction with the various components of the wash-coat critically affects the activity of the supported NMs. Notably, Pd has extensively been added to TWC formulations starting from the mid-1990s in an effort to produce less expensive, Rh-free materials. Better A/F control and modification of the support provided high NO_x conversion, comparable with the traditional Rh/Pt catalyst. With respect to the support, the replacement of CeO_2 with better-performing CeO_2-based materials, specifically CeO_2-ZrO_2 mixed oxides, in the formulations of the TWCs has significantly increased their performance. The following positive effects have been attributed to the ceria-based components:

Promotion of the noble metal dispersion
Enhancement of the thermal stability of the Al_2O_3 support
Promotion of the water–gas-shift (WGS) and steam-reforming reactions
Promotion of catalytic activity at the interfacial metal-support sites
Promotion of CO removal through oxidation, employing lattice oxygen
Storage and release of oxygen under respectively lean and rich conditions; the quantity of oxygen stored or released is known as the oxygen storage capacity (OSC)

The last ability is particularly important to minimize the effect of inevitable oscillations in the A/F ratio. In fact, the A/F ratio must be kept within a narrow window near the stoichiometric composition, as indicated in Figure 11.1. This is achieved by monitoring the oxygen content with a λ-sensor to adjust the A/F ratio. A second λ-sensor placed after the converter is used as part of an onboard diagnostics (OBD) system to check the correct functioning of the TWC.

Once operational, TWCs convert more than 98% of the pollutants. The outstanding issue in relation to TWCs is emissions just after start-up, before the catalyst has reached operating temperature. In addition to the development of new TWCs that are active at low temperature, other possibilities being considered include the use of upstream HC adsorbers to trap initial HC emissions, which are subsequently converted after heat-up; electrical heating of the TWC; and the development of close-coupled catalysts (CCCs): TWCs with high thermal stability to be placed closer to the engine. In this way, the catalyst warm-up time is reduced and limits emissions immediately after start-up [15].

11.1.2.3.2   Lean DeNO$_x$ Catalysts

Lean engine operation (diesel or lean-burn gasoline) has the advantage of producing less NO, as the temperature of combustion is lower, and less CO and HCs, as the combustion is more complete due to the excess of oxygen. The improved fuel economy also means that overall CO_2 emissions are reduced. However, even though less primary NO$_x$ is formed, the oxidizing conditions (see Table 11.1) mean that TWCs are ineffective for NO$_x$ reduction. Extensive research has been conducted into possible catalysts to reduce NO$_x$ under the oxidizing conditions of the exhaust, using the HCs present (see Table 11.1). This research has been reviewed [15, 16]. The materials investigated can be grouped into the following categories: Pt/Al_2O_3 and related systems, Cu-ZSM5 and related systems, metal oxide catalysts, and Ag-based systems. Despite all of the research, it is true to say that all suffer from problems such as insufficient activity or low hydrothermal stability, which make them unsuitable for widespread application in transport.

A different approach to the problem of lean DeNO$_x$, that of the storage-reduction catalyst (SRC), has been developed by Toyota [17, 18]. Here, the NO$_x$ produced in lean operation is trapped or stored on a Pt-Ba catalyst. During short switches to rich operation, the stored NO$_x$ species pollutants are reduced. A disadvantage is the sulfur sensibility of the storage material, which adsorbs SO$_x$ species more strongly than NO$_x$ species. This means that it must be used only with low sulfur content fuels. However, the Toyota process is by far the most effective NO$_x$ removal method available, and its commercialization has probably largely contributed to a decrease in interest in direct catalytic solutions.

11.1.2.3.3   Diesel Applications

Particulate matter (PM) remains a particular problem for diesel engines, and two technologies are commonly used to control emissions: diesel oxidation catalysts for the liquid fraction of the PM [19] and particulate filters for the dry carbon (soot) fraction [20]. As indicated above, DeNO$_x$ is also an unresolved problem in the case of diesel engines, with the added complication that the amount of gaseous HCs to perform the reduction is small. In principle, the liquid PM could be used to reduce NO$_x$, but this is even more difficult to achieve. The idea of adding a reducing agent has also been tested. For example, urea is a potential solution that has been demonstrated for trucks [21]. However, there are a number of general problems associated with this approach that makes its application to personal automobiles doubtful. These include the space considerations of including an additional (urea) tank onboard, the risk of ammonia slip, and the absence of a urea distribution network.

11.1.2.3.4   Prevention

Pollution prevention in many ways appears to be a more obvious solution, and it is indeed an approach that is actively pursued. As may be inferred from the above, a whole range of measures aimed at pollution prevention are in fact adopted in conjunction with end-of-pipe technologies. These include engine modification (e.g., exhaust-gas recirculation), and the use of reformulated (with low aromatic content) and low-sulfur fuels pollution. For example, the problem of Pb was tackled by removing lead from fuels. The same approach is underway for SO_x emissions with the use of low-sulfur-content fuels. A more preventive approach, for example, is the use of alternative fuels such as liquefied petroleum gas (LPG) and natural gas (NG), which are intrinsically cleaner than diesel or gasoline, resulting in much lower values for the pollutants than those shown in Table 11.1 [22]. These could also be combined with postcombustion catalytic control. However, as will be discussed below, the transformation to a hydrogen economy would represent a switch in strategy from the combined approach currently used with end-of-pipe technologies to a purely preventive approach; indeed, it would eliminate the need for these technologies. In many ways, a transfer to a renewable hydrogen economy is the ultimate preventive measure.

11.1.3 Advantages/Disadvantages of H_2

While the advantages of H_2 as a fuel source are considerable and have led to much enthusiasm in some quarters, there are of course also drawbacks. Although there have been reports on the potentially negative environmental consequences of the release of large amounts of H_2 into the atmosphere [23, 24], there seems to be wider general agreement that when it is possible to use H_2 it would be a good thing. In most cases, therefore, the drawbacks highlighted relate not to the use of hydrogen but on how to arrive at the hydrogen economy, and these are thus more technical in nature [25]. The challenge of establishing a hydrogen economy can be divided into four categories: production, distribution, storage, and end use. Here, we will limit the comments mainly to the transport sector. These initial considerations will be discussed in more detail in Sections 11.2 to 11.4.

11.1.3.1 Production

A fundamental question in any strategy on how to establish H_2 as a viable energy source is how to produce H_2 in sufficient quantities to meet current energy demands. While hydrogen is the most abundant element on earth, less than 1% is present as H_2 [26]. Thus, H_2 must be produced from other sources. Industrial production of hydrogen is an established technology; however, the specific needs of the transport sector give rise to a number of problems that hinder wide-scale application of this technology. In all cases, the cost of H_2 is higher than equivalent amounts of currently used fuels and in some cases, such as the ideal scenario (electrolysis), considerably higher. The biggest advantage from a production point of view is flexibility, given the large number of options at least theoretically available.

11.1.3.2 Distribution

The absence of an adequate hydrogen distribution network perhaps represents the biggest barrier. This is especially true in the case of the personal automobile. It is clear that the transport sector is quite varied, and while it is possible to envisage economically more attainable solutions for fleet vehicles, massive investment is needed to establish a distribution infrastructure for automobiles.

11.1.3.3   Storage

This probably represents the second biggest barrier. While various storage technologies exist, the amount of hydrogen currently stored is rather limited. Of the best storage solutions available, safety is an issue. The development of an efficient, simple, and safe hydrogen storage method represents a goal of fundamental importance.

11.1.3.4   End Use

Here there seem to be only advantages. H_2 has a higher energy content and is cleaner than any other fuel [27]. However, while its environmental advantages are not in doubt, it must be remembered that H_2 is only as clean as the source used to produce it. As stated above, H_2 production from renewable sources is not considered immediately possible, and the transition requires the use of fossil-based fuels. In this situation, H_2 produces zero emissions only at the point of use. This could potentially resolve the problems of urban pollution discussed previously, but H_2 production would still result in undesirable emissions, which necessitates a sort of environmental accountancy in which the debit/credit columns are not always immediately obvious, above all in relation to CO_2 production. For example, the introduction of TWCs led to higher CO_2 emissions because of the need to operate engines at $\lambda = 1$. Legislation that envisaged zero-emission vehicles (ZEVs) did not take into account the fact that the source of electricity used to power vehicles may or may not produce airborne emissions. Reformulated fuels require more extensive processing and subtract H_2 from the H_2 pool. Biofuels from renewable sources are often considered as having a zero CO_2 impact. This is not true, as clearly demonstrated by examples of life-cycle-assessment (LCA) studies. In relation to CO_2 emissions, it is clear that complementary strategies, such as sequestration and forestation, are necessary if the commitments of the Kyoto Protocol are to be respected.

The existence of fuel cell technology, which uses H_2 very efficiently, is another apparent advantage. However, for automobiles, the only valid comparison between the efficiency of various engines is the so-called well-to-wheel efficiency, in which all stages from production to end use are considered. On this basis, the advantage of fuel cells over gasoline IC engines is less, as the H_2 has to be produced from other sources, and this requires energy. Cost is another very important factor. Although prices are decreasing in all areas relating to hydrogen use, any change from fossil-based fuels will have a greater or smaller added cost factor. In fact, it is unlikely that the transition be viable purely from an economic point of view and will only be acceptable if other factors are considered, such as environmental benefits.

The biggest imponderable relates to public acceptance. H_2 is perceived to be unsafe, which is usually traced back to the Hindenburg disaster. Although it has been reported that the problem related to the covering material used on the zeppelin [28], removing this perception of danger will not be easy. In addition, as a new technology, there will necessarily be a resistance to H_2, as history teaches, especially if any vehicle performance characteristics are compromised. Indeed, there have been many cases of new technologies that failed because of indifferent public reception.

11.2 HYDROGEN PRODUCTION METHODS

11.2.1 HYDROGEN PRODUCTION BY REFORMING

Figure 11.2 outlines the main options available for production of H_2 for use as an energy source. It should be noted that the hydrogen economy envisages combined

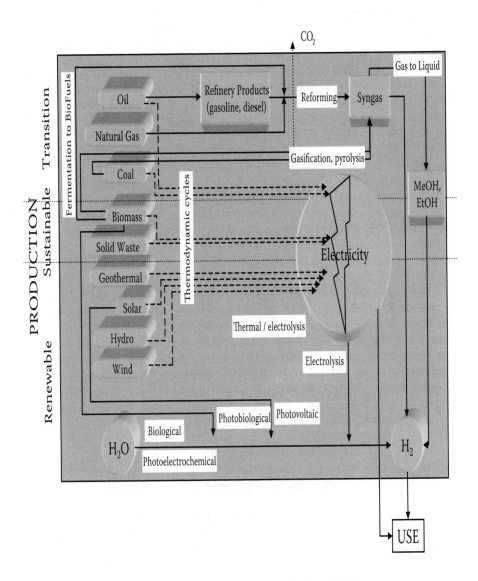

FIGURE 11.2 Scheme of the main options for the production of hydrogen.

H_2 and electricity generation. Electricity generation from nuclear energy could also be included in the scheme, but this is not an option available everywhere. Clearly, the choice is quite varied. The most realistic short-term option in the transition to a sustainable hydrogen economy is catalytic production from H_2 fossil fuels, although in principle this could be produced catalytically using any hydrogen-containing compound as a fuel. In general, for hydrocarbons or oxygenates, H_2 can be produced by one of three chemical reactions, or combinations thereof, collectively known as reforming reactions: steam reforming (SR), dry (or CO_2) reforming (DR), and partial oxidation (PO). These reactions can be represented as follows:

Steam reforming (SR)

$$C_xH_yO_z + (x - z)\, H_2O = x\, CO + (x{-}z + y/2)\, H_2 \tag{11.5}$$

$$C_xH_yO_z + (2x - z)\, H_2O = x\, CO_2 + (2x{-}z + y/2)\, H_2 \tag{11.6}$$

Dry reforming (DR):

$$C_xH_yO_z + (x{-}z)\, CO_2 = (2x{-}z)\, CO + y/2\, H_2 \tag{11.7}$$

Partial oxidation (PO):

$$C_xH_yO_z + (x+y/2{-}z)/2\, O_2 = x\, CO + y/2\, H_2O \tag{11.8}$$

$$C_xH_yO_z + (x{-}z)/2\, O_2 = x\, CO + y/2\, H_2 \tag{11.9}$$

Direct catalytic decomposition is another option, but this is realistically a possibility only for methanol and will not be further discussed here. Partial oxidation can also proceed noncatalytically (often denoted POX, to distinguish it from PO). Equations 11.5 through 11.9 represent the idealized reactions, and other reactions may occur at the catalyst surface:

Water–gas-shift reaction (WGSR):

$$CO + H_2O = CO_2 + H_2; \ \Delta H° = -41\ \text{kJ·mol}^{-1} \tag{11.10}$$

Reverse water–gas-shift reaction (rWGSR):

$$CO_2 + H_2 = CO + H_2O; \ \Delta H° = 41\ \text{kJ·mol}^{-1} \tag{11.11}$$

Oxidation reactions:

$$CO + \smallint O_2 = CO_2 \tag{11.12}$$

$$C_xH_yO_z + (2x +y/2 - z)/2\, O_2 = x\, CO_2 + y/2\, H_2O \tag{11.13}$$

$$H_2 + \int O_2 = H_2O \qquad (11.14)$$

Carbon formation reactions:

$$C_xH_y = a\ C + C_{x-a}H_{y-2a} + a\ H_2\ (a \leq x) \qquad (11.15)$$

$$2CO = C + CO_2 \qquad (11.16)$$

$$CO + H_2 = C + H_2O \qquad (11.17)$$

In reality, reforming reactions are mechanistically quite complicated and are composed of multiple steps and reactions. This makes it difficult to establish with certainty the exact mechanism. For example, while some reports claim a direct partial-oxidation mechanism (Equation 11.8), it has been established that, in the case of many Ni-based catalysts, the reaction takes places by complete combustion at the front of the catalytic bed (Equation 11.13), followed by steam and dry reforming (Equations 11.5 and 11.7) further along. In all cases, the mechanistic complexity of Equations 11.5 through 11.9 is indicated by the fact that the product stream produced invariably contains CO_2 and H_2O, along with unreacted fuel.

Before giving an overview of industrial-scale hydrogen production, a distinction should be made between the reforming reactions discussed above and industrial processes based on these reactions. SR, PO, and DR offer H_2/CO mixtures of various compositions (3, 2, and 1, respectively, in the case of methane). Industrial processes often use multiple reformers and variations of conditions to tailor the final composition to the required use. Some processes, such as autothermal reforming (ATR) (discussed below) are combinations of the reforming reactions. Differences between processes may be defined on the basis of the composition of the feed or modality of reaction and may thus be considered as engineering distinctions. However, due to the previously discussed complexity of the surface processes, from the point of view of the basic chemistry/catalysis taking place, distinctions become less clear, often with the same reactions taking place under different conditions.

11.2.2 Industrial Steam Reforming of Methane

Hydrogen production is of enormous importance in the chemical and refining industries. In mixtures with nitrogen, H_2 is used for ammonia synthesis. It is used in mixtures with CO for methanol and higher alcohol synthesis, for liquid hydrocarbons synthesis (Fischer Tropsch synthesis), and for synthesis of oxygenates (hydroformilation). It is also required for methanol-to-gasoline (MTG) processes. In fact, due to this importance, mixtures of CO and H_2 are commonly known as synthesis gas or syngas. In refineries, H_2 is used in a number of hydrotreating processes (hydrocracking, hydrotreating, hydroconversion, hydrodesulfurizing). The main source of H_2 in an oil refinery is catalytic reforming of naphtha, with some contribution from H_2 recovery from H_2-rich off-gases. However, the ever-decreasing hydrogen content of crude reserves (especially U.S. reserves), combined with the growing demand for hydrogen in processes such as fuel reformulating and desulfurization (to combat air pollution), has given rise to the so-called refinery hydrogen balance problem, where demand is approaching or has in some cases surpassed supply. The shortfall is being met by plants that produce H_2 by steam reforming of methane [29, 30].

Thus, H_2 production is a mature industrial technology, and processes based on catalytic reforming have been commercialized. In all cases, natural gas (predominantly CH_4) is the raw material. Two processes are mainly used to produce H_2 from methane:

Steam reforming (SR)
Autothermal reforming (ATR)

Steam reforming of methane (SRM) is the most widely employed process and was commercialized as early as 1926:

$$CH_4 + H_2O = CO + 3\ H_2; \quad \Delta H^\circ = 206\ kJ \cdot mol^{-1} \qquad (11.18)$$

Despite the fact that it is a highly endothermic and therefore an energy-intensive process, it is the preferred option for large-scale production due to the wide availability of natural gas and the fact that, unlike the other options, it does not require a costly O_2 plant. The reaction produces the highest theoretical CO/H_2 ratio, but the ratio can be tailored by changing the reaction conditions [30, 31]. Given its endothermicity, the reaction is favored at high temperature, and the increase in the total number of molecules means that it is also favored at low pressure.

A schematic illustration of a H_2 production process based on reforming is shown in Figure 11.3. Steps commonly used in SRM processes are highlighted in bold.

Ni-based catalysts (Ni/NiO, ca. 15 wt%) are used for the reforming step(s), typically supported on ceramic materials such as α-Al_2O_3, MgO, MgMAlO$_x$ spinel, and Zr_2O_3 [31]. A number of additives are included to enhance the overall performance characteristics of the basic formulation:

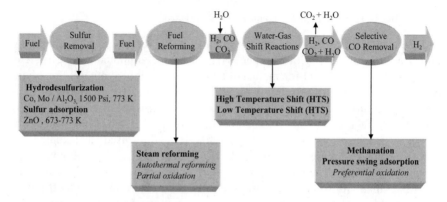

FIGURE 11.3 Schematic illustration of a hydrogen production process based on reforming.

Calcium aluminate (10 to 13 wt% CaO) as a binder to improve the mechanical
  strength of the material
Potassium oxide (up to 7 wt%) to inhibit coke formation
Silica (up to 7 wt%, as a silicate) to stabilize the potassium oxide

Such formulations have been developed to overcome the chief problem encountered under steam-reforming conditions: a strong tendency to deactivate through
various mechanisms. The severe hydrothermal conditions normally employed
strongly favor sintering and catalyst weakening (loss of mechanical strength), while
supports based on Al_2O_3 or MgO form spinels with Ni, thus removing the active Ni
phase. Calcium aluminate and the high Ni loadings can alleviate these problems.
Deactivation through coke formation is an insurmountable problem. In addition to
the potassium oxide additive, high steam/CH_4 ratios are used to minimize coke
deposits. However, high steam/CH_4 ratios put a further strain on the mechanical
integrity of the catalyst.

Autothermal reforming, which has also achieved widespread application, is a
combination of partial oxidation and steam reforming. The process was originally
developed by Haldor Topsoe in the late 1950s [30, 32] as a means of increasing the
H_2 content for ammonia plants. The process uses the exothermic partial oxidation
reaction to supply heat to the steam-reforming process and involves cofeeding natural
gas, steam, and oxygen. Original configurations were based on the idea of consec-
utive reaction zones. In the first, the thermal or combustion zone, noncatalytic partial
oxidation took place. The reaction was ignited externally with a burner:

$$CH_4 + 1.5\ O_2 = CO + 2\ H_2O; \quad \Delta H° = -35.7\ kJ·mol^{-1} \qquad (11.19)$$

This was followed by a catalytic zone, in which steam reforming and the WGSR
took place:

$$CH_4 + H_2O = CO + 3H_2; \quad \Delta H° = 206\ kJ·mol^{-1} \qquad (11.20)$$

$$CO + H_2O = CO_2 + H_2; \quad \Delta H° = -41\ kJ·mol^{-1} \qquad (11.21)$$

This technology has also seen considerable improvements. In later developments,
the two segments were combined into a single unit [33–36]. Not surprisingly, given
the similarity of the chemistry involved, Ni-based catalysts are also used in ATR
processes.

Industrial-scale catalytic partial oxidation has not yet been developed; however,
as it represents an attractive alternative to SR and ATR, it has been the subject of
intensive academic and industrial research. The advantages include the exothermicity
of the reaction, its greater selectivity to syngas, the requirement for high space
velocity, which offers the possibility of more compact design, and the fact that it
offers another option of syngas composition for industrial use. As mentioned earlier,
there have been claims for direct partial oxidation over some catalysts. However, in
most cases the mechanism involves total combustion of part of the methane at the

front of the catalyst, followed by steam and CO_2 reforming further along the bed, the partial oxidation reaction being a linear combination of these three reactions [37–41]. From this point of view, it has some similarities to ATR.

CO_2 reforming supplies syngas in the theoretical H_2/CO ratio of 1 and has been commercialized, for example, in the Calcor process [42, 43]. It is most widely used in secondary reforming processes to reduce the H_2/CO ratio obtained from steam reforming. Finally, a number of other processes have been described, including combined reforming (CR), gas-heated reforming (GHR), and combined autothermal reforming (CAR) [44]. These are various combinations of SR and ATR.

The WGSR serves two important purposes: the simultaneous reduction of the CO content of the primary reformate (up to 12% for steam reforming and 6 to 8% for ATR) and increase of the hydrogen content. A number of extensive reviews of the reaction have been published [45–49].

As a slightly exothermic process, the WGSR is favored at low temperature. The temperature at which activity is observed is therefore determined by the kinetics, which are slow, and the dew point [36]. In addition, the exothermicity is an important process consideration: the higher the shift conversion, the higher the bed temperature generated. This may result in catalyst sintering. In industrial applications, a compromise between high CO conversion and high temperature is achieved by dividing the water–gas-shift step into two adiabatic stages using separate high-temperature shift (HTS) and low-temperature shift (LTS) catalysts. Fe-based catalysts are used for HTS and Cu-Zn for LTS. On exiting from the reformer section, the gases are hot, which can be exploited for HT-WGSR. Further downstream, the again cooler gases can be treated in a second LTS unit .

The final stage shown in Figure 11.3, selective CO removal, is usually accomplished by selective CO methanation or pressure-swing adsorption. As will be discussed in more detail below, preferential CO oxidation (PrOx) is a more suitable option for automotive application.

11.2.3 Electrolysis of Water and Other Systems

Of the H_2 production methods shown in Figure 11.2, water electrolysis is also an established technique for H_2 production at an industrial level. However, because of its high cost, electrolysis is used only in specialized processes, where extremely high purity is required, and currently it represents <5% of total industrial production [29]. Of the other techniques, pyrolysis of coal is the closest to wide-scale industrial application.

11.2.4 Onboard Hydrogen Production and Purification

11.2.4.1   Requirements for Onboard Production

The process of conversion of a carbon-based fuel to H_2 for use in fuel cells is commonly called fuel processing. One option to supply H_2 for automotive applications is that of onboard reforming, in which the H_2 is obtained from a hydrocarbon or oxygenate fuel by reforming reactions onboard the vehicle. Because of the absence of a H_2 infrastructure, this was initially considered to be the most accessible option, and much early work was devoted to this approach, with a view to utilizing the existing gasoline network or another liquid fuel that could be adapted (usually methanol).

In principle, the industrial H_2 production methods discussed above can be applied to onboard reforming, with suitable modification. Thus, three options are described in most reports on onboard reforming: steam reforming (SR), autothermal reforming (ATR), and partial oxidation (PO).

It should also be noted that some reports, perhaps more correctly and more or less explicitly, divide the options into two: steam reforming and partial oxidation. Here, a note of caution should be sounded on the definitions involved in onboard reforming processes. As outlined by Ahmed, each process can be defined by the manner in which the reactants are introduced to the catalyst [50]. Thus, in steam reforming, a mixture of fuel and H_2O is fed to the reforming catalyst, while fuel and oxygen are used in partial oxidation (with H_2O added later for the WGSR), and in autothermal reforming fuel, H_2O and O_2 are co-fed. However, strictly speaking, the autothermal reforming process should only be considered so if the feed composition is such that the overall reaction is heat balanced, even though, in practice, an excess of air is often used (to compensate for heat losses and to obtain a reformer with a rapid response):

$$C_xH_yO_z + a\ H_2O + b\ O_2 = c\ H_2 + d\ CO_2; \Delta H° = 0 \qquad (11.22)$$

On the other hand, many applications use various amounts of H_2O and O_2 in the feed. As pointed out by Pettersson and Westerholm, such processes are better described as "combinatorial reforming" [51]. As discussed previously, the surface reactions are quite complex and mechanistic analysis is often not conducted. For these reasons, it is also common for reports to describe these systems as being based on one type of reforming reaction or another.

Given the established history of industrial reforming, it is not surprising that research into H_2 production for small-scale or mobile applications has taken an example from this knowledge. The concept of onboard H_2 production follows a strategy similar to that outlined in Figure 11.3, with the same basic aim: to produce H_2 with minimal CO content. As will be discussed below, the latter is particularly important, as CO is a poison for the polymer electrolyte membrane fuel cell (PEMFC) located downstream. If maximum H_2 content were the only consideration, then steam reforming would be the only option. However, onboard reformers must meet a different set of criteria than their industrial counterparts. These criteria include [50]:

An H_2 production capacity that is significantly lower than those in stationary
  plants
Significantly smaller size and weight
The ability to undergo multiple start-up/shutdown cycles
Variable and rapid response
Lower cost
High reliability and durability (although lower than stationary reformers)
Lower operating temperature

In addition, current industrial technology suffers from a number of drawbacks that must be overcome if reformers are to be candidates for onboard H_2 production. These include [35]:

The pyrophoric nature of Ni-based catalysts used for reforming and Cu- and Fe-based catalysts used for WGSR. This represents a potential risk (sintering and fire hazard) for onboard application.

The unsuitability of hydrodesulfurization and pressure-swing adsorption (PSA) for onboard application because they require high pressure. The absence of sulfur is necessary not only because Ni catalysts are extremely sulfur sensitive, but also because sulfur is a poison for PEMFCs (discussed below).

The unsuitability of the other method of CO removal commonly used. For methanation of CO, CO_2 must first be removed, as it can be also methanated consuming H_2. The inadaptability of both methods makes preferential oxidation (PrOx) the most viable option for this important step, although other options do exist (discussed below).

The endothermic nature of steam reforming, which requires complicated engineering for heat management.

The fact that industrial steam reforming is optimized for steady-state operation.

When all factors are taken into account, ATR/PO have considerable advantages over SR, although the hydrogen content of the primary reformate is lower. These processes can be achieved with more compact systems, and they are less energy intensive and have a more rapid response, thus making them much more suitable to the multiple start-up/shutdown operations expected in a mobile source. Despite this, it should be noted that steam reforming has also been extensively investigated, and indeed demonstrated, for onboard application [52]. The main thrust of current research is to meet the criteria by overcoming the limitations of current industrial methods.

11.2.4.2 Reforming Catalysts

Precious metals (Rh, Pt, Ru, Ir) supported on oxides have been shown to be highly active for reforming reactions [37, 39–41, 53]. They also generally show a higher tolerance to sulfur than Ni [35]. Their main drawback is their prohibitive cost, which has thus far precluded their industrial use. However, as these metals are intrinsically more active than Ni, their loading can be lowered by more than an order of magnitude. This high activity means that these metals have received attention as possible replacements for Ni.

Another significant field of research has been to find ways to overcome the limitations of Ni and the other first-row transition metals Fe through Zn. The effect of the support on activity has been investigated. With nonnoble metal catalysts (Ni, Co, Fe), supports with a low concentration of Lewis acid sites or the presence of basic sites (ZrO_2, MgO, La_2O_3) have been found to promote activity and offer a higher resistance to coking [54–56]. ZrO_2 and CeO_2–ZrO_2 mixed oxides have been shown to be particularly effective supports [57–59]. This approach is also used in the case of precious metal catalysts. For example, the higher activity of Pd supported on CeO_2 with respect to SiO_2 or Al_2O_3 was attributed to the ability of ceria to supply lattice oxygen to the steam-reforming reaction [60, 61]. Doped ceria supports (e.g., $Ce_{0.8}Sm_{0.15}Gd_{0.05}O_2$) have also been shown to be effective [62].

Due to the high thermal stability of hexa-aluminates [63], their use as supports has been investigated, especially for the partial oxidation reaction, which can give rise to very high bed temperatures [64–66].

The use of the support to promote reaction is often used in conjunction with novel preparation methods. Controlled synthesis of highly active metallic nanoparticles is an area of strong topicality in heterogeneous catalysis, and the synthesis of reforming catalysts is no exception. In the case of Ni, it has been demonstrated that coking is less of a problem with very small particles [67, 68]. In the case of precious metals, very active particles allow a decrease of the loading and therefore of cost. If sintering of these nanoparticles could be prevented, then such precious metal catalysts could become viable alternatives to Ni-based catalysts. One approach that has been successfully applied to achieve better and more-reproducible metal dispersion with respect to traditional synthesis methods is incorporation of the active phase within precursors such as perovskite oxides or hydrotalcite-type clays [42, 69–75]. As the active phase is present homogeneously within the precursor, very finely dispersed and highly active materials are formed upon heating the starting material to high temperature.

Another area that is being actively investigated is that of bimetallic catalyst, for which there is a considerable scientific literature. A common approach has been to promote the activity of Ni with small amounts of a precious metal.

Catalysts for so-called oxidative methanol reforming have also been widely investigated. Great attention has been devoted to the well-established commercial $Cu/ZnO/Al_2O_3$ systems [76–79]. Inclusion of zirconia in the formulation has been found to be beneficial for the methanol reforming reaction [80–82]. Noble, transition, and base metals (using various supports) have also been considered as the active phase [83–87]. More recently, there has been a growing interest in the reforming of ethanol to produce hydrogen [88–94]. This might be due to the fact that it is becoming generally accepted that ethanol, as a bioproduct of fermentation, can be one of the large-scale renewable hydrogen sources. It should be noted, however, that, with respect to the reforming of methanol, the presence of the C–C bond in the ethanol might favor coke formation on the surface of the catalyst and, therefore, catalyst deactivation. Indications that Co-based catalysts show good activity have made them a subject of particular attention for ethanol reforming [95, 96].

11.2.4.3   High-Temperature and Low-Temperature WGSR Catalysts

As an industrial process, the WGSR is efficiently achieved in two steps. Experimental and early-generation fuel processors use this type of arrangement with interstage cooling and industrial catalyst formulations. However, addition of an extra step increases the overall size, weight, and cost of the system. Thus, a single step (and nonpyrophoric) WGSR catalyst would be more desirable, although it should be noted that when methanol is used as a fuel, only one WGSR step is necessary. As this catalyst should ideally reduce the CO concentration to 1% or less, it must show high activity at a relatively low temperature (<600 K). In addition, it must satisfy the requirements common to all catalysts used in hydrogen production: high stability and durability, mechanical integrity and resistance to shock or temperature excursions, stability toward poisons such as H_2S and chlorine, and the absence of side reactions (particularly methanation activity). WGS catalysis is an area in which a breakthrough is necessary, and research into more suitable shift catalysts has mainly considered two categories of material: base metal catalysts and precious metal catalysts or gold catalysts [34, 35].

Molybdenum carbide catalysts have been reported to show higher WGSR activity than commercial Cu-Zn-Al LTS catalysts. In addition, the conditions employed were suitable for mobile applications (493 to 568 K and atmospheric pressure); there was no apparent deactivation or modification of the structure during 48 h on-stream; and there was negligible methanation activity [97].

A base metal catalyst with activity comparable with commercial LTS catalyst (and a wider operating temperature) has also been reported. The authors stress the nonpyrophoricity and the ability to withstand situations common in fuel processing, such as water condensation during start-up and shutdown [98, 99].

Cu- and Ni-doped (La)CeO_2 have been reported to show high WGS activity and stability at temperatures up to 900 K. A co-operative redox reaction mechanism was reported, involving oxidation of CO adsorbed on the metal cluster by oxygen supplied to the metal interface by ceria [100]. On the other hand, high-temperature stability of Cu was reported to be a problem, and even the most promising material investigated promoted CO methanation [101].

In the case of precious metals, ceria or promoted ceria is usually used as the support, and Pt is the most widely studied metal. The WGS activity of Pt/CeO_2 was recognized in the early 1980s, in connection with the use of CeO_2 as an additive in TWC formulations [102], and ceria-supported precious metal catalysts are considered potential candidates for fuel processing [34, 35, 103–106]. Strategies aimed at improving these materials include the use of promoters [104], nanostructuring of the support [106], and alloying of the metal [101]. Although the suitability of PGM catalysts has been questioned [107], Johnson Matthey (JM) has recently developed "non-pyrophoric PGM formulations" with suitable durability and no methanation activity across a wide temperature range (500 to 800 K) that can be incorporated into its own fuel processor in one WGS stage [34].

Following the report of the extraordinary low-temperature CO oxidation activity of gold-based catalysts [108], there has been an enormous interest in the use of supported gold as potential WGS catalysts. Au/CeO_2 [103, 109–113], Au/TiO_2 [114–118], and Au/Fe_2O_3 [113, 117, 119–123] have all been investigated. However, a general criticism of these catalysts is that reproducible catalyst synthesis has been a problem, with many papers stressing the sensitivity to synthesis. In addition, thermal stability is an issue: gold catalysts sinter too easily. For these reasons, gold-based catalysts are widely believed to be more suitable as PrOx candidates (discussed below), for which the lower operating temperatures allow them to maintain their sometimes remarkable activity.

11.2.4.4 Preferential Oxidation of CO (PrO$_x$)

After the WGSR step(s), the CO content of the reformate is on the order of 1%. CO must be reduced, as it is a poison for the electrodes of PEMFCs. For onboard application, there are a number of approaches that can, in principle, be considered to lower the final CO content of the stream to levels compatible for use in a fuel cell. These include preferential oxidation of CO (PrOx), CO methanation, Pd-membrane separation, and the development of electrodes with greater CO tolerance [124]. PrOx (also called selective oxidation) involves preferentially or selectively oxidizing the CO in the presence of the large excess of H$_2$, and is the most studied method, as it is a relatively advanced technology. The oxygen needed to oxidize the CO is added after the shift step(s). For onboard application, the catalyst should ideally operate between the temperature of the shift reactor and the operating temperature of the PEMFC, usually between ca. 473 and 353 K.

Commercial PrOx catalysts were first introduced in the 1960s, with the trade name Selectoxo™, to remove CO from H$_2$ in ammonia synthesis processes. The catalyst consists of Pt supported on alumina, promoted by a base metal. The platinum content is usually rather high, on the order of 5 wt%. These commercial catalysts are indeed capable of reducing CO levels to <10 ppm, but only under very carefully controlled conditions. Various parameters — selectivity, temperature control, space velocity — are of fundamental importance to the overall efficiency. The variations in conditions encountered with onboard applications are often incompatible with the ideal PrOx conditions. For example, a problem of existing Pt-based catalysts is that low space velocity favors CO production by the reverse WGSR. If a fuel reformer must operate with variable loads, this could be a problem.

Selectivity, which can be gauged from the O$_2$/CO ratio necessary to observe adequate removal of CO, is another important issue. If the catalyst is not 100% selective toward CO oxidation, which implies simultaneous conversion of H$_2$, an excess of air must be added. This results not only in a loss of H$_2$, but also in an overall increase of the heat generated through two highly exothermic reactions. This can have knock-on effects, as increased reaction temperature can lead to promotion of the reverse WGSR and therefore an increase in the outlet CO concentration.

Such problems can be overcome by: two-stage PrOx with interstage cooling; low inlet temperature to second stage; minimum air injection; fixed-flow operation; and second-stage preferential oxidation with catalyst immune to rWGSR. However, this would make an already complicated fuel reformer system even more complicated [35, 36].

Following the report by Haruta et al., the general properties of catalysts based on Au nanoparticles have become widely investigated [108]. Because they show higher activity toward CO oxidation than H$_2$ oxidation, they inevitably have attracted great interest as potential PrOx candidates [125–134]. Au-based catalysts show very high selectivity toward CO oxidation, even in the presence an excess of H$_2$ typical of reformate streams. However, deactivation has been reported to be a problem when, more realistically, CO$_2$ and H$_2$O are added to the feed [135]. This is a problem that must be overcome if such catalysts are to become viable alternatives to Pt-based catalysts.

Finally, Cu-based materials have recently emerged as promising candidates, in particular those containing ceria [135, 136].

11.2.4.5 Membrane Technology

In recent years, incorporation of membrane technology into hydrogen production processes has become an area of interest. To date, separation membranes, usually based on Pd-alloys, have been used most extensively [29]. Separation membranes can be incorporated into various steps of the H_2 production process. At the reforming step, oxygen-selective membranes can be used to deliver highly active oxygen to the catalyst and eliminate the need for gas-phase oxygen. At the WGS stage, hydrogen-selective membranes can be used to remove hydrogen from the stream and therefore overcome thermodynamic limitations. There have also been reports of CO or CO_2-selective membranes, which can be used to purify hydrogen and eliminate the need for PrOx [35]. Theses technologies are as yet at a preliminary stage, but they do offer interesting possibilities for the whole hydrogen production process.

11.2.4.6 Fuel Choice

Research has shown that methanol, ethanol, compressed natural gas, methane, biogas, naphtha, gasoline, diesel, kerosene, aviation fuels, marine fuels, hydrocarbons (pentane, hexane, octane, acetylene), propane, butane, dimethyl ether, and ammonia can all be used as feedstocks for onboard fuel reformers [51]. Thus, there is a high degree of flexibility from the point of view of fuels. Of these, methanol and gasoline are the favored options. Technically, methanol is the easiest option, and indeed methanol appears to be the most commonly investigated liquid fuel for onboard applications [52]. A liquid fuel also considerably simplifies handling problems. However, a methanol distribution network does not exist. A network for gasoline, on the other hand, does exist, but extremely high reforming ability is necessary for gasoline, especially in relation to coking of the catalyst. From the point of view of overall efficiency and CO_2 emissions, there is very little difference between methanol and gasoline if the synthesis of methanol is included in the comparison. The best approach would most likely be the development of fuel-flexible reformers, but in that case the reformer is unlikely to be optimized for any one fuel [29, 51, 137].

To meet the particular requirements of onboard hydrogen generation, highly complex, integrated systems are required. Despite this, many reformers have been produced commercially [52]. However, it should be noted that onboard reforming is now widely viewed as a first step toward the ultimate solution of onboard hydrogen storage.

11.3 HYDROGEN DISTRIBUTION AND ONBOARD STORAGE

11.3.1 HYDROGEN DISTRIBUTION SYSTEMS

It is arguable that in no other area is the scale of the challenge faced by a transition to a hydrogen economy greater than in the transport sector. This is because the

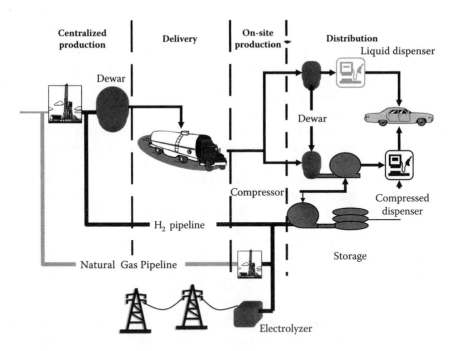

FIGURE 11.4 Scheme of an integrated model of a hydrogen distribution system.

difficulties posed are not only the technical problems of production, common to all sectors, but there is the added complication of establishing a hydrogen storage/distribution network. Figure 11.4 shows an integrated model of the possible solutions to such an H_2 distribution network. As can be inferred from the previous section, off-board hydrogen generation represents an easier solution to the problem of H_2 production. Steam reforming of methane represents the best option for off-board generation, utilizing the existing natural gas network.

Massive investment is needed to realize this model. The needs are especially acute in the case of the personal automobile. For fleet vehicles, centralized refueling could be used, which would limit the number of sites necessary. In fact, H_2/fuel-cell buses have already been introduced, albeit on a limited scale. Bigger vehicles are also more flexible in the choice of technology. In the case of automobiles, it would be necessary to replace all or most of the existing infrastructure, which gives rise to a classical chicken-or-egg problem: car manufacturers are unwilling to fully commit to H_2 until the network exists, yet it is unlikely that investment will be stimulated in the absence of a ready market. It seems likely that the resolution will be a combination of factors — gradual replacement, using niche markets to stimulate investment, and with a combination of public and private investment and consumer payment. This process has in part been started in the United States, Japan, and Europe, with the introduction of more than 70 hydrogen filling stations worldwide and the so-called Hydrogen HyWays programs.

11.3.2 Hydrogen Storage

It has been estimated that to have a driving autonomy of 400 km, an average personal car, "optimized for mobility and not for prestige," requires about 8 kg of H_2 for a hydrogen internal-combustion engine (H_2 ICE) or 4 kg for a fuel-cell-driven car [26]. The problems of storing such large amounts of H_2 are considerable.

There are numerous criteria that a hydrogen storage material must ideally satisfy before being considered suitable for use in automobiles. These include gravimetric storage density, volumetric storage density, safety, storage stability, and reproducibility (all as high as possible); price, rehydrogenation and equilibrium pressure, enthalpic effects, and toxicity (all as low as possible); hydrogenation rate (as quick as possible); and dehydrogenation rate (as quick as necessary). To make any discussion more manageable, these criteria can be distilled into four main categories:

Gravimetric storage density
Volumetric storage density
Operating temperature (which is linked to pressures, enthalpy, and rates)
Safety

The criteria set by the USDOE [138], which are generally accepted as the basic targets, include a gravimetric density of 6.5 wt%, a volumetric density of 65 $g \cdot l^{-1}$, and a decomposition temperature of between 333 and 393 K. Clearly, the storage method should be reversible, as safe as possible, and nontoxic. It should be noted that the densities refer to system characteristics, not material characteristics.

For the purpose of the present discussion, the potential methods of hydrogen storage will be subdivided into the following categories:

High-pressure gas cylinders
Liquid hydrogen in cryogenic tanks
Adsorbed hydrogen on high-surface-area sorbents (at T < 100 K)
Hydride systems

11.3.2.1 Compressed Hydrogen

In order to achieve the required storage densities, pressures must exceed 50 MPa, and a target of 70 MPa is generally considered to be the objective. At present, 35-MPa tanks are routinely available. For comparison, laboratory steel hydrogen cylinders are normally pressurized to ca. 20 MPa. To withstand such high pressures, advanced cylinder materials, constructed from carbon-fiber-reinforced composite materials, have been developed. However, these are quite costly. A further cost-related problem associated with this technology is that of compression, which also considerably diminishes the specific energy content [139]. Other costs include the necessity of advanced pressure-control mechanisms. Finally, there is the problem of safety both during gas compression and use. The potential hazards of such highly pressurized cylinders are enormous, and in fact they have been ruled out as an option for personal automobiles in Japan. From the point of view of structural stability, tanks should ideally be spherical. However, this is rather inconvenient for transport applications, where space is at a premium. Thus, cylinders are usually used, although this results in higher structural strain.

Advanced research into this technology is in the area of materials engineering. Research into new tank designs can be divided into two categories. One approach is to design new methods of using the carbon fiber more efficiently, thereby using less. More recently, the idea of conformable or replicant tanks has been proposed. In these, the tank is made from many repeatable structures and thus the tank shape can be chosen to suit space considerations. This approach is still at a very early stage of development [140, 141].

11.3.2.2   Liquefied Hydrogen

Cryogenic storage of H_2 in liquid form (21 K) is a relatively advanced technology [142], and it offers good storage H_2 density (70.8 kg·m^{-3}). However, as with the gas compression, the liquefaction process is energy consuming and costly. A second drawback is that of evaporation losses. The critical temperature of H_2 is very low (32 K). Even with the most advanced cryogenic tanks, some heat transfer occurs, which means that, to prevent very hazardous buildup of pressure, the system must be open or vented. This results in boil-off of the H_2. A potential solution to this loss of fuel is to capture the boil-off with hydrides.

As above, research in this area is focused on the cryogenic tank. Progress is being registered, and the latest tanks significantly reduce boil-off to less than 3% per day after the first 3 days, boil-off being negligible during the first 3 days [143].

Although they do not meet the ideal specifications, gas or cryogenic storage at present represent the most advanced hydrogen storage technologies.

11.3.2.3   High-Surface-Area Sorbents

The adsorption of a gas on a solid surface depends on the attractive forces between the two. These forces can in principle be divided into two types: weak van der Waals interactions that give rise to physical adsorption, or physisorption (binding energy ca. 0.1 eV); or stronger interaction that results in a chemical bond between the adsorbate (gas) and the adsorbent (surface), or chemisorption (binding energy ca. 2 to 3 eV). Both processes can potentially be exploited for onboard H_2 storage, and adsorption on high-surface-area sorbents is a hugely active research field. The adsorption capacity is dependent on the type of adsorption process, with chemisorption resulting in higher uptake.

In the specific case of H_2, it is important to note that, above the critical temperature of 32 K, only monolayer adsorption is possible during the physisorption process. This means that the physisorption capacity (intended as the maximum adsorption) is proportional to the surface area of the material. Within this framework, the adsorption observed depends on the temperature and pressure: at high temperature it decreases rapidly, while at a given temperature, the storage is a function of the pressure. This pressure dependence in turn gives rise to the reversibility, with hydrogen released as the pressure decreases and vice versa [144, 145].

The ability of various gases to adsorb on carbon-based materials is well known. As it is possible to obtain carbon as a very-high-surface-area, highly porous material (activated carbon), and as carbon is a lightweight solid at room temperature, carbon-based materials appear to have good characteristics as H_2 storage materials. Recently, reports of highly specific carbon materials, collectively known as nanostructured carbon materials, appear to have opened new possibilities in the field.

Various types of nanostructured carbon materials have been investigated [144–149]. Carbon nanotubes (CNT) were first reported by Ijima [150]. Single-walled nanotubes (SWNT) consist of a graphene sheet rolled into a cylinder with an inner diameter 0.4 to 3 nm a length of 10 to 100 nm. They typically bunch together to form so-called nanoropes (10 to 100 parallel tubes). Multiwalled nano-tubes (MWNT) consist of a series of concentric graphite cylinders (2 to 50 tubes, 30 to 50 nm). The interlay distance is similar to that of graphite. Graphite nanofibers (GNF) are stacks of graphite platelets (length: 5 to 100 μm, diameter: 5 to 200 nm) in various formations, with an interlayer spacing similar to that of graphite. Finally, carbon can also be nanostructured mechanically, usually by ball milling of graphite, which increases the surface area.

The issue of H_2 storage on carbon-based materials is rather controversial. Two reports in particular triggered a flurry of research activity into nanostructured carbon materials. The first claimed a storage capacity of 5 to 10 wt% for SWNTs under low-pressure/high-temperature adsorption (p = 40 kPa, T = 273 K) conditions [151]. Subsequently, under high-pressure/high-temperature conditions (P = 11,365 MPa, T = 298 K), storage values of 4.52 wt% for graphite, 11.26 wt% for CNTs, 53.68 wt% for platelet GNFs, and 67.55 wt% for herringbone GNFs were reported [152]. These results were hailed as the breakthrough that would usher in a new era of fuel-cell-driven cars.

Other groups have reported similar excellent, or at least very high, uptakes for carbon nanostructures or modified carbon nanostructures [153–156]. In some of these cases, rather high temperature was required to obtain maximum adsorption or desorption, which can be attributed to a chemisorption process. The applicability of such systems to hydrogen storage for mobile applications is doubtful. High storage capacities have also been reported for nanostructured graphite [157, 158]. It should also be noted that many of the groups that reported high uptakes later published downward revisions of the storage amounts.

Against these promising results, a number of groups have not been able to independently reproduce high H_2 uptakes. Uptake as low as 0.2 wt% [159], 1 wt% [160], 1 wt% [161], and 0.3 wt% [162] have been reported. Zuettel et al. concluded that adsorption of hydrogen at 77 K is due to physisorption and is therefore propor-tional to the specific surface area of the CNT and limited to 2 wt% for SWNT (theoretical surface area = 1315 $m^2 \cdot g^{-1}$) [145]. A similar conclusion was reached by Strobel et al. [162] and Schimmel et al. [163]. If physisorption were indeed respon-sible, then relatively high uptakes should be possible only at low temperature, and the comparative room temperature (RT) adsorption should be significantly smaller, or even negligible.

There are a number of reasons for the scatter in the data obtained. It has been rightly pointed out that the definition of "uptake" is sometimes rather vague and is not always used in the same way [144]. However, this is not the core of the problem, as large discrepancies still exist. It has been clearly shown that, in some cases, adsorption of water may give erroneous results [164]. Furthermore, large-scale production of nanotubes is notoriously difficult (although improving), and their purification is complicated. Impurities may contribute to or be responsible for the uptake observed. More specifically, it has been claimed that the presence of Ti-alloy, introduced as an impurity from a sonic probe habitually used to open SWNTs, is responsible for large H_2-uptake values at room temperature [165]. This has led to interest in the H_2-uptake properties of intentionally metal-decorated nanotubes and fullerenes.

Chemisorption as the adsorption process is also a possibility. However, like physisorption, this also creates a theoretical difficulty: if the adsorption energies associated with physisorption are too low to explain the high capacities reported for CNTs, then the adsorption energies associated with chemisorption are too high to explain reversible storage at ambient temperature. An intermediate state between physisorption and chemisorption has been hypothesized [166]. This suggestion has received support from theoretical calculations [167], but this has not been demonstrated. Another possibility is that an impurity catalyzes the chemisorption process.

Other high-surface-area sorbent materials have been investigated for their hydrogen storage capacities. Inorganic nanotubes, essentially boron and nitrogen analogues of carbon nanotubes, have received much attention [168, 169]. These are also referred to as collapsed nanotubes. Another class of materials that is beginning to attract attention as a hydrogen storage medium is that of metal-organic frameworks (MOFs) [170–172], which are distinguished by very high surface areas. In both cases, sufficient storage has not yet been reported. The reports on the latter materials have come under some criticism on the grounds that the isotherms reported are not consistent with any kind of adsorption process [173].

11.3.2.4   Hydride Systems

11.3.2.4.1   Reversible Metal Hydride Systems

As a highly reactive element, hydrogen can form hydrides or solid solutions with most elements as well as with thousands of metal alloys of various compositions (AB_5, AB_2, AB, A_2B, AB_3, A_2B_7, A_2B_{17}, A_6B_{23}, etc.) [26, 143, 174]. Although the rich chemistry of hydrides prevents easy subdivision of various hydride species, reversible hydrides for H_2 storage are conventionally divided into two categories: (a) interstitial hydrides, in which the hydrogen is absorbed on interstitial sites in a host metal, and (b) complex hydrides, in which the hydrogen is chemically bonded in covalent and ionic compounds. The complex elemental hydrides include mainly hydrides of elements in groups 1 to 3 and some transition metals (TM) hydrides. Reversible hydrides are charged and discharged by thermal means, and the storage behavior of a given material is described by a pressure-concentration-temperature (PCT) plot or, more simply, by van't Hoff plots ($\ln P$ vs. $1/T_{dec}$) [26, 143]. The reversibility of metallic hydrides can be influenced by appropriate alloying, usually by combining strong and weak hydride-forming elements to produce compounds with intermediate thermodynamic affinities for hydrogen. A classical example is

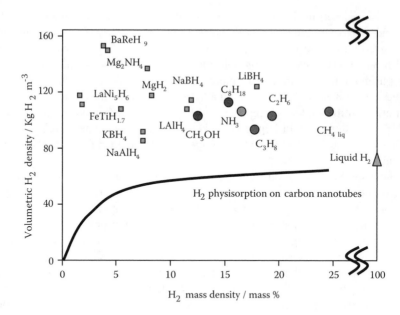

FIGURE 11.5 Comparison of the hydrogen stored in metal hydrides, carbon nanotubes, hydrocarbons, and other liquid fuels. (Adapted from Schlapbach, L. and Zuttel, A., *Nature*, 414, 353, 2001; Zuttel, A. et al., *J. Alloys Comp.*, 356, 515, 2003; and Zuttel, A. et al., *J. Power Sources*, 118, 1, 2003. With permission.)

$LaNi_5H_6$, which shows fast and reversible sorption at room temperature (RT), with a dissociation pressure higher than 0.1 MPa [143]. This material has already been commercialized in hydride-based batteries.

Although hydride chemistry offers a seemingly endless list of storage candidates, when the full set of requirements for automotive applications is considered, the number of suitable materials decreases dramatically (Figure 11.5). To meet the gravimetric density requirements, only light metal hydrides and complex hydrides can be considered. The potentially suitable complex hydrides include the transition metal complexes (e.g., $Mg_2NiH_4 = Mg^{2+} [NiH_4]^{4-} Mg^{2+}$) and nonmetal hydrides ($LiBH_4$, $NaBH_4$). However, not all of these are suitable thermodynamically. When reversibility (in general, complex hydrides have greater reversibility problems) or toxicity considerations are evaluated (e.g., Be hydrides, $Li_3Be_2H_7$), the list is even further reduced [175]. Indeed, when all factors are considered, it may be stated that no hydrogen storage material currently meets all the generally accepted criteria for a H_2 storage material [176, 177].

At present, complex metal hydrides are considered to be the most promising onboard storage systems. The alanate family (aluminum hydrides), especially sodium alanate ($NaAlH_4$), have received much attention. Despite the favorable thermodynamics, it was originally thought that $NaAlH_4$ was an unsuitable for hydride storage on the basis of its poor kinetic reversibility. In fact, it was considered impossible under practical conditions, necessitating a rehydrogenation temperature of 473 to 673 K (above the melting point) and a pressure of 10 to 40 MPa [143]. However, the observation that rehydrogenation could be catalyzed by Ti, under conditions acceptable for fuel cell use [178], has strongly renewed research interest in this material [173, 179–181].

Hydrogen is released in a two-step process:

$$3 \; NaAlH_4 = Na_3AlH_6 + 2 \; Al + 3 \; H_2 \qquad (11.23)$$

$$Na_3AlH_6 = 3 \; NaH + Al + 1.5 \; H_2 \qquad (11.24)$$

Both steps can be achieved near 373 K, while the first step has an equilibrium pressure of 0.1 MPa at 303 K [143]. The theoretical maximum gravimetric storage density is 5.5 wt%, which is below the target limit but represents a good compromise. Research into this system is mainly directed toward finding efficient dopants to catalyze the rehydrogenation, finding efficient methods of doping, and optimizing the temperature and pressure characteristics of decomposition [182]. Titanium remains the best catalyst thus far reported. Mechanical methods, usually high-energy ball-milling, to improve the kinetics have also been employed with some success. Apart from the specific interest in $NaAlH_4$, the discovery that catalysis can be used to overcome kinetic reversibility has opened up the general possibility that other hydrides can be similarly treated [183].

Magnesium-based systems are also potential hydrogen storage candidates. In many ways, Mg-based systems are a microcosm of the whole area of hydrogen storage materials: multiple possibilities exist, but always with at least one problem that makes them impractical for mobile application. Mg_2NiH_4 (3.6 wt%), Mg_2FeH_6 (5.5 wt%, volumetric density 150 kg·m^{-3}), and MgH_2 (7.6 wt%) have all been investigated. The first would be suitable but for its low gravimetric density; the second has reasonable gravimetric density and remarkable volumetric density, but unsuitable thermodynamic properties; the third shows high gravimetric density but poor kinetic reversibility. However, Mg_2FeH_6 is a strong candidate in applications other than hydrogen storage, such as thermal energy storage. Recently, a method of chemically activating MgH_2 has been reported [184], once more illustrating that kinetic problems can be overcome.

A novel hydrogen storage system is based on lithium nitride (LiN_3), which can be hydrogenated first to a mixture of lithium imide and lithium hydride and then, in a second step, to lithium amide [153]:

$$Li_3N + 2 \; H_2 \rightarrow Li_2NH + LiH + H_2 \leftrightharpoons LiNH_2 + 2 \; LiH \qquad (11.25)$$

The second step, with a theoretical gravimetric storage density (with respect to $LiNH_2$) of ca. 7%, was found to be almost completely reversible at 528 K and 558 K. However, it was also reported that addition of ca. 1% $TiCl_3$ substantially improved the kinetics of the system and eliminated the undesirable formation of ammonia [185], which is a poison for the PEMFC membrane. The analogous Mg and Ca systems have also been investigated. These are clearly complicated systems, containing mixtures at both "ends" of the hydrogenation/dehydrogenation reaction. However, this complexity also introduces an element of flexibility. Based on this idea, other composite systems, with mixtures of the light elements/metals, have also been investigated. The general aim of this strategy is to improve the kinetics of the system while paying a minimum penalty in storage capacity.

The complex boron hydrides $LiBH_4$ (18.4 wt%) and $NaBH_4$ have also received some attention as reversible storage materials [176, 177]. In their pure forms, these compounds do not show good promise for mobile application, as they are very stable and thus require high operating temperature. While some progress has been registered in improving the behavior, for example, using SiO_2 as a catalyst to lower the temperature of decomposition [177], it is true to say that their use in chemical storage systems at present offers a better alternative (discussed below).

Many other reversible hydride materials, even simple hydrides such as vanadium hydrides, have been investigated [26, 158, 174, 181, 183]. The strong interest is fueled by a widespread belief that hydride chemistry still offers the possibility of significant breakthroughs, either by synthesis of novel hydrides with suitable properties or by finding ways to catalyze the storage behavior of known materials.

11.3.2.4.2   *Chemical Hydride Storage*

Chemical hydride storage involves release of H_2 by chemical reaction of a hydride. The hydride is usually stored onboard as a solution or in a slurry. This might be considered as onboard production rather than storage; however, it is usually classified and discussed as chemical hydride storage, as hydrides are involved. The approach has the advantage of circumventing some of the reversibility problems of complex hydrides, but the disadvantage is that onboard regeneration is not possible. One option is hydrolysis of light hydrides or reactive metals [143]:

$$LiH + H_2O = LiOH + H_2 \qquad (11.26)$$

$$Li + H_2O = LiOH + 1/2\ H_2 \qquad (11.27)$$

$$NaBH_4 + 4H_2O = NaOH + H_3BO_3 + 4H_2 \qquad (11.28)$$

Alkaline solutions of alkali metal borohydrides ($NaBH_4$, $LiBH_4$) in H_2O have received particular attention. The decomposition of aqueous solutions of $NaBH_4$ to produce H_2 is a well-established process, especially in the presence of a suitable catalyst [186–190]. The Millenium company has commercialized a system for hydrogen generation based on this reaction, and it is the subject of ongoing research for use with automobiles [191, 192].

An interesting recent field of study is that of liquid hydrides, which can be reversibly hydrogenated/dehydrogenated [193, 194]. Such systems are at an early stage of development, but they seem promising.

11.4. END USE OF HYDROGEN IN AUTOMOBILES

11.4.1 DIRECT USE OF HYDROGEN

It is often stated that H_2 is not a primary source of fuel (like coal or petroleum) but is, rather, an energy vector that must be converted into electrical energy. However,

H_2 can also be used directly in an internal-combustion engine. Initial investigations into direct combustion were performed in the 1970s, where H_2 was mixed with gasoline to lower the combustion temperature and thus reduce NO_x emissions. The idea of a hydrogen-only internal-combustion engine (H_2 ICE) is now under serious investigation. The efficiency of a H_2 ICE is only marginally better than that of a gasoline-fueled ICE (and diesel), being limited by the Carnot efficiency. However, in compensation, there are no emissions except NO_x, which are formed at rich or near stoichiometric operation. Notably, such engines can be run as low as half-stoichiometric, in which case NO_x emissions can be reduced by more than 90% compared with stoichiometric operation [195]. In addition, a H_2 ICE offers more power than fuel-cell-powered engines, and it could, in fact, eliminate a costly and problematic technology: the fuel cell itself. A number of car manufacturers have developed or are developing H_2 ICE automobiles, often using engines that can also run on gasoline. The thinking on such vehicles with regard to widespread application is divided. They appear to be an ideal way to kick-start the transition to the hydrogen economy, as they are much more attainable. Some see them as a temporary measure to introduce hydrogen vehicles into the marketplace, start the transition to a hydrogen infrastructure, familiarize the general public with hydrogen technology, and give more time for the development of fuel cells for widespread use. In contrast, others see them as a viable alternative to fuel cells themselves.

11.4.2 Fuel Cells — Low-Temperature PEMFC

A fuel cell (FC) is an electrochemical device that converts chemical energy directly into electrical energy. As they are not limited by the Carnot cycle of heat engines, they are intrinsically more efficient than engines based on combustion processes. Comprehensive descriptions of various FCs are given in the literature [44]. Here, the discussion is limited to a brief overview. FCs are classified according to the electrolyte used. Figure 11.6 summarizes the types of H_2 FCs currently under investigation for both stationary and transport applications. Some of their operating characteristics or requirements are also included.

Alkaline fuel cell (AFC)
Polymer-electrolyte-membrane fuel cell (PEMFC)

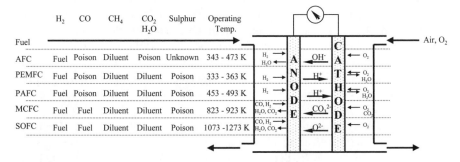

FIGURE 11.6 Scheme of the main types of H_2 fuel cells.

Phosphoric acid fuel cell (PAFC)
Molten carbonate fuel cell (MCFC)
Solid-oxide fuel cell (SOFC)
The direct methanol fuel cell (DMFC) uses methanol as a fuel

The first fuel cell was invented in 1839 by William Grove [196]; since then, interest in their development has been cyclical, with periods of intense activity, usually inspired by external factors, followed by periods of relative inactivity. The most important breakthrough in the development of FCs was provided by Francis T. Bacon, who demonstrated that, with properly engineered FC stacks, they could be practical as power sources. The U.S. space program resulted in another upsurge of interest. PEMFC fuel cells were developed for use on the Gemini missions; however, reliability problems led to their replacement by Bacon-type AFCs for life-support systems in the later Apollo missions. The oil crisis in the 1970s led to another bout of research activity, but interest subsided when FCs failed to meet acceptable costs. The latest surge of activity is due to the hydrogen economy.

A fuel cell can be operated either directly or indirectly. Direct operation involves feeding the pure fuel (H_2 or methanol), and is the most advantageous. Indirect operation means that the fuel must first be converted to H_2 through reforming processes, which considerably increases the overall complexity and cost of the system [27]. As discussed previously, this corresponds to onboard reforming in the case of automotive application. Indirect use also decreases the overall efficiency, although, even when operated indirectly, the efficiency is higher than for the ICE. For high-temperature FCs, the fuel-processing step can be incorporated into the FC itself, utilizing the high temperature of operation for the reforming reaction. This is known as internal fuel processing.

Of the various fuel cells shown in Figure 11.6, only PEMFCs, AFCs, and DMFCs are considered technically suitable for mobile application. Despite having been abandoned by the U.S. space program, advances in PEMFC, also referred to as solid polymer fuel cells (SPFCs), technology (Ballard, IFC), ironically during a period of general disinterest in the 1980s, have helped it emerge as the clear favorite for use in personal automobiles. The suitability of PEMFCs derives from compact design, fast start-up, suitable power density, and good load response. PEMFCs and advanced aerospace AFCs offer power density that is an order of magnitude higher than the other options [44]. Currently, the most commonly used membrane is Nafion® resin (Du Pont), which consists of a fluorocarbon polymer functionalized with sulfonic acid groups. This is an excellent proton conductor with good resistance to gas crossover and does not suffer from electrolyte loss, as the acid groups are fixed to the polymer backbone [197, 198]. In common with all fuel cells, the electrodes must be porous (to allow ion transfer) and usually are based on Pt (or other precious metals) supported on carbon. An excellent review on alternative membranes for PEMFC is given in the literature [199].

Although early demonstration model AFC-powered vehicles were developed [197], the AFC has received much less attention as a possible candidate. It is not entirely clear why this is so, but it is most probably because of the sensitivity of AFCs to both CO and CO_2. The early emphasis on onboard reforming perhaps overstressed this drawback. Nevertheless, the AFC has considerable advantages with respect to the PEMFC from the point of view of simplicity and cheapness. Base metals can be used for the electrodes. In situations where pure hydrogen can be delivered (such as expected with storage methods), there appears to be no reason why they should not offer a viable alternative.

The DMFC is also a potential candidate, but an outsider. It attracted much early attention, mainly because a liquid fuel in plentiful supply appeared to be an attractive option. However, among the disadvantages of methanol as a fuel are its toxicity and the absence of a distribution network. With regard to the latter point, the apparent commitment to establishing a H_2 distribution network significantly decreases the possibility that methanol could become a widely used fuel. Today, DMFCs are considered as very promising candidates for portable applications. Finally, interest in the development of PAFCs, MCFCs, and SOFCs is mainly for stationary application (domestic and commercial power generation, auxiliary/backup power), although PAFCs have been developed as power sources for buses [44].

11.4.2.1 Drawbacks of Existing PEM for Mobile Application

In addition to their current high cost, there are two main technical drawbacks to PEMFCs. The first relates to water management. The membrane must be kept hydrated at all times. Thus, the PEMFC must operate in conditions where the water evaporation rate is lower than its rate of production. This necessity determines the normal operating temperature of 343 to 363 K and an upper temperature limit of 393 K. The second problem relates to the extreme sensitivity of the electrodes to CO and SO_2. These poison the electrodes by blocking the active surface. In the case of CO, the poisoning is somewhat reversible. In practice, this means that CO levels must be less than 10 ppm, and the stream must be sulfur free. With electrodes that are more tolerant of CO, the upper limit can be 50 ppm, with some loss of performance, but even this level represents a significant barrier and illustrates how efficient the CO removal method needs to be.

11.4.3 COMMERCIAL PROTOTYPES OF EXISTING HYDROGEN FUEL-CELL-BASED CARS

Over the years, many example of hydrogen-powered vehicles have been produced [197]. However, most of these were isolated projects. In the 1990s, automobile manufacturers intensified their efforts, H_2 technology has advanced significantly. Table 11.2 lists some examples of vehicles that have been unveiled by many of the main automobile manufacturers in the last 15 years. In addition to these, there are many other examples of projects involving buses or specialty vehicles. In the case of buses, PAFCs are more common than PEMFCs.

Table 11.2, even if not exhaustive, clearly illustrates that the technology of hydrogen vehicles is at a very advanced state. Many of the previously discussed options are technically possible. The fundamental problem is one of cost. It should be noted that many examples in Table 11.2 are hybrid fuel-cell/battery vehicles, which might represent another transition step to fuel-cell-only powered vehicles.

TABLE 11.2
Commercial Prototypes of Hydrogen Vehicles

Organization	Vehicle	Engine Type	Fuel
Audi	A2	FC/battery	Gaseous H_2
BMW	Series 7 (745 h) (sedan)	ICE (FC APU)	Gasoline/liquid H_2
Daihatsu	MOVE EV-FC (micro van)	FC/battery	Methanol reformer
	MOVE FCV- K II (mini vehicle)	FC/battery	Compressed H_2 (3600 psi)
DaimlerChrysler	NECAR1 (180 van)	FC stack	Compressed H_2 (4300 psi)
	NECAR2 (V-class)	FC	Compressed H_2 (3600 psi)
	NECAR3 (A-class)	FC stack	Methanol reformer
	NECAR4 (A-class)	FC	Liquid H_2
	NECAR4A	FC	Compressed H_2 (5000 psi)
	NECAR5 (A-class)	FC	Methanol reformer
	NECAR5.2 (A-class)	FC/battery	Methanol reformer
	F-Cell (A-class)	FC/battery	Compressed H_2 (5000 psi)
	Jeep Commander 2 (SUV)	FC/battery	Methanol reformer
	Sprinter (van)		
	Natrium (minivan)	FC	Compressed H_2 (5000 psi)
	Jeep Treo	FC/battery	Catalyzed $NaBH_4$
ESORO	Hycar	FC/battery	Compressed H_2
Fiat	Seicento Elettra H_2 fuel cell	FC/battery	Compressed H_2
Ford Motor Co.	P2000 HFC (sedan)	FC	Compressed H_2
	Focus FCV	FC	Compressed H_2 (3600 psi)
	Think FC5	FC	Methanol reformer
	Advanced FCV	FC/battery	Compressed H_2 (5000 psi)
	Model U	ICE	
Opel / GM	Sintra (minivan)	FC	
	Zafira (minivan)	FC	Methanol reformer
	Precept FCEV	FC/battery	Metal hydride
	HydroGen 1 (zafira)	FC/battery	Liquid H_2
	HydroGen 3 (zafira)	FC	Liquid H_2
	Chevy S-10 (pickup truck)	FC/battery	Reformer CHF
	Advanced HydroGen3	FC	Compressed H_2 (5000 psi)
	Sequel	FC	Compressed H_2 (10000 psi)
		FC/battery	Compressed H_2 (10000 psi)
GM (Shanghai) PATAC	Phoenix	FC/battery	Compressed H_2

continued

TABLE 11.2 (CONTINUED)
Commercial Prototypes of Hydrogen Vehicles

Organization	Vehicle	Engine Type	Fuel
Honda Motor Co.	FCX-V1	FC/battery	Metal hydride (LaNi$_5$)
	FCX-V2	FC/battery	Methanol reformer (ATR)
	FCX-V3	FC/capacitors	Gas (250 atm, 3600 psi)
	FCX-V4	FC/capacitors	Gas (350 atm, 5000 psi)
	FCX	FC/capacitors	Gas (5000 psi)
	Kiwami concept	FC	Hydrogen
Hyundai	Santa Fe (SUV)	FC	Compressed H$_2$
	Tucson	FC	Compressed H$_2$
Kia	Sportage	FC	Compressed H$_2$
Mazda Motor Co.	Demio (passenger car)	FC/capacitors	Metal hydride
	Premacy FC-EV	FC	Methanol reformer
Mitsubishi	SpaceLiner	FC/battery	Methanol reformer
	Grandis FCV	FC/battery	Compressed H$_2$
Nissan	R'nessa (SUV)	FC/battery	Methanol
	Xterra (SUV)	FC/battery	Compressed H$_2$
	X-TAIL (SUV)	FC/battery	Compressed H$_2$ (5000 psi)
	Effis	FC/battery	
Peugeot/Citroen	Peugeot Hydro-Gen	FC/battery	Compressed H$_2$
	Peugeot Fuel Cell Cab	FC/battery	Compressed H$_2$ (4300 psi)
	H$_2$O fire fighting	Battery/FC APU	Catalyzed NaBH$_4$
Renault	EU-FEVER Project (Laguna Wagon)	FC/battery	Liquid H$_2$
Suzuki	Covie	FC	—
	Mobile Terrace	FC	
Toyota	RAV 4 FCEV (SUV)	FC/battery	Metal hydride
	RAV 4 FCEV (SUV)	FC/battery	Methanol reformer
	FCHV-3 (SUV)	FC/battery	Metal hydride
	FCHV-4 (SUV)	FC/battery	Compressed H$_2$ (3600 psi)
	FCHV-5 (SUV)	FC/battery	Reformer (CHF)
	FCHV (SUV)	FC/battery	Compressed H$_2$ (5000 psi)
Volkswagen	EU Capri Project (VW Estate)	FC/battery	Methanol reformer
		FC	Liquid H$_2$
	HyPower	FC/capacitors	Compressed H$_2$

11.4.4 HYDROGEN VERSUS COMPETING TECHNOLOGIES

Hydrogen-powered vehicles have to contend with two main competing technologies. The first is current TWC technology. Advances being made in postcombustion emission control mean that it is possible to envisage that gasoline-fueled IC engines fitted with superior TWCs will become zero emitting at the point of use (with the exception of CO_2). For car manufacturers, this would involve only minor changes in production methods: simply changing the type of TWC or modifying the arrangement of the converter in the exhaust stream. The same cannot be said for changing production to hydrogen-powered vehicles. The second is electric vehicles (EVs), either as battery-powered EVs or as hybrid electric vehicles (HEVs), which involve the combination of a combustion engine and a battery. In HEVs, the two power sources may be connected to a single drive train (in series), or two separate power trains may be used (parallel), so that the vehicle can be operated by either or both. Various EVs have been developed over the years, only to be met by consumer indifference, and they have only found niche markets. They are unattractive mainly due to long recharging times and rather limited range. In addition, there are technical problems associated with the batteries. However, these issues have been confronted, and two HEV designs have recently been commercialized: Toyota's Prius and Honda's Civic (both parallel designs with nickel-metal-hydride batteries). More importantly, these vehicles have achieved a degree of commercial success, which augurs well for more widespread production. In a recent report, a comparison was made between the well-to-wheel efficiencies of an automobile powered by a gasoline ICE, a hybrid HEV, and a fuel-cell-powered vehicle, in which the hydrogen is derived from onboard gasoline reforming. In agreement with Honda's own findings, the results indicated that there is no significant difference in efficiency or CO_2 emissions between the last two options [200]. This prompted the authors to conclude that efforts to develop HEVs should be given highest priority in the transition toward the hydrogen economy, given the advanced state of the technology.

11.5 CONCLUSIONS AND PERSPECTIVES

It is clear that many of the technologies necessary for the transition to the hydrogen economy are technically possible. The main problem is their high cost. Indeed, the possibility must be accepted that equivalent or lower costs than the current situation may never be achieved. However, cost is not the only issue, and the need to replace fossil fuels as well as environmental benefits must also be considered. Furthermore, as there now appears to be a firm commitment to research and development in this direction, costs will certainly come down, either through research breakthroughs or through the economies of scale that come into play as any technology becomes more widespread. For the developing world, the cost issue is of course a more acute problem; however, it may be possible for developing countries to take advantage of the worldwide research effort, perhaps skipping the necessity to establish the fossil fuel infrastructures present in the developed world. Further along the line, in the case of establishing a renewable hydrogen economy, the natural resources of any country will play an important role. For example, Iceland is usually cited as the country closest to achieving a renewable hydrogen economy. There, the enormous geothermal energy resources available are harnessed for hydrogen production, and its widespread distribution is made easier by the relatively small population. The potential of any other country to develop a renewable hydrogen economy will depend in a similar way on its natural resources.

It is difficult to predict which of the technological options discussed in this chapter will become the standard technology for automotive application. It is possible that more than one system will be widely used. However, at present, the general consensus seems to be that the long-term option will be fuel-cell vehicles with onboard hydrogen storage linked to the hydrogen distribution network discussed above. With regard to the storage material, the situation of carbon-based materials is delicately balanced, and if the reproducibility problems are not resolved soon, this area will almost certainly experience a decline of interest. This leaves hydrides, although it is an open question as to which hydride or form of hydride storage system will prevail.

ACKNOWLEDGMENT

University of Trieste, Centre of Excellence for Nanostructured Material, Consortium INSTM, MIUR FIRB2001, and FISR2002 are acknowledged for financial support.

REFERENCES

1. International Partnership for the Hydrogen Economy; available on-line at http://www.iphe.net, 2005.
2. Painter, D.E., *Air Pollution Technology*, Reston Publishing Co., Reston, VA, 1974.
3. De Nevers, N., *Air Pollution Control Engineering*, 2nd ed., McGraw-Hill, Boston, 2000.
4. Obuchi, A. et al., A practical scale evaluation of catalysts for the selective reduction of NO_x with organic substances using a diesel exhaust, *Appl. Catal. B, Environ.*, 15, 37, 1998.
5. Ciambelli, P. et al., Lean NO_x reduction CuZSM5 catalysts: evaluation of performance at the spark ignition engine exhaust, *Catal. Today*, 26, 33, 1995.
6. Degobert, P., *Automobiles and Pollution*, Society of Automotive Engineers, Warrendale, PA, 1995.
7. Iwamoto, M. et al., Performance and durability of Pt-MFI zeolite catalyst for selective reduction of nitrogen monoxide in actual diesel engine exhaust, *Appl. Catal. B, Environ.*, 5, L1, 1994.
8. The European Environmental Agency, Annual Report 2005. Available on-line at http://themes.eea.europa.eu.
9. Antia, A.N. et al., Basin-wide particulate carbon flux in the Atlantic Ocean: regional export patterns and potential for atmospheric CO_2 sequestration, *Global Biogeochemical Cycles*, 15, 845, 2001.
10. Conte, M. et al., Hydrogen economy for a sustainable development: state-of-the-art and technological perspectives, *J. Power Sources*, 100, 171, 2001.

11. Kyoto Protocol to the United Nations Framework Convention on Climate Change. Available on-line at http://unfccc.int/resource/docs/convkp/kpeng.html, 2005.

12. Church, M.L., Cooper, B.J., and Willson, P.J., *Catalyst Formulations 1960 to Present*, SAE 890815, Society of Automotive Engineers, Warrendale, PA, 1989.

13. McCabe, R.W. and Kisenyi, J.M., Advances in automotive catalyst technology, *Chem. Ind. (London)*, 15, 605, 1995.

14. Kaspar, J., Graziani, M., and Fornasiero, P., Ceria-containing three way catalysts, in *Handbook on the Physics and Chemistry of Rare Earths: The Role of Rare Earths in Catalysis*, Vol. 29, Gschneidner, Jr., K.A. and Eyring, L., Eds., Elsevier Science B.V., Amsterdam, 2000, chap. 184, pp. 159–267.

15. Kaspar, J., Fornasiero, P., and Hickey, N., Automotive catalytic converters: current status and some perspectives, *Catal. Today*, 77, 419, 2003.

16. Burch, R., Knowledge and know-how in emission control for mobile applications, *Catalysis Reviews: Science Engineering*, 46, 271, 2004.

17. Takahashi, N. et al., The new concept 3-way catalyst for automotive lean-burn engine: NO_x storage and reduction catalyst, *Catal. Today*, 27, 63, 1996.

18. Shinjoh, H. et al., Effect of periodic operation over Pt catalysts in simulated oxidizing exhaust gas, *Appl. Catal. B, Environ.*, 15, 189, 1998.

19. Farrauto, R.J. and Voss, K.E., Monolithic diesel oxidation catalysts, *Appl. Catal. B, Environ.*, 10, 29, 1996.

20. van Setten, B.A.A.L., Makkee, M., and Moulijn, J.A., Science and technology of catalytic diesel particulate filters, *Catal. Rev., Sci. Eng.*, 43, 489, 2001.

21. Amon, B. et al., The $SINO_x$ system for trucks to fulfil the future emission regulations, *Top. Catal.*, 16, 187, 2001.

22. Guibet, J.C., *Fuels and Engines*, Editions Technip, Paris, 1999.

23. Tromp, T.K. et al., Potential environmental impact of a hydrogen economy on the stratosphere, *Science*, 300, 1740, 2003.

24. Prather, M.J., An environmental experiment with H_2? *Science*, 302, 581, 2003.

25. Service, R.F., The hydrogen backlash, *Science*, 305, 958, 2004.

26. Schlapbach, L. and Zuttel, A., Hydrogen-storage materials for mobile applications, *Nature*, 414, 353, 2001.

27. McNicol, B.D., Rand, D.A.J., and Williams, K.R., Fuel cells for road transportation purposes: yes or no?, *J. Power Sources*, 100, 47, 2001.

28. Bain, A. and Van Vorst, W.D., The Hindenburg tragedy revisited: the fatal flaw found, *Int. J. Hydrogen Energy*, 24, 399, 1999.

29. Rostrup-Nielsen, J.R. and Rostrup-Nielsen T., Large-scale hydrogen production, *CATTECH*, 6, 150, 2002.

30. Pena, M.A., Gomez, J.P., and Fierro, J.L.G., New catalytic routes for syngas and hydrogen production, *Appl. Catal. A Gen.*, 144, 7, 1996.

31. Freni, S., Calogero, G., and Cavallaro, S., Hydrogen production from methane through catalytic partial oxidation reactions, *J. Power Sources*, 87, 28, 2000.

32. Reyniers, M.F. et al., Catalytic partial oxidation, Part I: catalytic processes to convert methane: partial or total oxidation, *CATTECH*, 6, 140, 2002.

33. Edwards, N. et al., On-board hydrogen generation for transport applications: the HotSpot™ methanol processor, *J. Power Sources*, 71, 123, 1998.

34. Ghenciu, A.F., Review of fuel processing catalysts for hydrogen production in PEM fuel cell systems, *Curr. Opin. Solid St. M.*, 6, 389, 2002.

35. Farrauto, R. et al., New material needs for hydrocarbon fuel processing: generating hydrogen for the PEM fuel cell, *Ann. Rev. Mater. Res.*, 33, 1, 2003.

36. Giroux, T. et al., Monolithic structures as alternatives to particulate catalysts for the reforming of hydrocarbons for hydrogen generation, *Appl. Catal. B, Environ.*, 56, 95, 2005.

37. Wei, J.M. and Iglesia, E., Isotopic and kinetic assessment of the mechanism of methane reforming and decomposition reactions on supported iridium catalysts, *Phys. Chem. Chem. Phys.*, 6, 3754, 2004.

38. Wei, J.M. and Iglesia, E., Isotopic and kinetic assessment of the mechanism of reactions of CH_4 with CO_2 or H_2O to form synthesis gas and carbon on nickel catalysts, *J. Catal.*, 224, 370, 2004.

39. Wei, J.M. and Iglesia, E., Mechanism and site requirements for activation and chemical conversion of methane on supported Pt clusters and turnover rate comparisons among noble metals, *J. Phys. Chem. B*, 108, 4094, 2004.

40. Wei, J.M. and Iglesia, E., Reaction pathways and site requirements for the activation and chemical conversion of methane on Ru-based catalysts, *J. Phys. Chem. B*, 108, 7253, 2004.

41. Wei, J.M. and Iglesia, E., Structural requirements and reaction pathways in methane activation and chemical conversion catalyzed by rhodium, *J. Catal.*, 225, 116, 2004.

42. Hayakawa, T. et al., CO_2 reforming of CH_4 over Ni perovskite catalysts prepared by solid phase crystallization method, *Appl. Catal. A, Gen.*, 183, 273, 1999.

43. Ashcroft, A.T. et al., Partial oxidation of methane to synthesis gas using carbon dioxide, *Nature*, 352, 225, 1991.

44. Song, C.S., Fuel processing for low-temperature and high-temperature fuel cells: challenges and opportunities for sustainable development in the 21st century, *Catal. Today*, 77, 17, 2002.

45. Liu, Q.S. et al., Progress in water-gas-shift catalysts, *Prog. Chem.*, 17, 389, 2005.

46. Armor, J.N., The multiple roles for catalysis in the production of H_2, *Appl. Catal. A, Gen.*, 176, 159, 1999.

47. Gunardson, H., *Industrial Gases in Petrochemical Processing*, Marcel Dekker, New York, 1998.

48. Newsome, D.S., The water-gas shift reaction, *Catal. Rev., Sci. Eng.*, 21, 275, 1980.

49. Liu, Q.S. et al., Progress in water-gas-shift catalysts, *Prog. Chem.*, 17, 389, 2005.

50. Ahmed, S. and Krumpelt, M., Hydrogen from hydrocarbon fuels for fuel cells, *Int. J. Hydrogen Energy*, 26, 291, 2001.

51. Pettersson, L.J. and Westerholm, R., State of the art of multi-fuel reformers for fuel cell vehicles: problem identification and research needs, *Int. J. Hydrogen Energy*, 26, 243, 2001.

52. Moon, D.J. et al., Studies on gasoline fuel processor system for fuel-cell powered vehicles application, *Appl. Catal. A, Gen.*, 215, 1, 2001.

53. York, A.P.E., Xiao, T.C., and Green, M.L.H., Brief overview of the partial oxidation of methane to synthesis gas, *Top. Catal.*, 22, 345, 2003.

54. Tomishige, K., Chen, Y.G., and Fujimoto, K., Studies on carbon deposition in CO_2 reforming of CH_4 over nickel-magnesia solid solution catalysts, *J. Catal.*, 181, 91, 1999.

55. Tsipouriari, V.A. and Verykios, X.E., Carbon and oxygen reaction pathways of CO_2 reforming of methane over Ni/La_2O_3 and Ni/Al_2O_3 catalysts studied by isotopic tracing techniques, *J. Catal.*, 187, 85, 1999.

56. Slagtern, A. et al., Specific features concerning the mechanism of methane reforming by carbon dioxide over Ni/La_2O_3 catalyst, *J. Catal.*, 172, 118, 1997.

57. Hegarty, M.E.S., O'Connor, A.M., and Ross, J.R.H., Syngas production from natural gas using ZrO_2-supported metals, *Catal. Today*, 42, 225, 1998.

58. Montoya, J.A. et al., Methane reforming with CO_2 over Ni/ZrO_2-CeO_2 catalysts prepared by sol-gel, *Catal. Today*, 63, 71, 2000.

59. Roh, H.S. et al., Highly stable Ni catalyst supported on Ce-ZrO_2 for oxy-steam reforming of methane, *Catal. Lett.*, 74, 31, 2001.

60. Sharma, S. et al., Evidence for oxidation of ceria by CO_2, *J. Catal.*, 190, 199, 2000.

61. Wang, X. and Gorte, R.J., A study of steam reforming of hydrocarbon fuels on Pd/ceria, *Appl. Catal. A, Gen.*, 224, 209, 2002.

62. Krumpelt, M. et al., *Catalytic Autothermal Reforming*, annual progress report, USDOE, Energy Efficiency and Renewable Energy, Office of Transportation Technologies, Washington, D.C., 2000, pp. 66–70.

63. Zarur, A.J. and Ying, J.Y., Reverse microemulsion synthesis of nanostructured complex oxides for catalytic combustion, *Nature*, 403, 65, 2000.

64. Groppi, G., Cristiani, C., and Forzatti, P., Preparation, characterisation and catalytic activity of pure and substituted La-hexaaluminate systems for high temperature catalytic combustion, *Appl. Catal. B, Environ.*, 35, 137, 2001.

65. Schicks, J. et al., Nanoengineered catalysts for high-temperature methane partial oxidation, *Catal. Today*, 81, 287, 2003.

66. Chu, W.L., Yang, W.S., and Lin, L.W., The partial oxidation of methane to syngas over the nickel-modified hexaaluminate catalysts $BaNi_yAl_{12-y}O_{19}$-delta, *Appl. Catal. A, Gen.*, 235, 39, 2002.

67. Bartholomew, C.H., Carbon deposition in steam reforming and methanation, *Catal. Rev., Sci. Eng.*, 1982, 67, 1982.

68. Swaan, H.M. et al., Deactivation of supported nickel-catalysts during the reforming of methane by carbon-dioxide, *Catal. Today*, 21, 571, 1994.

69. Basile, F. et al., New hydrotalcite-type anionic clays containing noble metals, *Chem. Commun.*, 2435, 1996.

70. Basile, F. et al., Catalytic partial oxidation and CO_2-reforming on Rh- and Ni-based catalysts obtained from hydrotalcite-type precursors, *Appl. Clay Sci.*, 13, 329, 1998.

71. Shiozaki, R. et al., Partial oxidation of methane over a $Ni/BaTiO_3$ catalyst prepared by solid phase crystallization, *J. Chem. Soc., Faraday Trans.*, 93, 3235, 1997.

72. Takehira, K. et al., Autothermal reforming of CH_4 over supported Ni catalysts prepared from Mg-Al hydrotalcite-like anionic clay, *J. Catal.*, 221, 43, 2004.

73. Shishido, T. et al., CO_2 reforming of CH_4 over Ni/Mg-Al oxide catalysts prepared by solid phase crystallization method from Mg-Al hydrotalcite-like precursors, *Catal. Lett.*, 73, 21, 2001.

74. Shishido, T. et al., Partial oxidation of methane over Ni/Mg-Al oxide catalysts prepared by solid phase crystallization method from Mg-Al hydrotalcite-like precursors, *Appl. Catal. A, Gen.*, 223, 35, 2002.

75. Takehira, K., Shishido, T., and Kondo, M., Partial oxidation of CH_4 over $Ni/SrTiO_3$ catalysts prepared by a solid-phase crystallization method, *J. Catal.*, 207, 307, 2002.

76. Raimondi, F. et al., Hydrogen production by methanol reforming: post-reaction characterisation of a $Cu/ZnO/Al_2O_3$ catalyst by XPS and TPD, *Appl. Surf. Sci.*, 189, 59, 2002.

77. Peppley, B.A. et al., Methanol-steam reforming on $Cu/ZnO/Al_2O_3$, Part 1: the reaction network, *Appl. Catal. A, Gen.*, 179, 21, 1999.

78. Turco, M. et al., Production of hydrogen from oxidative steam reforming of methanol, I: preparation and characterization of $Cu/ZnO/Al_2O_3$ catalysts from a hydrotalcite-like LDH precursor, *J. Catal.*, 228, 43, 2004.

79. Turco, M. et al., Production of hydrogen from oxidative steam reforming of methanol, II: catalytic activity and reaction mechanism on $Cu/ZnO/Al_2O_3$ hydrotalcite-derived catalysts, *J. Catal.*, 228, 56, 2004.

80. Breen, J.P. and Ross, J.R.H., Methanol reforming for fuel-cell applications: development of zirconia-containing Cu-Zn-Al catalysts, *Catal. Today*, 51, 521, 1999.

81. Velu, S. et al., Selective production of hydrogen for fuel cells via oxidative steam reforming of methanol over CuZnAl(Zr)-oxide catalysts, *Appl. Catal. A, Gen.*, 213, 47, 2001.

82. Velu, S., Suzuki, K., and Osaki, T., Oxidative steam reforming of methanol over CuZnAl(Zr)-oxide catalysts; a new and efficient method for the production of CO-free hydrogen for fuel cells, *Chem. Commun.*, 2341, 1999.

83. Liu, S.T. et al., Hydrogen production by oxidative methanol reforming on Pd/ZnO, *Appl. Catal. A, Gen.*, 283, 125, 2005.

84. Shabaker, J.W. et al., Aqueous-phase reforming of methanol and ethylene glycol over alumina-supported platinum catalysts, *J. Catal.*, 215, 344, 2003.

85. Bi, Y.P. et al., Correlation of activity and size of Pt nanoparticles for methanol steam reforming over $Pt/(Y)-Al_2O_3$ catalysts, *Acta Chimica Sinica*, 63, 802, 2005.

86. Tsai, A.P. and Yoshimura, M., Highly active quasicrystalline Al-Cu-Fe catalyst for steam reforming of methanol, *Appl. Catal. A, Gen.*, 214, 237, 2001.

87. Segal, S.R. et al., Low temperature steam reforming of methanol over layered double hydroxide-derived catalysts, *Appl. Catal. A, Gen.*, 231, 215, 2002.

88. Akande, A., Idem, R.O., and Dalai, A.K., Synthesis, characterization and performance evaluation of Ni/Al_2O_3 catalysts for reforming of crude ethanol for hydrogen production, *Appl. Catal. A, Gen.*, 287, 159, 2005.

89. Comas, J. et al., Bio-ethanol steam reforming on Ni/Al_2O_3 catalyst, *Chem. Eng. J.*, 98, 61, 2004.

90. Diagne, C. et al., Efficient hydrogen production by ethanol reforming over Rh catalysts: effect of addition of Zr on CeO_2 for the oxidation of CO to CO_2, *Comptes Rendus Chimie*, 7, 617, 2004.

91. Fatsikostas, A.N. and Verykios, X.E., Reaction network of steam reforming of ethanol over Ni-based catalysts, *J. Catal.*, 225, 439, 2004.

92. Fatsikostas, A.N., Kondarides, D.I., and Verykios, X.E., Steam reforming of biomass-derived ethanol for the production of hydrogen for fuel cell applications, *Chem. Commun.*, 851, 2001.

93. Klouz, V. et al., Ethanol reforming for hydrogen production in a hybrid electric vehicle: process optimisation, *J. Power Sources*, 105, 26, 2002.

94. Navarro, R.M. et al., Production of hydrogen by oxidative reforming of ethanol over Pt catalysts supported on Al_2O_3 modified with Ce and La, *Appl. Catal. B, Environ.*, 55, 229, 2005.

95. Haga, F. et al., Catalytic properties of supported cobalt catalysts for steam reforming of ethanol, *Catal. Lett.*, 48, 223, 1997.

96. Haga, F. et al., Effect of particle size on steam reforming of ethanol over alumina-supported cobalt catalyst, *Nippon Kagaku Kaishi*, 11, 758, 1997.

97. Patt, J. et al., Molybdenum carbide catalysts for water-gas shift, *Catal. Lett.*, 65, 193, 2000.

98. Ruettinger, W., Ilinich, O., and Farrauto, R.J., A new generation of water gas shift catalysts for fuel cell applications, *J. Power Sources*, 118, 61, 2003.

99. Ruettinger, W. et al., Non-pyrophoric water-gas shift catalysts for hydrogen generation in fuel cell applications, *Abstr. Pap. Am. Chem. Soc.*, 221, U494, 2001.

100. Li, K., Fu, Q., and Flytzani-Slephanopoulos, M., Low-temperature water-gas shift reaction over Cu- and Ni-loaded cerium oxide catalysts, *Appl. Catal. B, Environ.*, 27, 179, 2000.

101. Krause, T. et al., *Water Gas Shift Catalysis*, DOE Hydrogen Program, FY 2004 Progress Report, U.S. Department of Energy, Washington, D.C., 2004.

102. Mendelovici, L. and Steinberg M., *J. Catal.*, 96, 285, 1985.

103. Fu, Q., Saltsburg, H., and Flytzani-Stephanopoulos, M., Active nonmetallic Au and Pt species on ceria-based water-gas shift catalysts, *Science*, 301, 935, 2003.

104. Gorte, R.J. and Zhao, S., Studies of the water-gas-shift reaction with ceria-supported precious metals, *Catal. Today*, 104, 18, 2005.

105. Bunluesin, T., Gorte, R.J., and Graham, G.W., Studies of the water-gas-shift reaction on ceria-supported Pt, Pd, and Rh: implications for oxygen-storage properties, *Appl. Catal. B, Environ.*, 15, 107, 1998.

106. Swartz, S.L. et al., Fuel processing catalysts based on nanoscale ceria, *Fuel Cell Bull.*, 30, 7, 2001.

107. Zalc, J.M., Sokolovskii, V., and Loffler, D.G., Are noble metal-based water-gas shift catalysts practical for automotive fuel processing? *J. Catal.*, 206, 169, 2002.

108. Haruta, M., Yamada, N., Kobayashi, T., and Ijima, S., Gold catalysts prepared by coprecipitation for low-temperature oxidation of hydrogen and of carbon monoxide, *J. Catal.*, 115, 301, 1989.

109. Fu, Q. et al., Activity and stability of low-content gold-cerium oxide catalysts for the water-gas shift reaction, *Appl. Catal. B, Environ.*, 56, 57, 2005.

110. Kim, C.H. and Thompson, L.T., Deactivation of Au/CeO_x water gas shift catalysts, *J. Catal.*, 230, 66, 2005.

111. Tabakova, T. et al., FTIR study of low-temperature water-gas shift reaction on gold/ceria catalyst, *Appl. Catal. A, Gen.*, 252, 385, 2003.

112. Andreeva, D. et al., Low-temperature water-gas shift reaction over Au/CeO_2 catalysts, *Catal. Today*, 72, 51, 2002.

113. Luengnaruemitchai, A., Osuwan, S., and Gulari, E., Comparative studies of low-temperature water-gas shift reaction over Pt/CeO_2, Au/CeO_2, and Au/Fe_2O_3 catalysts, *Catal. Commun.*, 4, 215, 2003.

114. Idakiev, V. et al., Gold catalysts supported on mesoporous titania for low-temperature water-gas shift reaction, *Appl. Catal. A, Gen.*, 270, 135, 2004.

115. Panagiotopoulou, P. and Kondarides, D.I., Effect of morphological characteristics of TiO_2-supported noble metal catalysts on their activity for the water-gas shift reaction, *J. Catal.*, 225, 327, 2004.

116. Boccuzzi, F. et al., Gold, silver and copper catalysts supported on TiO_2 for pure hydrogen production, *Catal. Today*, 75, 169, 2002.

117. Boccuzzi, F. et al., FTIR study of the low-temperature water-gas shift reaction on Au/Fe_2O_3 and Au/TiO_2 catalysts, *J. Catal.*, 188, 176, 1999.

118. Sakurai, H. et al., Low-temperature water-gas shift reaction over gold deposited on TiO_2, *Chem. Commun.*, 271, 1997.

119. Andreeva, D. et al., $Au/alpha-Fe_2O_3$ catalyst for water-gas shift reaction prepared by deposition-precipitation, *Appl. Catal. A, Gen.*, 169, 9, 1998.

120. Andreeva, D. et al., Low-temperature water-gas shift reaction on $Au/alpha-Fe_2O_3$ catalyst, *Appl. Catal. A, Gen.*, 134, 275, 1996.

121. Andreeva, D. et al., Low-temperature water-gas shift reaction over $Au/alpha-Fe_2O_3$, *J. Catal.*, 158, 354, 1996.

122. Venugopal, A. and Scurrell, M.S., Low temperature reductive pretreatment of Au/Fe_2O_3 catalysts, TPR/TPO studies and behaviour in the water-gas shift reaction, *Appl. Catal. A, Gen.*, 258, 241, 2004.

123. Daniells, S.T., Makkee, M., and Moulijn, J.A., The effect of high-temperature pretreatment and water on the low temperature CO oxidation with Au/Fe_2O_3 catalysts, *Catal. Lett.*, 100, 39, 2005.

124. Farrauto, R.J., Introduction to solid polymer membrane fuel cells and reforming natural gas for production of hydrogen, *Appl. Catal. B, Environ.*, 56, 3, 2005.

125. Rossignol, C. et al., Selective oxidation of CO over model gold-based catalysts in the presence of H_2, *J. Catal.*, 230, 476, 2005.

126. Marino, F., Descorme, C., and Duprez, D., Noble metal catalysts for the preferential oxidation of carbon monoxide in the presence of hydrogen (PROX), *Appl. Catal. B, Environ.*, 54, 59, 2004.

127. Panzera, G. et al., CO selective oxidation on ceria-supported Au catalysts for fuel cell application, *J. Power Sources*, 135, 177, 2004.

128. Schumacher, B. et al., Kinetics, mechanism, and the influence of H_2 on the CO oxidation reaction on a Au/TiO_2 catalyst, *J. Catal.*, 224, 449, 2004.

129. Shin, W.S. et al., Development of Au/Ce catalysts for preferential oxidation of CO in PEMFC, *J. Ind. Eng. Chem.*, 10, 302, 2004.

130. Luengnaruemitchai, A., Osuwan, S., and Gulari, E., Selective catalytic oxidation of CO in the presence of H_2 over gold catalyst, *Int. J. Hydrogen Energy*, 29, 429, 2004.

131. Schubert, M.M. et al., Influence of H_2O and CO_2 on the selective CO oxidation in H_2-rich gases over Au/alpha-Fe_2O_3, *J. Catal.*, 222, 32, 2004.

132. Cameron, D., Holliday, R., and Thompson, D., Gold's future role in fuel cell systems, *J. Power Sources*, 118, 298, 2003.

133. Choudhary, T.V. et al., CO oxidation on supported nano-Au catalysts synthesized from a $[Au_6(PPh_3)(6)](BF_4)(2)$ complex, *J. Catal.*, 207, 247, 2002.

134. Schubert, M.M. et al., Activity, selectivity, and long-term stability of different metal oxide supported gold catalysts for the preferential CO oxidation in H_2-rich gas, *Catal. Lett.*, 76, 143, 2001.

135. Avgouropoulos G. et al., CuO-CeO_2 mixed oxide catalysts for the selective oxidation of carbon monoxide in excess hydrogen, *Catal. Lett.*, 73, 33, 2001.

136. Avgouropoulos, G. et al., A comparative study of $Pt/gamma$-Al_2O_3, Au/alpha-Fe_2O_3 and CuO-CeO_2 catalysts for the selective oxidation of carbon monoxide in excess hydrogen, *Catal. Today*, 75, 157, 2002.

137. Joensen, F. and Rostrup-Nielsen, J.R., Conversion of hydrocarbons and alcohols for fuel cells, *J. Power Sources*, 105, 195, 2002.

138. The U.S. Department of Energy, Website. U.S. Department of Energy; available online at www.doe.gov, 2005.

139. Wolf, A. and Schuth, F., A systematic study of the synthesis conditions for the preparation of highly active gold catalysts, *Appl. Catal A, Gen.*, 226, 1, 2002.

140. Aceves, S.M., Perfect, S., and Weisberg, A., *Optimum utilization of available space in a vehicle through conformable hydrogen vessels*, DOE Hydrogen Program, FY 2004 Progress Report, U.S. Department of Energy, Washington, D.C., 2004.

141. Weisberg, A., *Next-generation physical hydrogen*, DOE Hydrogen Program, FY 2004 Progress Report, U.S. Department of Energy, Washington, D.C., 2004.

142. Wolf, J., Liquid-hydrogen technology for vehicles, *MRS Bull.*, 27, 684, 2002.

143. Schuth, F., Bogdanovic, B., and Felderhoff, M., Light metal hydrides and complex hydrides for hydrogen storage, *Chem. Commun.*, 2249, 2004.

144. Darkrim, F.L., Malbrunot, P., and Tartaglia, G.P., Review of hydrogen storage by adsorption in carbon nanotubes, *Int. J. Hydrogen Energy*, 27, 193, 2002.

145. Zuttel, A. et al., Hydrogen storage in carbon nanostructures, *Int. J. Hydrogen Energy*, 27, 203, 2002.

146. Ajayan, P.M. and Ebbesen, T.W., Nanometre-size tubes of carbon, *Rep. Prog. Phys.*, 60, 1025, 1997.

147. Ajayan, P.M., Carbon nanotubes: novel architecture in nanometer space, *Prog. Crystal Growth Characterization Mater.*, 34, 37, 1997.

148. Dresselhaus, M.S. et al., *Graphite Fibres and Filaments*, Springer, Berlin, 1998.

149. Hirscher, M. et al., Hydrogen storage in carbon nanostructures, *J. Alloys Comp.*, 330, 654, 2002.

150. Iijima, S., Helical microtubules of graphitic carbon, *Nature*, 354, 56, 1991.

151. Dillon, A.C. et al., Storage of hydrogen in single-walled carbon nanotubes, *Nature*, 386, 377, 1997.

152. Chambers, A. et al., Hydrogen storage in graphite nanofibers, *J. Phys. Chem. B*, 102, 4253, 1998.

153. Chen, P. et al., Interaction of hydrogen with metal nitrides and imides, *Nature*, 420, 302, 2002.

154. Gupta, B.K. and Srivastava, O.N., Further studies on microstructural characterization and hydrogenation behaviour of graphitic nanofibres, *Int. J. Hydrogen Energy*, 26, 857, 2001.

155. Ye, Y. et al., Hydrogen adsorption and cohesive energy of single-walled carbon nanotubes, *Appl. Phys. Lett.*, 74, 2307, 1999.

156. Fan, Y.Y. et al., Hydrogen uptake in vapor-grown carbon nanofibers, *Carbon*, 37, 1649, 1999.

157. Orimo, S. et al., Hydrogen desorption property of mechanically prepared nanostructured graphite, *J. Appl. Phys.*, 90, 1545, 2001.

158. Orimo, S. et al., Hydrogen in the mechanically prepared nanostructured graphite, *Appl. Phys. Lett.*, 75, 3093, 1999.

159. Ahn, C.C. et al., Hydrogen desorption and adsorption measurements on graphite nanofibers, *Appl. Phys. Lett.*, 73, 3378, 1998.

160. Tibbetts, G.G., Meisner, G.P., and Olk, C.H., Hydrogen storage capacity of carbon nanotubes, filaments, and vapor-grown fibers, *Carbon*, 39, 2291, 2001.

161. Shiraishi, M., Takenobu, T., and Ata, M., Gas-solid interactions in the hydrogen/single-walled carbon nanotube system, *Chem. Phys. Lett.*, 367, 633, 2003.

162. Strobel, R. et al., Hydrogen adsorption on carbon materials, *J. Power Sources*, 84, 221, 1999.

163. Schimmel, H.G. et al., Hydrogen adsorption in carbon nanostructures: comparison of nanotubes, fibers, and coals, *Chem-Eur. J.*, 9, 4764, 2003.

164. Yang, R.T., Hydrogen storage by alkali-doped carbon nanotubes-revisited, *Carbon*, 38, 623, 2000.

165. Hirscher, M. et al., Hydrogen storage in sonicated carbon materials, *Appl. Phys. A, Mater. Sci. Process.*, 72, 129, 2001.

166. Dillon, A.C. and Heben, M.J., Hydrogen storage using carbon adsorbents: past, present and future, *Appl. Phys. A, Mater. Sci. Process.*, 72, 133, 2001.

167. Cheng, H., Pez, G.P., and Cooper, A.C., Mechanism of hydrogen sorption in single-walled carbon nanotubes, *J. Am. Chem. Soc.*, 123, 5845, 2001.

168. Fakioglu, E., Yurum, Y., and Veziroglu, T.N., A review of hydrogen storage systems based on boron and its compounds, *Int. J. Hydrogen Energy*, 29, 1371, 2004.

169. Chen, J. and Wu, F., Review of hydrogen storage in inorganic fullerene-like nanotubes, *Appl. Phys. A, Mater. Sci. Process.*, 78, 989, 2004.

170. Rosi, N.L. et al., Hydrogen storage in microporous metal-organic frameworks, *Science*, 300, 1127, 2003.

171. Kesanli, B. et al., Highly interpenetrated metal-organic frameworks for hydrogen storage, *Angewandte Chemie-International Edition*, 44, 72, 2005.

172. Panella, B. and Hirscher, M., Hydrogen physisorption in metal-organic porous crystals, *Adv. Mater.*, 17, 538, 2005.

173. Zhou, L., Progress and problems in hydrogen storage methods, *Renew. Sust. Energy Rev.*, 9, 395, 2005.

174. Sandrock, G., A panoramic overview of hydrogen storage alloys from a gas reaction point of view, *J. Alloys Comp.*, 295, 877, 1999.

175. Grochala, W. and Edwards, P.P., Thermal decomposition of the non-interstitial hydrides for the storage and production of hydrogen, *Chem. Rev.*, 104, 1283, 2004.

176. Zuttel, A. et al., Hydrogen storage properties of $LiBH_4$, *J. Alloys Comp.*, 356, 515, 2003.

177. Zuttel, A. et al., $LiBH_4$: a new hydrogen storage material, *J. Power Sources*, 118, 1, 2003.

178. Bogdanovic, B. and Schwickardi, M., Ti-doped alkali metal aluminium hydrides as potential novel reversible hydrogen storage materials, *J. Alloys Comp.*, 253, 1, 1997.

179. Zaluska, A., Zaluski, L., and Strom-Olsen, J.O., Sodium alanates for reversible hydrogen storage, *J. Alloys Comp.*, 298, 125, 2000.

180. Gross, K.J., Thomas, G.J., and Jensen, C.M., Catalyzed alanates for hydrogen storage, *J. Alloys Comp.*, 330, 683, 2002.

181. Sandrock, G., Gross, K., and Thomas, G., Effect of Ti-catalyst content on the reversible hydrogen storage properties of the sodium alanates, *J. Alloys Comp.*, 339, 299, 2002.

182. Anton, D.L., Hydrogen desorption kinetics in transition metal modified $NaAlH_4$, *J. Alloys Comp.*, 356, 400, 2003.

183. Bogdanovic, B. and Sandrock, G., Catalyzed complex metal hydrides, *MRS Bull.*, 27, 712, 2002.

184. Johnson, S.R. et al., Chemical activation of MgH_2: a new route to superior hydrogen storage materials, *Chem. Commun.*, 2823, 2005.

185. Ichikawa, T. et al., Lithium nitride for reversible hydrogen storage, *J. Alloys Comp.*, 365, 271, 2004.

186. Kojima, Y. et al., Compressed hydrogen generation using chemical hydride, *J. Power Sources*, 135, 36, 2004.

187. Kojima, Y. et al., Hydrogen generation by hydrolysis reaction of lithium borohydride, *Int. J. Hydrogen Energy*, 29, 1213, 2004.

188. Kojima, Y. et al., Development of 10 kW-scale hydrogen generator using chemical hydride, *J. Power Sources*, 125, 22, 2004.

189. Kojima, Y. et al., Hydrogen generation using sodium borohydride solution and metal catalyst coated on metal oxide, *Int. J. Hydrogen Energy*, 27, 1029, 2002.

190. Schlesinger, H.I. et al., Sodium borohydride, its hydrolysis and its use as a reducing agent and in the generation of hydrogen, *J. Am. Chem. Soc.*, 75, 215, 1953.

191. Wu, Y., Kelly, M.T., and Ortega, J.V., *Low-Cost, off-board regeneration of sodium borohydride*, DOE Hydrogen Program, FY 2004 Progress Report, U.S. Department of Energy, Washington, DC, 2005.

192. Wilding, B. et al., *Hydrogen storage: radiolysis for borate regeneration*, DOE Hydrogen Program, FY 2004 Progress Report, U.S. Department of Energy, Washington, DC, 2005.

193. Cooper, A. et al., *Design and development of new carbon-based sorbent systems for an effective containment of hydrogen*, DOE Hydrogen Program, FY 2004 Progress Report, U.S. Department of Energy, Washington, DC, 2004.

194. Jensen, C.M. et al., Catalytically enhanced systems for hydrogen storage, in *Proceedings of the 2002 U.S. DOE Hydrogen Program Review*, U.S. Department of Energy, Washington, DC, 2002.

195. Cho, A., Fire and ICE: revving up for H_2, *Science*, 305, 964, 2004.

196. Grove, W.R., On voltaic series and the combination of gases by platinum, *Philos. Mag.*, 14, 127, 1839.

197. Kordesch, K. and Simader, G., Fuel cell powered electric vehicles, chap. 7 in *Fuel Cells and Their Applications*, VCH, Weinheim, 1996.

198. Mauritz, K.A. and Moore, R.B., State of understanding of Nafion®, *Chem. Rev.*, 104, 4535, 2004.

199. Hickner, M.A. et al., Alternative polymer systems for proton exchange membranes (PEMs), *Chem. Rev.*, 104, 4587, 2004.

200. Demirdoven, N. and Deutch, J., Hybrid cars now, fuel cell cars later, *Science*, 305, 974, 2004.

12 Efficiently Distributed Power Supply with Molten Carbonate Fuel Cells

Peter Heidebrecht and Kai Sundmacher

CONTENTS

12.1 INTRODUCTION

Fuel cells are devices for electric power production. Similar to their classical counterparts, which are based on a cycle process, they convert chemical energy (usually the enthalpy of combustion of a combustible substance) into electric energy. The substantial difference between these two classes of processes is that cycle processes usually use three energy transformation steps — from chemical enthalpy to thermal, then kinetic, and finally electric energy — whereas fuel cells directly convert chemical into electrical energy and thereby offer a chance to obtain higher degrees of efficiency [1–4].

The environmental impact of fuel cells strongly depends on what the fuel the cell is fed with. Fuel production based on fossil resources is neither free of greenhouse gases, because carbon dioxide is emitted, nor is it sustainable. Fuel based on renewable resources has overall zero net carbon dioxide emissions, but the combination of it with fuel cells is not compulsory. Thus, fuel cells are not a sustainable technique by themselves, but they promise a more efficient use of available fuels, be they based on fossil or renewable resources.

Among the known fuel cell types, the two high-temperature fuel cells, namely, the molten carbonate fuel cell (MCFC) and the solid oxide fuel cell (SOFC), do not require hydrogen as their primary fuel gas, but they can be fed with any fuel gases containing short-chained hydrocarbons, carbon monoxide, and hydrogen. While a dominant part of low-temperature fuel cell systems is occupied by the reforming process, which transforms fuel gas into hydrogen, this process can simply be integrated into high-temperature fuel cells. This so-called internal reforming concept not only offers a simpler system design compared with that of low-temperature fuel cells with their external reforming units, but it also significantly increases the overall electric system efficiency. In addition, their insensitivity with respect to carbon monoxide allows a wide spectrum of fuels to be used in high-temperature fuel cells.

However, they not only provide electric power. Due to their operating temperature, these concepts combine high electric efficiency with a wide spectrum of heat utilization, for example, steam production, cold chillers, or a downstream cycle process. Compared with classical concepts, they offer a higher electricity/heat ratio, which is preferable in many stationary applications. Due to their combined heat and power production together with their fuel flexibility, high-temperature fuel cells are attractive for several areas of application. They can be used in small power plants based on natural gas, where they accommodate residential areas or a single larger building, for example, a hospital or an office building.

Beyond the replacement of classical units in today's stationary applications, where they offer superior efficiency, high-temperature fuel cells are also strong candidates for sustainable energy systems. The most prominent example is the use of biogases from fermentation processes (e.g., from wastewater treatment or from cattle-breeding farms). Further applications are the consumption of lean waste gases from processes in food or chemical industries, landfill gases, or mining gases. All these applications use fuels for on-site production of electricity in combination with heat or steam, depending on the specific demands of each application. High-temperature fuel cells are more suitable for these applications than a cycle process with low efficiency in combination with a boiler. They offer a chance to alter today's centralized energy supply system toward a distributed system where numerous opportunities for renewable energy supply can be exploited.

Although these properties apply to both known high-temperature fuel cell types, namely, the MCFC and the SOFC, we will focus on the first mentioned in the following sections. This is mainly because the MCFC is technically better developed, with most vital questions about material stability, system reliability, and production procedures already solved and the first commercial series actually available.

In the following, after a technical introduction into MCFC, its general advantages and drawbacks are discussed. Afterwards, several existing technical realizations are described and compared with each other. Several actual examples of application of MCFCs are discussed, and some future development trends are indicated.

12.2 TECHNICAL BACKGROUND

12.2.1 WORKING PRINCIPLE

Like any other fuel cell, the MCFC consists of several layers (Figure 12.1). The electrolyte layer of the MCFC consists of a eutectic carbonate melt (38% K_2CO_3, 62% Li_2CO_3), which is immobilized in a porous aluminium oxide structure (γ-$LiAlO_2/\alpha$-Al_2O_3). It serves as a semipermeable layer that only allows carbonate ions (CO_3^{2-}) to pass through the layer. Other substances, especially dissolved nonionic gases, cannot pass through.

On each side of the electrolyte layer, a porous catalyst layer, an electrode, is placed. They consist of an electron-conducting solid material, which also serves as a catalytic promoter for the reactions occurring at the respective electrode. Alternatively, the functionality of electron conduction and reaction promotion can be separated by using a carrier material, which mainly serves as the conductor, and placing the catalyst in a thin layer upon the surface of the carrier material. In the case of MCFC, nickel and nickel oxide are preferred materials for the anode and the cathode electrodes, respectively. A part of the molten carbonate is also located in the electrodes' pores, held in place by capillary forces. The remainder of the pores is filled with gas through which the educts and products of the reactions inside the electrodes are transported. The electrochemical reactions (see below) basically happen at the three-phase boundary between gas, liquid, and catalyst, so a large interfacial area is required in the pores' structure.

At each electrode, on the opposite side of the electrolyte layer, a gas channel is located. The anode channel is fed with a mixture of steam and the fuel gas, for example, methane. Prior to the reaction at the electrode, this gas has to be converted to hydrogen in the reforming process. The major chemical reactions in this process are the steam-reforming and the water–gas-shift reaction:

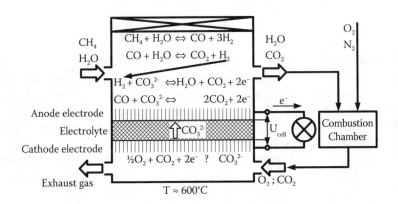

FIGURE 12.1 Working principle of an MCFC.

$$CH_4 + H_2O \leftrightarrow CO + 3\,H_2$$

$$CO + H_2O \leftrightarrow CO_2 + H_2$$

This process requires two things: heat, because it is endothermic, and high temperatures, because its conversion is severely limited by its chemical equilibrium at low temperatures. Both are available in the MCFC. The catalyst commonly used to promote these reactions is based on nickel.

The reforming products, mainly hydrogen and carbon monoxide, diffuse into the anode electrode and dissolve in the electrolyte inside the electrode pores. There they react at the electrode catalyst surface, thereby consuming carbonate ions from the electrolyte and producing free electrons, which are located on the electron-conducting solid phase of the electrode after the reaction. In addition, carbon dioxide and water are produced. Generally, the hydrogen oxidation is intrinsically faster than the carbon monoxide oxidation. But, especially with carbon monoxide-rich fuel gases, this reaction becomes important:

$$H_2 + CO_3^{2-} \leftrightarrow H_2O + CO_2 + 2e^-$$

$$CO + CO_3^{2-} \leftrightarrow 2\,CO_2 + 2e^-$$

Because full conversion of hydrogen and carbon monoxide is not possible for thermodynamic and energetic reasons, the anode exhaust gas contains significant amounts of hydrogen and carbon monoxide, as well as a small portion of unreformed fuel gas. This gas has to be oxidized completely, so it is mixed with air and fed into a combustion unit. Along with the heat-releasing electrochemical reactions, this combustion is the main heat source within the MCFC system, and it is used to heat up the fresh air to the process temperature.

The completely oxidized gas is then fed into the cathode channel. Here the carbon dioxide is consumed together with some oxygen to form new carbonate ions, closing the carbonate ion loop. In the same reaction, two electrons are taken out of the electron-conducting solid phase of the cathode:

$$CO_3^{2-} \leftrightarrow 1/2\,O_2 + CO_2 + 2e^-$$

The cathode exhaust gas leaves the system.

Between the extra electrons at the anode and the "missing" electrons, i.e., the positive electron holes at the cathode electrode, an electric voltage occurs. Connecting the anode and cathode via an electric load, for example, an electric engine or a light bulb, allows the electrons to move from the anode to the cathode and do electrical work on that device.

Obviously, the MCFC exhausts carbon dioxide. This is contrary to the common wisdom that fuel cells do not emit greenhouse gases. In fact, this fuel cell even requires carbon dioxide in its cathode reaction; otherwise, the carbonate ions that are consumed at the anode reaction could not be replaced and the cell would quickly run out of electrolyte. It also is not sufficient to exclusively recycle the carbon dioxide that is produced at the anode electrode. This would require that every molecule of carbon dioxide fed into the cathode channel be converted to carbonate ions, which is not possible for thermodynamic reasons. Thus, a continuous feed of carbon to the system is necessary, causing a continuous exhaust of carbon dioxide. A global "zero emission" operation can only be achieved by using biofuels.

12.2.2 FEATURES OF THE MCFC

One of the major aspects in the MCFC is temperature. With a typical operating temperature of about 550 to 650°C, relatively inexpensive metal materials can be used, in contrast to the SOFC, which operates at 800 to 1000°C and consequently needs expensive ceramic materials. On the other hand, the MCFC temperature is high enough to obtain sufficiently high reaction rates with inexpensive and less active catalysts like nickel. For the reforming process, a temperature of 700 to 800°C would be preferable. At this temperature not only is the reaction rate high, but also the chemical equilibrium, which limits the conversion in this process, is significantly more favorable than at lower temperatures. In the internal reforming concept shown in Figure 12.1, the continuous removal of the reforming products, i.e., hydrogen and carbon monoxide by the oxidation process, helps to obtain high degrees of reforming conversion, although the MCFC temperature is relatively low.

Because of the absence of highly precious metals like platinum, high-temperature fuel cells are tolerant with respect to carbon monoxide. This is what makes the MCFC suitable for a wide range of different fuels. In principle, the MCFC would even operate on carbon monoxide only. Sulfur is a catalyst poison in the MCFC, so it must be removed from the feed gas.

A further advantage of the MCFC is its potential in the combined production of heat and electric power. Even if the exhaust gas is used to preheat the feed gas of the system, it still has a temperature of about 400°C, which is sufficient to generate pressurized steam in an industrial application or hot water for a residential building. Such coproductions are also possible with classical apparatuses, but oftentimes the electric power demand is equal or even higher than the demand for heat, a ratio that cannot be satisfied by engines or turbines alone. Due to their high efficiency, high-temperature fuel cells can meet these requirements. A hybrid system consisting of an MCFC and a downstream turbine can further increase the portion of electric energy produced by the system. Because of the low efficiency and high costs of very small turbines, this is economically useful for systems above 1 MW.

Like most other fuel cells, the MCFC principle promises low maintenance costs. Except for the blowers, which move the gases through the channels with comparably low pressure drop, there are no moving parts in the system. The major part of maintenance effort is the replacement of the cell stack, which has to take place after a certain degradation of the electrode catalysts. Today's stack lifetime expectancy is about 2 to 4 years, during which the system can continuously deliver heat and power.

12.3 TECHNICAL REALIZATIONS

12.3.1 CELL STACK

A single fuel cell, as depicted in Figure 12.1, is capable of delivering a voltage of about 0.7 to 0.9 V under operating conditions. According to today's standards, a typical current density of an MCFC is at about 150 mA/cm^2, so a cell of 1 m^2 size delivers approximately 1 kW electric power. To obtain systems with higher power, several cells are combined in a cell stack (Figure 12.2). From the chemical engineering point of view, these are parallel reactors; from the electric point of view, this is a series of current sources.

Channel structures between the cells distribute the gases across the electrode area and simultaneously collect the produced charges at the electrodes and transfer them to the neighboring cell. Because they connect the anode of a cell to the cathode of the next one, they are referred to as bipolar plates. Cell stacks can contain up to several hundred cells.

One practical issue with the stacking of fuel cells is the sealing. Due to temperature gradients in the system, and because of different expansion coefficients of the materials, the layered structure of the stack has to be pressed with a pressure of several bar.

The plates at either end of the stack are oftentimes massive structures. This is not only because they have to take up the pressure the stack is under, but also because here the electric current is collected in one point. While in bipolar plates the electric current flows mainly directly through the plate in a direction perpendicular to the cell plane, the current in the end plate has to flow along the plane. To minimize resistance, a high cross-sectional area is required; thus the end plates are comparably thick.

12.3.2 PERIPHERAL DEVICES

In addition to the fuel cell itself, a fuel cell system requires several additional peripheral process units. These are mainly: a desulfurization unit for the feed gas, an evaporator for the steam necessary for the reforming process, an upstream prereformer, a combustion unit between the anode exhaust and the cathode inlet, and eventually one or more heat exchangers to recover the thermal energy of the exhaust stream.

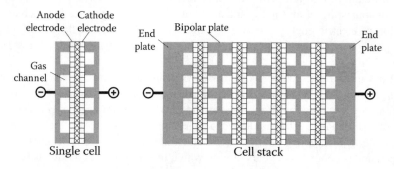

FIGURE 12.2 Fuel cell stack, bipolar plates, and end plates.

Especially for fuel gases from fermentation processes, sulfur plays an important role. For a MCFC — as for any other fuel cell — sulfur is a strong catalyst poison and has to be removed from the feed gas. Depending on the fuel gas in the respective application, this unit has to fulfill different requirements. Processed natural gas contains a well-defined, constant portion of sulfuric components. In this case, active coal filters are applied, but this choice is not optimal due to the high replacement frequency and its costs. As an alternative, an adsorption bed with zinc oxide under cyclic operation is being discussed. It transforms the sulfur-containing components to hydrosulfide and then adsorbs it during the first cycle period. Once the catalyst bed is loaded to a certain extent, the feed gas is switched to a second identical adsorption bed in parallel, and air is fed into the first one. With this, the sulfur is oxidized and removed from the bed as sulfur dioxide, leaving a cleaned adsorption bed behind.

Gases containing larger portions of sulfur require additional measures. Biogas, for example, often contains significant amounts of sulfur with frequently changing concentrations and varying molecular composition. Here, biological anaerobic processes have been proposed, which produce elementary sulfur. The advantage of this concept is that the sulfur is not emitted as sulfur oxide, but it is available in a pure and less toxic form. On the other hand, these processes cannot clean the gas sufficiently for application in MCFC. Consequently, a combination of biological and adsorptive desulfurization methods seems advisable here [5–7].

In addition to hydrocarbons or carbon monoxide, steam is an important component in the feed gas. The amount of water in the feed gas is usually described by the so-called steam-to-carbon ratio, S/C. In most applications, this ratio is between 2 to 3. Mixtures with significantly less steam tend to reversibly deposit carbon on the surfaces of the pipes and electrodes. This leads to the blocking of the anode electrode and the reforming catalyst in the fuel cell and quickly decreases the performance of the system [8–12]. On the other hand, too high a steam dosage only dilutes the feed gas and leads to decreased system efficiency, especially because of the energy costs of the water evaporation. Prior to usage in the MCFC, the water has to be cleaned. Reverse osmosis is a preferred principle here to remove most of the undesired ions from the water, which could possibly act aggressively toward the piping, the catalysts, and the electrolyte at the high temperature inside the MCFC. Furthermore, an evaporator is obviously required in state-of-the-art MCFC systems.

While in principle the reforming process can take place exclusively inside the anode channel — as depicted in Figure 12.1 — a part of it usually is relocated to an external, separate reactor outside the cell. This concept, which is commonly used with low-temperature fuel cells, is known as external reforming (ER) or prereforming (PR). Another concept is the indirect internal reforming (IIR), in which the reforming takes place inside a stack or is thermally directly attached to it, but not within the cells themselves. Consequently, the concept of performing the reforming process inside the anode channel is called direct internal reforming (DIR). All three concepts are shown schematically in Figure 12.3 [13–21].

The advantage of the ER is that the design of the reactor and its operating parameters are largely independent of the fuel cell itself, giving several degrees of freedom with respect to geometry, heat management, fuel flexibility, and control. On

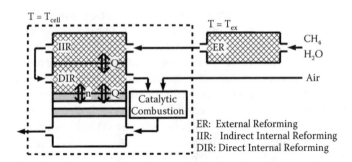

FIGURE 12.3 Reforming concepts: external reforming (ER), indirect internal reforming (IIR), and direct internal reforming (DIR).

the other hand, as already mentioned, the reforming process requires heat and high temperature, which in the case of ER have to be provided from the outside of the reactor. This can either be done by using the heat of the exhaust gas, accepting a limit process temperature equal to that of the cells exhaust gas, or by direct oxidation of a portion of the fuel with air inside or outside of the reforming reactor. The choice of direct oxidation not only contradicts the fuel cell principle of turning reaction enthalpy into electric energy, but it also tends to lower the overall electric efficiency of the system. Furthermore, the ER is subject to the chemical equilibrium of the reforming process, which at an operating temperature of 400°C is at about 50% conversion, and for 800°C approaches full conversion, depending on the feed gas composition and pressure. The application of the ER is still advisable in MCFC systems for two reasons. First, it increases the fuel flexibility. If the feed gas contains short-chained hydrocarbons like ethane or propane, these have to be converted to methane before being reformed to hydrogen. This cracking of carbon-carbon bonds is significantly slower than the steam reforming process and thus requires larger amounts of catalysts. These can be provided by a larger external reforming reactor. Thus the external reformer makes it possible to adapt the MCFC system to the specific fuel of each individual application without the need to manipulate the design parameters of the fuel cell itself. Secondly, the external reformer serves as a kind of warning system in applications with strongly alternating sulfur content. In the event of a sulfur breakthrough, the external reformer is the first unit to be poisoned by sulfur. This can easily be detected by a set of thermocouples that show the position of the reaction zone moving through the catalyst bed. In this case, a shutdown or the fix of the desulfurization unit can save the fuel cell from harm. Even in the worst case, that is, if the external reforming unit is irreversibly damaged, it is cheaper to replace than an irreversibly damaged fuel cell stack. Thus, application of the ER concept is advisable, especially in systems with an unreliable desulfurization unit or with strongly alternating sulfur content.

The IIR concept is based on the idea to provide the required heat for the reforming process by direct coupling to the heat-releasing electrochemical reactions in the fuel cells. This can be realized either by a separate reactor inside the hot housing of the cell stack or by inserting a reforming reactor between several fuel cells. This promotes the heat exchange toward the reforming process, so it takes place at about cell temperature.

The highest degree of process integration is the DIR concept, in which the reforming process is located directly inside the anode channel. This not only leads to an intense energetic exchange between heat-releasing electrochemical reactions and the endothermic reforming process, but it also includes the mass coupling of the production of hydrogen and carbon monoxide and their consumption at the anode electrode. This direct consumption of the reforming products accelerates the reforming process and shifts its chemical equilibrium toward an extent of conversion to almost 100%, even at lower temperature of about 600°C.

As already mentioned in the technical introduction, the anode exhaust and the cathode inlet are coupled via a combustion device. In the MCFC, a catalytic unit is preferred because it allows oxidization of diluted gas/air mixture over the complete cell stack without a pilot light. In addition, it helps to avoid extreme temperatures and thus suppresses the formation of nitrogen oxides, NO_x. In fact, nitrogen oxide concentration in the exhaust gas is below the detection limit.

12.3.3 ACTUAL SYSTEM DESIGN

MCFCs are developed by different companies around the world. Although their products are working on the same basic principle, they differ in system design, power class, and intended applications. In the following, the technical solutions of the most prominent developers are discussed.

12.3.3.1 MTU CFC Solutions

In Germany, the MTU CFC Solutions company has developed the so-called Hotmodule MCFC. MTU and Fuel Cell Energy are partners to a technology and supply exchange contract. Development started in 1990, and the functionality of the concept was first publicly demonstrated in 1997 at a prototype plant in Dorsten, Germany [22, 23]. This was followed by a series of about 25 field-test plants, which where installed in various applications in Germany, Europe, and other parts of the world. Exemplary applications are power supply in telecommunications, combined heat and power supply in hospitals and a university, and combined steam and power supply at a tire manufacturer. Currently, preparations for a series production of the Hotmodule are ongoing.

The Hotmodule is a complete MCFC system with a nominal power of 250 kW. The stack consists of 343 cells, in which the anode and cathode gas channels are arranged in a cross-flow design. The stack is in horizontal position so that the cells are standing upright, with the anode channels running from the bottom to the top of the stack and the cathode channels going from one side to the other. The stack is located in a cylindrical vessel in which all the hot compartments of the system are located — thus the name Hotmodule (Figure 12.4).

As shown in the flow scheme (Figure 12.5) the Hotmodule includes all three reforming concepts. The IIR is realized by inserting a flat reforming reactor after each package of eight fuel cells. Before the external reformer, the feed gas is heated up and mixed with steam in a combined heat exchanger and humidifier.

The Hotmodule has all the general advantages of the MCFC concept and combines them with a simple, efficient design. The cylindrical housing of the rectangular

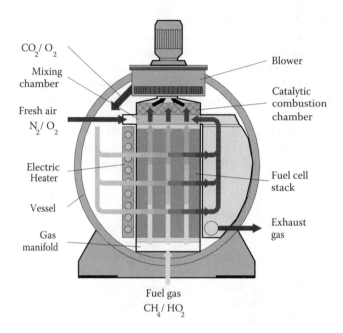

FIGURE 12.4 Cross-section of the Hotmodule.

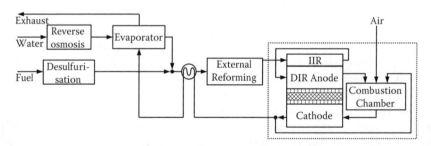

FIGURE 12.5 Flow scheme of the Hotmodule. The dotted line indicates the cylindrical vessel inside which the hot compartments are located.

cell stack automatically segments the surrounding space into four compartments. One of them contains the gas manifold, which distributes the preformed feed gas to the anode channels, and another one contains the combustion chamber. The other two volumes connect the combustion chamber with the cathode inlet, thereby splitting the cathode exhaust gas into the cathode recycle stream and the exhaust gas stream. The fact that the cell stack is completely surrounded by hot gases reduces the mechanical stress due to thermal gradients. As the stack is mounted on rails, it can be replaced by simply opening the vessel front cover, pulling the old stack out and moving the new one in. A complete replacement procedure takes about 3 to 4 days. This is mostly due to the long time required to heat up the new stack and cool down the old one.

12.3.3.2   Ansaldo Fuel Cells

In Trieste, Italy, Ansaldo Fuel Cells has successfully demonstrated the feasibility of a 100-kW MCFC system and is working on its "2TW" model, a 500-kW system including four stacks [24–26]. In contrast to the Hotmodule by MTU, the 2TW system is operated at an elevated pressure of several (3 to 5) bar. The compressor for the feed gas is coupled with a turbine in the exhaust gas stream, and the energy balance of these two is positive, meaning that the 2TW provides electrical power not only from the cell stack, but also from the combined turbine/compressor.

Another remarkable characteristic trait of the 2TW is the focus on a single reforming unit (Figure 12.6). This unit is combined with the combustion chamber to provide the energy required by the endothermic reforming process and is located inside the containment vessel of the hot system parts. The second advantage of this combination is that high operating temperatures and consequently high degrees of conversion can be obtained in the reformer. Although this reforming unit is located in the hot cell housing, it must be considered as an ER because its temperature is dominated by the combustion temperature instead of the cell temperature. The air is inserted at the cathode inlet, so it is heated up by the hot combustion exhaust gas, and a high oxygen concentration is obtained at the cathode electrode. In the Hotmodule, the cool air is fed into the combustion chamber, which reduces the temperatures inside the chamber, but in the 2TW system, a high temperature in the combustion and consequently in the reforming unit is favorable.

The MCFC systems by Ansaldo are also realized within a cylindrical vessel, which is advantageous for pressurized operating conditions.

12.3.3.3   Ishikawajima-Harima Heavy Industries

In Japan, MCFC development is conducted by a consortium of companies led by Ishikawajima-Harima Heavy Industries (IHI) [27, 28]. The MCFC system they are focusing on is a 300-kW stack with features similar to those of the 2TW by Ansaldo. It is operated under pressurized conditions and only contains a single ER unit combined with the combustion chamber. An important addition in this system is the water-recovery system, which by condensation of the steam in the exhaust gas provides the water required for the reforming process, so under regular operating

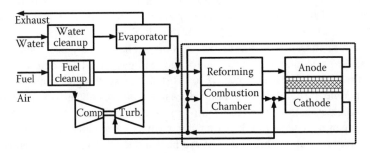

FIGURE 12.6 Flow scheme of the MCFC system by Ansaldo.

conditions, no external water supply and cleanup is required. A first plant has been demonstrated at the Kawagoe test site, and plans also include the development and demonstration of a plant of the size of several MW.

12.3.3.4   Fuel Cell Energy

Fuel Cell Energy (FCE), located in Danbury, CT, is one of the earliest suppliers of MCFC components and complete systems. They offer a cargo-container-sized 250-kW system as well as 1- and 2-MW plants based on MCFC. Just like the Hotmodule of the MTU, FCE's fuel cells feature both internal reforming concepts (IIR and DIR), using the external reformer solely to crack longer-chained hydrocarbons. In fact, early MTU systems were based on FCE cell design. A large number of demonstration plants inside the United States in different areas of application prove the technical feasibility of the FCE MCFC systems [29–31].

12.3.4   ACTUAL APPLICATIONS

MCFC systems have proven their potential in a large number of trial plants in quite different applications. They demonstrate the full functionality of the fuel cell and its superior electric system efficiency in comparison with classical units of the same size. The efficiency reaches up to 48%, depending on the system size and the individual application.

One typical application is the implementation of an MCFC system in a power plant delivering electricity and heat to a larger building complex, as is the case at the university hospital in Magdeburg, Germany. The specific Hotmodule system in Magdeburg is fed with natural gas, and it is combined with a tube-bundle heat exchanger utilizing the exhaust gas for the heat generators and cold chillers of the plant (Figure 12.7). While the system has an overall electric efficiency of 47%, its combined electric and thermal efficiency is at about 70%.

A similar MCFC system operated with natural gas is installed at a technical center of a telecommunications provider in Munich, Germany. The highly complex electronic devices in telecommunications require direct current, which is produced by the fuel cell, so AC/DC converters can be omitted and their energetic losses can be avoided. Secondly, the MCFC serves as an on-site uninterruptible power supply, which helps to avoid costly power failures in critical electronic systems. Finally, the heat produced by the Hotmodule is used to operate absorption chillers for the air-conditioning of the offices and technical rooms of the facility. In this application, the full range of the potential products of the MCFC is unfolded [32].

The application of the Hotmodule with biogas has already been demonstrated on a smaller scale, and specifications for the feed gas have been determined [33]. The sulfur content must be lowered to 20 ppm before the gas is convenient for the Hotmodule. It seems as if ammonia does not spoil the cell performance; indeed, it is electrochemically converted to nitrogen, water, and electric energy. No nitrogen oxide can be detected in the exhaust gas whatsoever. Only siloxane is a major issue, as it is a catalyst poison and has to be removed from the feed gas. In 2005, the installation of a Hotmodule at a wastewater treatment plant near Ahlen,

FIGURE 12.7 The Hotmodule MCFC system made by MTU in the IPF power plant at the University Hospital in Magdeburg, Germany.

Germany, is planned. It will mainly be fed with the sewage gases from the fermentation process and will supply the plant with electricity and heat for the offices and operations buildings.

One of the main fields of future applications of MCFC is seen in the use of secondary gases, i.e., waste gases from biological or industrial processes. Currently, only a small part of these gases are used for electric power supply; another part is burned to provide heat; and the rest is simply burned without any beneficial use. Here, the MCFC could exploit additional attractive applications due to its potential in combined heat and power supply, even with lean gases.

Other intended applications include the use of gasified coal as a fuel for MCFC, which is currently favored by IHI in Japan. While this is not a renewable technique in itself, it offers a highly efficient alternative to classical coal power plants and helps to preserve limited resources.

12.4 FUTURE CHALLENGES AND DEVELOPMENTS

The current developments in the field of MCFC aim at adapting it to new applications and lowering its production costs. Investment costs per installed nominal power are expected to remain slightly higher than those of classical units, but the higher system efficiency, in the form of lower fuel consumption, should overcompensate that. Especially with fuel prices expected to rise in the future, fuel efficiency and fuel costs will become increasingly important.

To further increase the area of possible applications, the MCFC must be able to operate under quickly changing conditions, as is the case in a stand-alone operating mode or in standby power systems. At the moment, this is not possible with an MCFC. This is mainly because its dynamic behavior is not yet fully understood. Even at steady state, the crucial temperature profile is not well known. As temperature measurements inside the stack are problematic from a technical point of view, other approaches are required to observe the system and provide appropriate control systems. These issues raise several different questions in the field of modeling, thermal management, reactor design, and process control.

While the adaptation of the Hotmodule MCFC system to a sewage gas application is currently under way, several other potential applications of economic and ecological interest will have to be investigated in the future. Although in principle it is suitable for a wide range of fuels, its sophisticated design requires some adjustments to each new situation. Many future applications are traditional ones, where the MCFC will replace cycle processes due to its higher efficiency, but in addition, the MCFC offers new opportunities to use renewable energy resources in an economically attractive way that is not available with classical units of power production.

REFERENCES

1. Kordesch, K. and Simader, G., *Fuel Cells and Their Applications*, 1st ed., Wiley VCH, Weinheim, Germany, 1996.
2. Larminie, J. and Dicks, A., *Fuel Cell Systems Explained*, 2nd ed., John Wiley & Sons, Chichester, U.K., 2003.
3. Vielstich, W., Lamm, A., and Gasteiger, H., Eds., *Handbook of Fuel Cells: Fundamentals, Technology, Applications*, John Wiley & Sons, Chichester, U.K., 2003.
4. Carrette, L., Friedrich, K.A., and Stimming, U., Fuel cells: fundamentals and applications, *Fuel Cells*, 1, 5, 2001.
5. Shennan, J.L., Microbial attack on sulphur-containing hydrocarbons: implications for the biodesulphurisation of oils and coals, *J. Chem. Tech Biotech.*, 67, 109, 1996.
6. Maxwell, S. and Yu, J., Selective desulphurisation of dibenzothiophene by a soil bacterium: microbial DBT desulphurisation, *Proc. Biochem.*, 35, 551, 2000.
7. Bagreev, A. et al., Desulfurization of digester gas: prediction of activated carbon bed performance at low concentration of hydrogen sulfide, *Catalysis Today*, 99, 329, 2005.
8. Trimm, D.L., Catalysts for the control of coking during steam reforming, *Catalysis Today*, 49, 3, 1999.
9. Tomishige, K., Chen, Y.G., and Fujimoto, K., Studies on carbon deposition in CO_2 reforming of CH_4 over nickel-magnesia solid solution catalysts, *J. Catalysis*, 181, 91, 1999.
10. Swaan, H.M. et al., Deactivation of supported nickel-catalysts during the reforming of methane by carbon-dioxide, *Catalysis Today*, 21, 571, 1994.
11. Sperle, T. et al., Pre-reforming of natural gas on a Ni catalyst: criteria for carbon free operation, *Appl. Cat. A*, 282, 195, 2005.
12. Shamsi, A., Carbon formation on Ni-MgO catalyst during reaction of methane in the presence of CO_2 and CO, *Appl. Cat A*, 277, 23, 2004.

13. Dicks, A.L., Hydrogen generation from natural gas for the fuel cell systems of tomorrow, *J. Power Sources*, 61, 113, 1996.
14. Rostrup-Nielsen, J.R. and Christiansen, L.J., Internal steam reforming in fuel-cells and alkali poisoning, *Appl. Cat. A*, 126, 381, 1995.
15. Clarke, S.H. et al., Catalytic aspects of the steam reforming of hydrocarbons in internal reforming fuel cells, *Catalysis Today*, 38, 411, 1997.
16. Freni, S., Rh based catalysts for indirect internal reforming ethanol applications in molten carbonate fuel cells, *J. Power Sources*, 94, 14, 2001.
17. Berger, R.J. et al., Nickel catalysts for internal reforming in molten carbonate fuel cells, *Appl. Cat. A*, 143, 343, 1996.
18. Bradford, M.C.J. and Vannice, M.A., Catalytic reforming of methane with carbon dioxide over nickel catalysts, 1: catalyst characterization and activity, *Appl. Cat. A*, 142, 73, 1996.
19. Effendi, A. et al., Steam reforming of a clean model biogas over Ni/Al_2O_3 in fluidised- and fixed-bed reactors, *Catalysis Today*, 77, 181, 2002.
20. Marquevich, M., Coll, R., and Montane, D., Steam reforming of sunflower oil for hydrogen production, *Ind. Eng. Chem. Res.*, 39, 2140, 2000.
21. Comas, J. et al., Bio-ethanol steam reforming on Ni/Al_2O_3 catalyst, *Chem. Eng. J.*, 98, 61, 2004.
22. Bischoff, M. and Huppman, G., Operating experience with a 250 kWel molten carbonate fuel cell (MCFC) power plant, *J. Power Sources*, 105, 216, 2002.
23. Available on-line at www.mtu-cfc-solutions.de.
24. De Simon, G. et al., Simulation of process for electrical energy production based on molten carbonate fuel cells, *J. Power Sources*, 115, 210, 2003.
25. Bosio, B. et al., Industrial experience on the development of the molten carbonate fuel cell technology, *J. Power Sources*, 74, 175, 1998.
26. Available on-line at www.ansaldofuelcells.com.
27. Morita, H. et al., Performance analysis of molten carbonate fuel cell using a Li/Na electrolyte, *J. Power Sources*, 112, 509, 2002.
28. Available on-line at www.ihi.co.jp/index-e.html.
29. Doyon, J., Farooque, M., and Maru, H., The Direct FuelCell™ stack engineering, *J. Power Sources*, 118, 8, 2003.
30. Lukas, M.D., Lee, K.Y., and Ghezel-Ayagh, H., Modeling and cycling control of carbonate fuel cell power plants, *Control Eng. Practice*, 10, 197, 2002.
31. Available on-line at www.fce.com.
32. Available on-line at www.deteimmobilien.de/Projekt_Brennstoffzelle.102.0.html.
33. Schmack, D., Einsatz von landwirtschaftlich erzeugtem Biogas zur Gewinnung elektrischer Energie mittels Brennstoffzellen, paper presented at PRO3-Symp. Schmelzkarbonat-brennstoffzellen, Magdeburg, Germany, 27 April 2004, p. 9.

Part IV

Trends, Needs,
and Opportunities
in Developing Countries

13 Renewable Resources and Energy in the Asia Pacific Region

Olivia L. Castillo

CONTENTS

The Asia Pacific region holds about one-third of the world's population. Poverty alleviation or eradication should be the centerpiece and platform of governments in the region. The majority of the donor organizations, like the World Bank, the Asian Development Bank, InWent of Germany, UNIDO (United Nations Industrial Development Organization), UNEP (United Nations Environment Program), UNDP (United Nations Development Program), and many others have this issue as their main objective. Due to rapid population growth rates, there has to be simultaneous, immediate economic development to arrest the widespread growth of poverty in the region.

Water and energy need to be priority issues in the economic development. According to a World Bank report (Asia Alternative Energy Programme, Status Report No. 9, Washington, D.C., 2001), there are about 2 billion people around the world who lack access to modern energy services, and of these, about 1.2 billion are found in Asia. The per capita energy consumption across the region is rising. Most countries have a large gap between the demand for and the supply of energy. Rural electrification remains a problem, and a sizeable segment of the population depends on traditional biomass-based energy sources.

This chapter focuses on the energy situation, particularly the renewable resources and energy that are being utilized and implemented in the less-developed countries such as the Philippines, Vietnam, Sri Lanka, Bangladesh, Laos PDR, Cambodia, Indonesia, and Nepal. The status of the renewable resources and energy in other developed countries like Malaysia and Thailand is also reviewed.

The role of renewable resources and energy is not yet substantial enough to replace fossil fuel. Countries such as India, China, Japan, and Australia are more advanced than the others.

This chapter reviews the main constraints and problems encountered, the renewable energy programs being developed, the policies governments are adopting, as well as the challenges and potential opportunities that abound and that could be implemented. The initiatives that are being undertaken now and for the future are also discussed.

13.1 INTRODUCTION

The Asia Pacific region needs a coherent and effective framework for sustainable development — which inevitably has to mandate the rapid deployment of renewable energy and energy efficiency policies and practices.

**2004 Renewable Energy Foundation Limited/World Council
for Renewable Energy Asia Pacific**

The framework should be supported and implemented by governments and inter-governmental organizations, with a focus on the appropriate energy policy formulation and adoption of specific sustainable energy strategies. Present development programs are fossil-fuel- or nuclear-based projects. Renewable energy and energy efficiency seem to be difficult tasks to do.

According to a World Bank report (Asia Alternative Energy Programme, Status Report No. 9, Washington, D.C. 2001), the Asia Pacific region holds about one-third of the world's population. Poverty alleviation or eradication should be the centerpiece and platform of governments in the region. Many donor organizations like the World Bank, the Asian Development Bank, InWent of Germany, UNIDO (United Nations Industrial Development Organization), UNEP (United

Nations Environment Program), UNDP (United Nations Development Program), and others have this issue as their main agenda or objective. The cited World Bank report notes the rapid population growth rates and emphasizes the need for simultaneous, immediate economic development to arrest the widespread growth of poverty in the region.

Now is the time to have renewable energy applications in the Asia Pacific region, and the sooner, the better. This is in line with the United Nations' Millennium Development Goals. With climate change and greenhouse gases emerging, the only reasonable approach will be the concept of renewable energy.

13.2 FOCUS

This chapter presents the energy situation in the Asia Pacific region, particularly the renewable resources and energy that are being utilized and implemented in the less-developed countries today.

Some Asian countries such as Malaysia and Thailand have a more advanced status regarding the use of renewable resources and energy. The chapter will merely enumerate some of the activities that have been carried out with regard to renewable energy and energy efficiency.

Asia has a great diversity that makes it difficult to undertake a unified project. It becomes more challenging because this diversity comes in terms of physical and geographical location, language, culture, religion, and indigenous communities.

Another dimension that compounds this problem is the economic status of these countries — from the poorest to the wealthiest — and their adaptation to environmental, economic, and social pressures. The poorer countries have to deal primarily with their basic needs of food, clothing, and shelter. However, to address these needs, it is necessary first to address the issue of energy poverty — the lack of ready access to locally generated, affordable, and easily replenished power and fuel . The wealthier nations, however, are threatened with dangerous and massive greenhouse-gas emissions due to their dependence on fossil fuels. As such, there is the threat of global climate change.

13.3 MAIN CONSTRAINTS/PROBLEMS

Studies have demonstrated that the following are the main constraints in adopting renewable energy programs:

Government policies
Lack of funds
Raising awareness
Capacity building
Capital intensive
Integrated programs
Transfer of technologies

13.4 RENEWABLE ENERGY PROGRAMS

The following are the renewable energy programs being developed:

Solar thermal
Wind energy
Photovoltaic
Biomass
Geothermal
Small hydro power
Energy from waste
Ocean energy

13.5 ASIAN DEVELOPMENT BANK'S PREGA PROJECT

PREGA is an acronym for the Netherlands Cooperation Fund on Promotion of Renewable Energy, Energy Efficiency and Greenhouse Gas Abatement.

13.5.1 OVERVIEW

The main objective of PREGA is to promote investments in renewable energy, energy efficiency, and greenhouse-gas abatement technologies in developing member countries. Such investments will:

Increase access to energy services by the poor
Realize other strategic development objectives
Help reduce greenhouse-gas emissions

To achieve this objective, PREGA will:

Develop capacities of national policy makers, technical experts, and staff of financing institutions to promote renewable energy, energy efficiency, and greenhouse-gas abatement
Support policy, regulatory, and institutional reforms, including removal of energy-pricing distortions
Facilitate access to private-sector financing

13.5.2 COUNTRIES

PREGA covers these 15 developing member countries:

Bangladesh
Cambodia
India
Indonesia
Kazakhstan
Kyrgyz Republic

Mongolia
Nepal
Pakistan
People's Republic of China
Philippines
Samoa
Sri Lanka
Uzbekistan
Vietnam

13.6 CHALLENGES

The major energy challenges facing the Asia Pacific region are to:

Provide energy to large populations in the developing countries
Reduce greenhouse-gas emissions while simultaneously addressing energy
  barriers to impoverished populations

13.7 STATUS OF RENEWABLE RESOURCES IN DEVELOPED COUNTRIES

13.7.1 MALAYSIA

Limited potential for wind energy; just pilot projects; total installed capacity
  of 150 kW in 1999.
Small hydropower accounts for about 23 MW; possibilities for future expan-
  sion.
Biomass-based energy offers opportunities due to large palm-oil/wood-pro-
  cessing industries and crop residues.
The whole of peninsular Malaysia is electrified.
Potential for solar photovoltaic systems limited to only special applications
  such as telecommunications, lighthouses, and remote locations.
Solar thermal energy offers opportunities.

13.7.2 THAILAND

Solar energy: solar water heating and solar photovoltaic.
Wind energy: used for water pumping in remote areas and for electricity
  generation on some islands.
Small hydro: small hydro power plants with installed capacities of 200 to
  6000 kW have been established by the Department of Energy Development
  and Promotion (DEDP), now the Department of Energy Development and
  Efficiency (DEDE): since 1980. To date, 25 projects with a total installed
  capacity of 132 MW have been completed. DEDE and the Energy Policy
  Office are actively pursuing energy conservation activities such as consumer

education, industrial energy audits, thermal/electricity efficiency demonstration projects, and end-use studies.
Biomass: accounted for 24% of final energy consumption in 1995.

Thailand has a five-year plan (1997–2001) to promote new and renewable energy
technologies (Demonstration and Promotion of Alternative Energy Production and
Utilization [DPAEPU] Plan).

13.8 STATUS OF RENEWABLE RESOURCES IN LESS-DEVELOPED COUNTRIES

13.8.1 BANGLADESH

The Rural Electrification Board and Grameen and Shakti, renewable energy companies,
are working on commercializing solar photovoltaic systems for rural electrification.
    The Asian Development Bank (ADB) PREGA will undertake feasibility studies for:

Cogeneration in sugar industries
Solar-wind diesel

13.8.2 NEPAL

The government of Nepal is very keen on developing renewable energy. The main
source of energy in the country is hydropower, and the country uses only 15% of
the potential energy supply. There are plans for developing the biomass potential of
the country. The government's initiated project on energy efficiency targets the
industrial sector. The government needs to institutionalize capacity building and to
formulate a national energy policy. The country is inviting foreign investors to
support the country's need for energy efficiency projects. Government officials also
want to know if there has been successful implementation on energy efficiency, what
kind of technology is appropriate for the country, and the investment cost. The
PREGA project will be helpful for capacity building in Nepal, especially in the
industrial sector.

13.8.3 SRI LANKA

Sri Lanka is implementing its "Village Hydro Project" to use off-grid microhydro
systems for rural electrification. (Note that only 38% of rural households have access
to electricity.)
    The ADB PREGA will also undertake feasibility studies and capacity-building
projects to ensure the success of the country's renewable energy projects.

13.8.4 INDONESIA

The current major source of energy in Indonesia is oil. The demand for energy is
high since only 53% of the household population has access to electricity. The
use of PV was previously introduced, but due to high cost, and as an effect of the

recent economic crisis, its operation stopped. Its use is being promoted again. Generally, the use of renewable energy in the country cannot compete commercially with conventional energy because of its high cost. Indonesia has already initiated activities in energy conservation by conducting training, education campaigns, and research. With the country's natural reserves, the opportunity for renewable energy is significant. However, there are barriers to successful implementation and these include lack of policy, finances, technology, and awareness. A development strategy to address the barriers, which aims to formulate policies and identify financing schemes to increase research and development in the energy sector, has been designed. The PREGA project is anticipated to assist in the implementation of Indonesia's energy projects.

13.8.5 VIETNAM

The application of renewable energy is taken seriously in Vietnam. In fact, the country has great potential in new and renewable energy, which includes solar, wind, biomass, and geothermal power. Vietnam is currently drafting an Energy Policy for the approval of the Prime Minister. The policy gives emphasis to the promotion of renewable energy. Activities undertaken on climate change are formulation of the Renewable Energy Masterplan, development of a microhydro project, and rural electrification. The systems loss in the distribution of energy is high due to obsolete powerplants and distribution network. The country expects that the PREGA project will provide solutions to the difficulties encountered in the areas of financial, technical and management capability. It is also expected that a work plan would be formulated during the workshop for implementation in Vietnam.

13.8.6 PHILIPPINES

The Philippine energy sector is moving towards privatization as reflected in the Energy Plan. Currently, 44% of energy supply is from hydro and geothermal plants. Some projects undertaken by the government in relation to climate change include retirement of oil-fired plants, deregulation of the oil industry, and use of renewable energy in rural electrification projects. Other projects are now being developed. The Department of Energy has already identified a strategy for the effective implementation of these projects through partnership with the private sector. The country hopes for a future committed to a renewable energy program. The PREGA project is in line with the thrust of the Energy Department. The Philippines joins other countries in the PREGA project in helping to solve the climate change problem and achieving sustainable development.

13.8.7 CAMBODIA

Efficient and sustainable energy and power supply is a priority in Cambodia. One of the thrusts of the government is to ensure reliable electricity for the people to promote national economic growth. The main source of energy in the country is biomass, which supplies 93% of the total energy demand. The hydropower potential is high, although electricity derived is still insignificant. Development of solar energy has started but due to high investment cost, its use is insignificant. The country has discussed plans to improve its energy status. Recently, the World Bank instituted training and capacity building to strengthen the energy sector. Lighting audits were also performed in some of the factories. The country has identified baseline and abatement scenarios and GHG abatement plans. It is expected that the participating countries would benefit from this PREGA project.

13.8.8 PEOPLE'S REPUBLIC OF CHINA (PRC)

The Government of PRC believes that the PREGA project has similar objectives and strategies as outlined and integrated in their sustainable development strategies. Renewable energy in the country cannot compete with traditional energy, as there is no market due to its high price compared to fossil fuels. The country is looking for financial institutions, which will assist in the promotion of renewable energy. There is also a plan to formulate a policy to push and encourage provinces to use renewable energy. Currently, the World Bank has a project in the country on creating and managing market share. PRC has initiated projects to mitigate climate change through capacity building, institutional arrangement, and participation in international fora and programs. A number of energy projects have already been implemented by funding agencies such as GEF and ADB. PREGA is timely for working together with the PRC government. The wide scope and work plan of PREGA may be integrated with other energy efficiency projects of PRC such as urbanization and construction.

13.8.9 INDIA

Coal is the major source of energy in India supplying about 54% of the total commercial energy requirement. The supply of coal in the country can last more than a hundred years but in spite of this, India is dependent on imported oil, and more than 60% of the rural population has no access to commercial energy. The oil and gas reserves in the country are nearing depletion, which is a cause for alarm. Continuing the use of coal on the other hand would increase greenhouse gas (GHG) emissions. Given this scenario, India recognizes the need to shift its energy approach from increasing supply to efficiency enhancement. The energy conservation potential for the different sectors in India is high. However, despite efforts to benefit from energy efficiency activities, barriers such as lack of awareness, financing, technology, and national coordination have restrained the effective implementation of the projects. To address these barriers, an Energy Conservation Bill was formulated for enactment. The PREGA project will directly address the problems hindering the implementation of energy efficiency projects in India.

13.9 CONCLUSIONS AND RECOMMENDATIONS

Increase access to energy services by the poor
Realize other strategic development objectives

Encourage investments in renewable energy, energy efficiency, and green-house-gas abatement technologies in developing member countries

Help reduce greenhouse-gas emissions

Develop capacities of national policy makers, technical experts, and staff of financial institutions to promote renewable energy, energy efficiency, and greenhouse-gas abatement

Support policy, regulatory, and institutional reforms, including removal of energy-pricing distortions

Facilitate access to private-sector financing

14 Development of Renewable Energy in Malaysia

Ah Ngan Ma, Yuen May Choo, Kien Yo Cheah, and Yusof Basiron

CONTENTS

14.1 INTRODUCTION

Over the last 40 years, the Malaysian palm oil industry has grown by leaps and bounds to become the world's largest producer and exporter of palm oil and its products. In 2004, Malaysia had about 3.87 million hectares (Ha) of land under oil palm cultivation. There were also 380 palm oil mills processing about 70 million tonnes (t) of fresh fruit bunch (FFB) to produce 13.98 million t of crude palm oil (CPO). The total export earnings of palm oil products, constituting refined palm oil, palm kernel oil, palm kernel cake, oleochemicals, and finished products amounted to RM 30.41 billion (RM 3.80 = US $1).

In 2003, Malaysia had an installed energy-generating capacity of about 14,296 MW, which was sufficient to meet the peak demand of 12,449 MW (Ministry of Energy, Water and Communication, 2004). The generating capacity was expected to expand to 16,834 MW by the year 2005 (8th Malaysian Plan, 2000) to meet the anticipated increase in demand for power due to rapid industrialization and urbanization.

Malaysia is fortunate to have plentiful supplies of petroleum and natural gas. These two sources of energy are expected to account for most of the nation's commercial energy requirements. However, the fast diminishing energy reserves, greater environmental awareness, and increasing energy consumption as a result of rapid industrialization make it necessary to examine alternative energy resources.

Alternative energy resources such as silviculture, solar, and wind appear to be promising from the environmental and renewable-resource perspectives. However, in the Malaysian context, there are several technical and socioeconomic uncertainties involved in their utilization. Their long-term feasibility, nevertheless, is an important aspect of the nation's research and development efforts, but their immediate application on a commercial scale is of a limited nature. What is needed is a renewable energy resource whose utilization system is presently proven and operational as well as amenable to expansion without much difficulty.

14.2 ENERGY FROM FIBER, SHELL, AND EMPTY FRUIT BUNCHES

Oil palm is a perennial crop. It has an economic life span of about 25 years. Oil palm is grown for its oils. Palm oil and palm kernel oil are extracted from the mesocarp and kernels of the fruits, respectively. In general, the fresh fruit bunches (FFB) contains about 20 to 22% palm oil, 6 to 7% palm kernel, 14% fiber, 7% shell, and 23% empty fruit bunch (EFB) (Ma, 2002). Table 14.1 shows the type and amount of biomass generated as well as their heat values.

14.2.1 FIBER AND SHELL

All of the palm oil mills in Malaysia use fiber and shell as the boiler fuel to produce steam and electricity for palm oil and kernel production processes. The fiber and shell alone can supply more than enough electricity to meet the energy demand of a palm oil mill. It is estimated that 20 kW·h (lower kW·h for higher capacity mill) of electrical energy is required to process 1 t of FFB. Thus in 2004, about 1400 million kW·h of electricity was generated and consumed by the palm oil mills. Assuming that each mill operates on an average of 393.35 h per month, the palm oil mills together will have a generating capacity of 296 MW. This constitutes about 2.5% of the energy demand of the country. It must be mentioned here that the palm oil mills generally have excess fiber and shell that are not used and have to be

---

TABLE 14.1
Biomass Generated by Palm Oil Mills in 2004

Biomass	Quantity (million tones)	Moisture Content (%)	Oil Content (%)	Heat Value (dry) (kJ/kg)
EFB	16.1	65	5	18,883
Fiber	9.8	35	5	19,114
Shell	4.9	12	1	20,156
POME	49.0	93	1	—

Note: EFB = empty fruit bunch; POME = palm oil mill effluent.

disposed of separately. In other words, the palm oil mills still have excess capacity to produce more renewable energy.

Assuming that a diesel power generator consumes 0.34 l of diesel for every kW·h of electricity output, the palm oil industry in 2004 is estimated to have saved the country about 476 million l of diesel, which amounted to about RM 395 million (price of diesel @ RM 0.83/l). The energy requirement for palm oil mills is mounting, as palm oil production is expected to reach 14 million t by the year 2005. Furthermore, the fuel cost could have been more if fuel oil were used as boiler fuel to generate steam separately for the milling processes.

The palm oil industry is indeed fortunate that the fiber and shell can be used directly as the boiler fuel without any further treatment. With proper control of combustion, black smoke emission usually associated with the burning of solid fuel can be controlled. Another intangible advantage of using both these residues as fuel is that it helps to dispose of these bulky materials, which otherwise would contribute to environmental pollution. Unless these materials can be more beneficially utilized, it is envisaged that they will remain as boiler fuel for the foreseeable future. It has generally been considered that energy is free in the palm oil mills. This has undoubtedly contributed greatly to the success of the palm oil industry.

14.2.2 Empty Fruit Bunch

Apart from fiber and shell, EFB is another valuable source of biomass that can be readily converted into energy. However, this material has only been utilized to a very limited extent, mainly because there is already enough energy available from fiber and shell. Also, due to its physical nature and high moisture content of 65%, the EFB has to be pretreated to reduce its bulkiness and moisture content to below 50% to render it more easily combustible (Jorgensen, 1985; Chua, 1991).

The EFB has a heat value of 18,883 kJ/kg on dry weight. Thus the total heat energy obtainable from the EFB in 2004 would be 106×10^{12} kJ. This is sufficient to generate about 26.5 million t of steam (at 65% boiler efficiency and 2604 kJ per kg of steam) and 980 million kW·h of electricity, saving the country 333 million l of diesel or RM 276 million. The above calculation was based on a standard noncondensing turbo-alternator working against a backpressure of 3 bars gauge. More than double the energy could be obtained if condensing turbines working at a vacuum of 0.25 bar (absolute) were used for power generation (Chua, 1991).

The above estimate represents the total obtainable energy from all of the 380 palm oil mills distributed throughout the country. Thus it can be said that the energy generated from a single palm oil mill will not be significant in volume, and it may not be viable for commercial consideration or to supply the electricity to the national grid. However, the EFB, unlike fiber, can be easily collected and transported. The possibility of producing electricity at a central power generating plant could be a viable proposition. The central power plants could be sited at locations where there are high concentrations of palm oil mills so that the EFB and the surplus fiber and shell from the mills could be transported at a reasonable distance and cost to the various central power plants. Also, since the power plants can be independent entities, they can be

TABLE 14.2
Energy Database for Palm Biomass

Sample	Caloric Value (kJ/kg)	Ash (%)	Volatile Matter (%)	Moisture (%)	Hexane Extractable (%)
Empty fruit bunch (EFB)	18,795	4.60	87.04	67.00	11.25
Fibers	19,055	6.10	84.91	37.00	7.60
Shell	20,093	3.00	83.45	12.00	3.26
Palm kernel cake	18,884	3.94	88.54	0.28	9.35
Nut	24,545	4.05	84.03	15.46	4.43
Crude palm oil	39,360	0.91	1.07	1.07	95.84
Kernel oil	38,025	0.79	0.02	0.02	95.06
Liquor from EFB	20,748	11.63	78.50	88.75	3.85
Palm oil mill effluent	16,992	15.20	77.09	93.00	12.55
Trunk	17,471	3.39	86.73	76.00	0.80
Petiole	15,719	3.37	85.10	71.00	0.62
Root	15,548	5.92	86.30	36.00	0.20

Source: Chow et al., Energy Database of the Oil Palm, in *Proceedings of 2003 MPOB International Palm Oil Congress*, Putrajaya, Malaysia, 24 Aug. 2003.

operated throughout the year. Table 14.2 shows the energy data for various types of palm biomass when they are used as boiler fuels to generate electricity.

14.3 BIOGAS FROM POME

Besides the solid residues, palm oil mills also generate large quantities of liquid waste in the form of palm oil mill effluent (POME), which, due to its high biochemical oxygen demand (BOD), is required by law to be treated to acceptable levels before it can be discharged into watercourses or onto land. In a conventional palm oil mill, about 0.7 m³ of POME is generated for every tonne of FFB processed. Hence in 2004, about 49 million m³ of POME was generated in this country. The palm oil mills use anaerobic processing to treat their POME. The biogas produced during the decomposition is a valuable energy source. It contains about 60 to 70% methane, 30 to 40% carbon dioxide, and trace amount of hydrogen sulfide. Its fuel properties are shown in Table 14.3 together with other gaseous fuels.

About 28 m³ of biogas is generated for every cubic meter of POME treated. However, most of the biogas is not recovered. So far, only a few palm oil mills harness the biogas for heat and electricity generation (Chua, 1991; Gillies and Quah, 1984; Quah et al., 1982). It has been reported that about 1.8 kW·h of electricity could be generated in a gas engine from 1 m³ of biogas (Quah et al., 1982). The potential energy from biogas generated by POME is shown in Table 14.4. Again,

TABLE 14.3
Some Properties of Gaseous Fuels

	Biogas	Natural Gas	LPG [a]
Gross calorific value (kJ/Nm³)	19,908–25,830	3,797	100,500
Specific gravity	0.847–1.002	0.584	1.5
Ignition temperature (°C)	650–750	650–750	450–500
Inflammable limits (%)	7.5–21	5–15	2–10
Combustion air required (m³/m³)	9.6	9.6	13.8

Note: All gases evaluated at 15.5°C, atmospheric pressure, and saturated with water vapor.

[a] LPG = liquefied petroleum gas.

Source: Quah, S.K. and Gillies, D., *Practical Experience in Production Use of Biogas*, in Proceedings of National Workshop on Oil Palm By-Product Utilization, Palm Oil Research Institute of Malaysia, Kuala Lumpur, 1981, pp. 119–125.

TABLE 14.4
Potential Energy from Biogas

Year	Palm oil production (million tonnes)	POME (million m³)	Biogas (million m³)	Electricity (million kW·h)
2004	13.98	49	1372	2470

as all the palm oil mills have enough energy from fiber and shell, there is no outlet for this surplus energy. Considering the costs of storage and transportation of the biogas, perhaps the most viable proposition is to encourage the setting up of industries in the vicinity of the palm oil mills where the biogas energy can be directly utilized. This could result in a substantial saving in energy bills (Chua, 1991).

It is estimated that 1 m³ of biogas is equivalent to 0.65 l of diesel for electricity generation. Hence the total biogas energy could substitute for as much as 892 million l of diesel in 2004. This would amount to RM 740 million. Again, the amount of biogas generated by an individual palm oil mill is not significant for commercial exploitation. However, the economic viability might be attractive if the palm oil mills could utilize all of their fiber, shell, EFB, and biogas for steam and electricity generation.

14.4 PALM OIL METHYL ESTERS AS DIESEL SUBSTITUTE

Biodiesel has gained much attention in recent years due to increasing environmental awareness. Biodiesel is produced from renewable plant resources and thus does not contribute to the net increase of carbon dioxide. From 1996 to 2004, the biodiesel production capacity in the European Union (EU) has increased by a factor of four from 591,000 t to a total of 2.355 million t (Bockey, 2002; Bockey, 2004).

Further utilization of biodiesel is anticipated due to the initiative of the respective authorities to promote biodiesel and the high cost of petroleum diesel. For example, by the end of 2005, at least 2% (about 3.1 million t) of fossil fuels will be replaced by biofuels (biodiesel, bioethanol, biogas, biomethanol, etc.) in all EU countries. This minimum target quantity has been set out in the EU commission action plan, and the proportion will be increased annually by 0.75% to reach 5.75% (about 17.5 million t) in the year 2010 (Bockey and Körbitz, 2002; Markolvitz, 2002; Schöpe and Britschkat, 2002). Biodiesel will take up about 10 million t. This proposal also envisages that by 2020, the proportion of biofuels will be 20% and obligatory blending of 1% of biofuels will be introduced in 2009 (1.75% from 2010 onward). The current trend and legislation will set a momentum for greater biodiesel production and consumption worldwide. Thus, there will be an upward course and new market opportunities for biodiesel.

Methyl esters of vegetable oils have been successfully evaluated as a diesel substitute worldwide (Choo and Ma, 2000; Choo et al., 1997). For example, rapeseed methyl esters in Europe, soybean oil methyl esters in the United States, sunflower oil methyl esters in both Europe and the United States, and palm oil methyl esters in Malaysia. As the choice of vegetable oil depends on the cost of production and reliability of supply, palm oil would be the preferred choice, as it is the highest oil-yielding crop (4 to 5 t/Ha/yr) among all the vegetable oils and the cheapest vegetable oil traded in the world market.

Malaysia has embarked on an extensive biodiesel program since 1982. The biodiesel program includes development of production technology to convert palm oil to palm oil methyl esters (palm diesel), a pilot-plant study of palm diesel production, as well as exhaustive evaluation of palm diesel as a diesel substitute in conventional diesel engines (both stationary engines and exhaustive field trials).

Crude palm oil can be readily converted to their methyl esters. The MPOB/PETRONAS patented palm diesel technology (Choo et al., 1992) has been successfully demonstrated in a 3000 t/yr pilot plant (Choo et al., 1995; Choo et al., 1997; Choo and Cheah, 2000). The novel aspect of this patented process is the use of solid acid catalysts for the esterification. The resultant of the reaction mixture, which is neutral, is then transesterified in the presence of an alkaline catalyst. The conventional washing stage or neutralization step after the esterification process is obviated, and this is an economic advantage.

Crude palm oil methyl esters (palm diesel) were systematically and exhaustively evaluated as a diesel fuel substitute from 1983 to 1994 (Choo et al., 1995; Choo et al., 2002a). These included laboratory evaluation, stationary engine testing, and field trials on a large number of vehicles, including taxis, trucks, passenger cars, and buses. All of these tests have been successfully completed. It is worth mentioning that the tests also covered field trials with 36 Mercedes Benz engines mounted onto passenger buses running on three types of fuels, namely, 100% petroleum diesel, blends of palm diesel and petroleum diesel (50:50), and 100% palm diesel. Each bus ran for 300,000 km, the expected life of the engines. (Total mileage of the ten buses running on 100% palm diesel was 3.7 million km.) Very promising results have been obtained from the exhaustive field trial. Fuel consumption by volume was comparable with that of the diesel fuel. Differences in engine performance were so

TABLE 14.5
Fuel Characteristics of Malaysian Diesel, Palm Diesel, and Palm Diesel with Low Pour Point

Property	Malaysian Diesel	Palm Oil Methyl Esters (Palm Diesel)	Palm Diesel with Low Pour Point
Specific gravity	0.8330	0.8700	0.8803
ASTM D1298	@15.5°C	@ 23.6°C	@ 15.5°C
Sulfur content (wt%)	0.10	<0.04	< 0.04
IP 242			
Viscosity at 40°C (cSt)	4.0	4.5	4.5
ASTM D445			
Pour point (°C)	15.0	16.0	−15.0
ASTM D97			
Cetane index	53	50	NA [a]
ASTM D976			
Gross heat of combustion (kJ/kg)	45,800	40,135	39,160
ASTM D 2382			
Flash point (°C)	98	174	153
ASTM D 93			
Conradson carbon residue (wt%)	0.14	0.02	0.01
ASTM D 189			

[a] NA: not available.

small that an operator would not be able to detect it. The exhaust gas was found to be much cleaner, as it contained comparable NO_x and less hydrocarbon, CO, and CO_2. The very obvious advantage is the absence of black smoke and sulfur dioxide from the exhaust. This is a truly environmentally benign fuel substitute.

Palm diesel has fuel properties that are very similar to those of petroleum diesel (Table 14.5). It also has a higher cetane number (63) than diesel (<40) (Table 14.6). A higher cetane number indicates shorter ignition time-delay characteristics and, generally, a better fuel. Palm diesel can be used directly in unmodified diesel engines, and obviously it can also be used as a diesel improver. Compared with crude palm oil, the palm diesel has very much improved viscosity and volatility properties. It does not contain gummy substances. However, it has a pour point of 15°C, and this has confined its utilization to tropical countries.

In recent years, palm diesel with low pour point (without additives) has been developed to meet seasonal pour-point requirements, for example spring (−10°C), summer (0°C), autumn (−10°C), and winter (−20°C). The MPOB patented technology (Choo et al., 2002b) has overcome the pour-point problem of palm diesel. With the improved pour point, palm diesel can be utilized in temperate countries. Besides having good low-temperature flow characteristics, the palm diesel with low pour point also exhibits comparable fuel properties as petroleum diesel (Table 14.5).

The storage properties of the palm diesel are very good. After storing for more than 6 months in a 50-m³ storage tank, it was found that there was little deterioration in the

TABLE 14.6
Cetane Numbers of Palm Diesel, Petroleum Diesel, and Their Blends

Blends		
CPO Methyl Esters (%)	Petroleum Diesel (%)	Cetane Number [a]
100	0	62.4
0	100	37.7
5	95	39.2
10	90	40.3
15	85	42.3
20	80	44.3
30	70	47.4
40	60	50.0
50	50	52.0
70	30	57.1

[a] Calculated according to ASTM D613.

fuel quality parameters except for the color, which had changed from orange to light yellow. This was due to the breakdown of the high-value colored carotene compounds.

The main benefit derived from such a renewable source of energy is the reduction of emission of greenhouse gases (GHG) such as CO_2. The production and consumption of palm diesel has a closed carbon cycle. This closed carbon cycle recycles the carbon dioxide, so there is no net accumulation of carbon dioxide in the atmosphere. Consequently, production of palm diesel, with its lower emissions, is in line with the Clean Development Mechanism (CDM) of the 1997 Kyoto Protocol.

Under the terms of the 1997 Kyoto Protocol (a major international initiative established to reduce the threat of global warming), there is a potential financial gain to transact these GHG benefits to the palm oil industry under the CDM. This mechanism allows emission-reduction projects to be implemented and credits to be awarded to the investing parties. Financial incentives like an attractive carbon-credit scheme should further enhance the economic viability of these renewable fuels.

In 2003, Malaysia consumed 8.91 million t of petroleum diesel (Ministry of Energy, Water and Communication, 2004). The transport sector alone consumed 4.941 million t and generated 19.32 million t of carbon dioxide. The transport sector has also been identified as one of the chief contributors to air pollution, particularly black smoke (due to diesel) and carbon dioxide. If 10% of the diesel (0.4941 million t) were replaced by palm diesel, the industry would be entitled to enjoy 1.932 million t of carbon credit, which amounts to US \$19.32 million at a rate of US \$10/t of carbon dioxide.

14.5 PALM OIL AS DIESEL SUBSTITUTE

Many researchers have investigated the possibility of using vegetable oils (straight or blended) as a diesel substitute. A good account of their attempts was reported in the 1983 JAOCS Symposium on Vegetable Oils as Diesel Fuels (Klopfenstein and Walker,

1983; Pryde 1983; Strayer et al., 1983). The symposium revealed that vegetable oils have good potential as alternative fuels if some problems could be overcome satisfactorily. These include high viscosity, low volatility, and the reactivity (polymerization) of the unsaturated hydrocarbon chains if the oil is highly unsaturated. These will give rise to coking on the fuel injectors, carbon deposits, oil ring sticking, and thickening and gelling of the lubricating oil as a result of contamination with vegetable oil.

It is possible to reduce the viscosity of the vegetable oil by incorporating a heating device with the diesel engine, as has been successfully demonstrated by the engine manufacturer Elsbett (Basiron and Hitam, 1992). Other factors that may have long-term effects on the engine are free fatty acids and gummy substances, which are found in the crude vegetable oils. The incomplete combustion residues may contribute undesirable deposits on the engine components, and the gummy substances may cause filters to plug. This will call for more regular and frequent servicing and maintenance of the engine.

Various blends of crude palm oil and palm oil products such as refined, bleached, and deodorized palm olein with medium fuel oil (MFO) and petroleum diesel, respectively, have been evaluated as boiler fuels and diesel substitutes (Hitam et al., 2001). Crude palm oil (CPO) and refined, bleached, and deodorized palm olein (RBDPOo) were blended with MFO and petroleum diesel, respectively, at various ratios by volume. The resultant fuel blends, CPO/MFO and RBDPOo/petroleum diesel, exhibit advantages and fuel characteristics that are better compared with those when the individual CPO, RBDPO, RBDPOo, MFO, and petroleum diesel are used solely as fuel (Table 14.7, Table 14.8, and Table 14.9) (Basiron, 2002). Currently, field trials using MPOB's in-house vehicles are being conducted to evaluate blends of RBDPOo/petroleum diesel (up to 10% of the former) as a diesel substitute. No technical problems have been reported so far.

TABLE 14.7
Fuel Characteristics of Crude Palm Oil (CPO), Medium Fuel Oil (MFO), and Blends of Crude Palm Oil/Medium Fuel Oil (CPO/MFO)

Property	Method	Unit	MFO	CPO	CPO/MFO (50:50)
Gross heat of combustion	D 240	Btu/lb	18,350 Min	17,064	17,692
		kJ/kg	42,680 Min	39,690	41,150
Sulfur	D 4294	wt%	3.5 Max	0.03	1.55
Viscosity @ 50°C	D 445	cSt	180 Max	25.6	67.3
Flash point	D 93	°C	66 Min	268	99
Ash	D 482	wt%	0.1 Max	NA [a]	0.012
Pour point	D 97	°C	21 Max	21.0	−6
Carbon residue	D 4530	wt%	13.0 Max	8.5	7.0
Density @ 15°C	D 1298	kg/l	0.98 Max	0.9140	0.9408
Sediment by extraction	D473	wt%	0.10 Max	NA [a]	0.02
Water by distillation	D 95	vol%	0.5 Max	NA [a]	0.25

[a] NA: not available.

TABLE 14.8
Fuel Characteristics of RBD Palm Olein (RBDPOo), Petroleum Diesel, and Blends of RBD Palm Olein/Petroleum Diesel (RBDPOo/Diesel)

Test Conducted	RBD Palm Olein (RBDPOo)	Blends of RBDPOo/Diesel					Petroleum Diesel
		90:10	70:30	50:50	30:70	10:90	
Density @ 40°C (kg/l) ASTM D1298	0.9150	0.8940	0.8770	0.8600	0.8435	0.8275	0.8190
Sulfur content (wt%) IP 242	0.035	0.035	0.055	0.060	0.080	0.090	0.100
Viscosity @ 40°C (cSt) ASTM D445	39.2	29.5	14.8	8.6	7.0	3.8	3.7
Pour point (°C) ASTM D97	9	9	12	12	12	15	15
Gross heat of combustion (kJ/kg) ASTM D240	38,975	39,800	40,625	41,450	42,275	43,100	45,000
Flash point (°C) PM cc D93	326	142	110	99	93	90	89

14.6 CONCLUSION

The progressive escalation of energy shortages and fuel prices in recent times has led to an intensified global search for viable alternative sources of energy. As conventional energy resources become more difficult to obtain, efforts must be directed toward development of alternative energy sources.

The palm oil industry is bestowed with a plentiful supply of by-products that can be readily used as energy resources. When EFB and biogas are properly processed using proven and innovative techniques, a considerable amount of energy can be economically recovered. The utilization of these by-products from the palm oil mills, if accepted by the authorities, will, to some extent, help in lowering the escalation of energy shortages. The production and application technologies have been fully demonstrated.

Energy is considered free by the palm oil mills. Fiber and shell together can supply more than enough energy to meet their energy demands. The electricity generated indirectly from fiber and shell represents about 2% of the national electricity demand. Energy from biogas and empty fruit bunch has so far been ignored, although they represent a hefty 4% of the national energy demand in terms of electricity. Efforts are being made to encourage palm oil mills to sell this excess energy in the form of electricity to the national grid.

TABLE 14.9
Fuel Characteristics of RBD Palm Oil (RBDPO), Petroleum Diesel, and Blends of RBD Palm Oil/Diesel (RBDPO/Diesel)

Test Conducted	Petroleum Diesel	Blends of RBDPO/Diesel					RBDPO
		2:98	3:97	5:95	6:94	7:93	
Density @ 15°C (kg/l) ASTM D1298	0.8479	0.8492	0.8499	0.8502	0.8521	0.8525	0.9151
Sulfur content (wt%) IP 242	0.16	0.13	0.11	0.11	0.11	0.11	0.12
Viscosity @ 40°C (cSt) ASTM D445	0.4248	4.895	4.576	4.656	5.010	5.021	40.68
Pour point (°C) ASTM D97	9	9	9	9	9	12	24
Gross heat of combustion (kJ/kg) ASTM D240	45,050	45,340	45,160	45,095	45,085	45,015	39,260
Flash point (°C) ASTM D93 ASTM D92	84.0	84.0	84.0	84.0	85.0	86.0	322.0

Palm diesel has been fully evaluated as a potential diesel substitute and diesel/cetane improver. Low-pour-point palm diesel (−21°C) without any additives has been produced to meet stringent winter diesel specifications. The palm diesel is an environmentally benign fuel substitute in terms of exhaust gas emission. Blends of CPO/MFO and RBDPOo/diesel have also been evaluated as potential fuels for boilers and diesel engines.

All of these energy sources are renewable and their supply is readily available and assured. Currently, burning of the biomass residues is often considered as a way to dispose of the product rather than as an energy source. These residues should be commercially exploited. This will make the palm oil industry more environmentally sustainable.

REFERENCES

8th Malaysian Plan, Economy Planning Unit, Prime Minister's Department, 2001, Kuala Lumpur, Malaysia.

Basiron, Y. and Hitam, A., *Cost Effectiveness of the CPO Fuel in the Mercedes Elsbett Engine Car*, PORIM Information Series, No. 4, July 1992.

Basiron, Y., *Palm Oil and Palm Oil Products as Fuel Improver*, Malaysian Patent PI 20020396, 2002.

Bockey, D. and Körbitz, W., Situation and Development Potential for the Production of Biodiesel: An International Study, Union for Promoting Oilseeds and Protein Plants, 2002 (www.ufop.de).

Bockey, D., Policy initiative schemes and benefits of biofuel promotion in germany: current status of legislation and production, paper presented at *Conference on Biofuels: Challenges for Asian Future*, Bangkok, 30–31 Aug. 2004.

Choo, Y.M., Ong, A.S.H., Cheah, K.Y., and Bakar, A., Production of Methyl Esters from Oils and Fats, Australian Patent 626014, 1992.

Choo, Y.M., Ma, A.N., and Basiron, Y., Production and evaluation of palm oil methyl esters as diesel substitute, *Elaeis*, November 5, 1995.

Choo, Y.M., Ma, A.N., and Ong, A.S.H., Biofuel, in *Lipids: Industrial Applications and Technology*, Gunstone, F.D. and Padley, F.B., Eds., Marcel Dekker, New York, 1997, pp. 771–785.

Choo, Y.M. and Cheah, K.Y., Biofuel, in *Advances of Oil Palm Research*, Vol. II, Yusof, B., Jalani, B.S., and Chan, K.W., Eds., Malaysian Palm Oil Board, Selangor, Malaysia, 2000, pp. 1293–1345.

Choo, Y.M. and Ma, A.N., Plant power, *Chemistry & Industry*, August, 530, 2000.

Choo, Y.M., Ma, A.N., and Basiron, Y., Palm diesel, paper presented at *2002 Oils and Fats International Congress* (OFIC), Kuala Lumpur, Malaysia, 7–10 Oct. 2002a.

Choo, Y.M., Cheng, S.F., Yung, C.L., Lau, H.L.N., Ma, A.N., and Basiron, Y., Low Pour Point Palm Diesel, Malaysian Patent PI 20021157, 2002b.

Chow, M.C., Subramaniam, V., and Ma, A.N., Energy Database of the Oil Palm, in *Proceedings of 2003 MPOB International Palm Oil Congress*, Putrajaya, Malaysia, 24 Aug. 2003.

Chua, N.S., Optimal utilization of energy sources in a palm oil processing complex, paper presented at *Seminar on Developments in Palm Oil Milling Technology and Environment Management*, Genting Highlands, Pahang, Malaysia, 16–17 May 1991.

Gillies, D. and Quah, S.K., Tennmaran biogas project, paper presented at the *Second Asean Workshop on Biogas Technology*, Kuala Trengganu, Trengganu, Malaysia, 8–13 October 1984.

Hitam, A., Choo, Y.M., Hasamuddin, W.H., and Yusof B., Palm Oil as Fuel, in *Proceedings of 2001 MPOB International Palm Oil Congress*, Kuala Lumpur, Malaysia, 20–23 Aug. 2001.

Jorgensen, H.K., Treatment of empty bunches for recovery of residues oil and additional steam production, *JAOCS*, 62, 282, 1985.

Klopfenstein, W.E. and Walker, H.S., Efficiencies of various esters of fatty acids as diesel fuels, *JAOCS*, 60, 1596, 1983.

Ma, A.N., Carbon credit from palm: biomass, biogas and biodiesel, *Palm Oil Engineering Bulletin*, 65, 24, 2002.

Malaysian Palm Oil Board, *Malaysian Oil Palm Statistics 2003*, Ministry of Plantation Industries and Commodities, Selangor, Malaysia, 2004.

Markolvitz, M., *The European Biodiesel Market: Biodiesel Status Report*, Degussa AG, Niederkassel, Germany, 2002.

Ministry of Energy, Water and Communication, *National Energy Balance 2003*, Malaysia Energy Centre, Selangor, Malaysia, 2004.

Pryde, E.H., Vegetable oils as diesel fuels: overview, *JAOCS*, 60, 1557, 1983.

Quah, S.K. and Gillies, D., Practical experience in production use of biogas, in *Proceedings of National Workshop on Oil Palm By-Product Utilization*, Palm Oil Research Institute of Malaysia, Kuala Lumpur, 1981, pp. 119–125.

Quah, S.K., Lim, K.H., Gillies, D., Wood, B.J., and Kanagaratnam, K., Sime Darby POME Treatment and Land Application System, in *Proc. of Reg. Workshop on Palm Oil Mill. Techy. Effl. Treat.*, Palm Oil Research Institute of Malaysia, Kuala Lumpur, 1982, pp. 193–200.

Schöpe, M. and Britschkat, G., Macroeconomic evaluation of rape cultivation for biodiesel production in Germany, *Munich*, March 2002 (www.ufop.de).

Strayer, R.C., Blake, J.A., and Craig, W.K., Canola and high erucic rapeseed oil as substitutes for diesel fuel: preliminary tests, *JAOCS*, 60, 1587, 1983.

15 Outlook on Catalytic Technologies for Sustainable Development: The Argentina Case

Carlos R. Apesteguía

CONTENTS

15.1 INTRODUCTION

Sustainable industrial development requires a balance between economic, technological, and environmental aspects. In particular, the development of cleaner and renewable technologies, as well as being fundamental to process optimization, waste reduction, and pollution prevention, are becoming imperative for developing countries, such as the Latin American transition-economy countries. The use of renewable resources in the chemical industry also represents an opportunity for the growth for agricultural production while reducing the impact of oil-based processes on the environment. However, in the transition toward a market economy, Latin American countries encounter significant difficulties that are in part related to the presence of old and environmentally unfriendly technologies.

Catalysis is a powerful tool for the development of clean catalytic technologies in chemical manufacturing and the use of renewable resources, particularly crop-derived raw materials, and thereby meets the economic and social interests of Latin American countries. Expected benefits of catalytic routes are higher process efficiency, decreased waste production, elimination of hazardous reagents, and development of valuable by-products. Thus, increasing research work has lately been focused on the development of novel catalyst formulations and catalytic processes in sectors including the polymer, agricultural, petrochemical, pharmaceutical, and fine-chemicals industries. However, the industrial development of catalysts is an expensive and labor-intensive activity because research in catalysis is still dominated largely by experimental studies. The development and transfer of highly advanced and sophisticated catalytic technologies and processes, and their subsequent industrial application and adoption, are therefore crucial for the industrialization programs of Latin American countries.

Argentina is a developing country requiring sustainable economic growth based in industrial development to improve income for the nation's population. However, industrial development cannot be achieved without sustainable chemistry and protection of the environment and the quality of life. Advanced research in catalysis is needed to develop novel clean technologies for chemical design and to achieve these economic and environmental goals. This chapter presents three examples of innovative catalytic processes developed by the Argentinean academic community in the field of renewable resources, clean technologies, and sustainable chemistry. Two examples deal with the use of solid catalysts for the synthesis of fine chemicals. Traditionally, fine and specialty chemicals have been produced predominantly by noncatalytic or homogeneously catalyzed synthesis. Unfortunately, these processes are characterized by the coproduction of large amounts of unwanted products, the use of toxic or corrosive reagents and solvents, and the use of harmful liquid catalysts. In particular, the use of strong liquid acids (HF, H_2SO_4, HCl) and Friedel-Crafts catalysts ($AlCl_3$, $TiCl_4$, $FeCl_3$) in homogeneous commercial processes poses problems of high toxicity, corrosion, and disposal of spent acid.

New industrial strategies for fine-chemical synthesis demand the use of renewable raw materials and the replacement of liquid acids or bases by solid catalysts.

The first example describes the use of acid zeolites for replacing $AlCl_3$ Friedel-Crafts catalyst in the gas-phase synthesis of aromatic ketones from acylation of phenol. The second example reports, for the first time, the development of bifunctional solid catalysts for producing menthols from citral, a renewable raw material, in a one-step process. Finally, the third example illustrates the use of catalytic technologies to promote a larger use of renewable low-value raw materials for production of an environmentally acceptable fuel. This involves the use of a two-step catalytic process for efficiently obtaining biodiesel from inedible acid materials such as frying oil, greases, tallow, and lard.

15.2 GREEN CATALYTIC PROCESSES IN THE SYNTHESIS OF FINE CHEMICALS

15.2.1 INTRODUCTION

Aromatic ketones are valuable intermediate compounds in the synthesis of important fragrances and pharmaceuticals. In many cases, these compounds are currently obtained in homogeneous processes via Friedel-Crafts acylations. However, Friedel-Crafts acylations are real and alarming examples of very widely used acid-catalyzed reactions that are based on 100-year-old chemistry and are extremely wasteful. Replacement of $AlCl_3$ Friedel-Crafts catalysts by solid acids for acylation reactions would drastically reduce the aqueous discharge and solid waste.

In particular, hydroxyacetophenones are useful intermediate compounds for the synthesis of pharmaceuticals. For example, *para*-hydroxyacetophenone (*p*-HAP) is used for the synthesis of paracetamol, a well-known antipyretic drug, and *ortho*-hydroxyacetophenone (*o*-HAP) is a key intermediate for producing 4-hydroxycoumarin and warfarin, both of which are used as anticoagulant drugs in the therapy of thrombotic disease [1, 2]. In the classical commercial process, *p*-HAP is obtained via the Fries rearrangement of phenyl acetate in a liquid-phase process involving the use of homogeneous catalysts such as $AlCl_3$, $TiCl_4$, $FeCl_3$, and HF, which pose problems of high toxicity, corrosion, and spent-acid disposal [3]. In an attempt to develop a suitable and environmentally benign process for producing *p*-HAP, strong solid acids such as ion-exchange resins, zeolites, Nafion, and heteropolyacids have been tested in liquid phase for the Fries rearrangement of phenyl acetate. However, solid acids form significant amounts of phenol together with *p*-HAP, and these are, in general, rapidly deactivated [4, 5].

Hydroxyacetophenones can also be obtained by the acylation of phenol in liquid or gas phases by employing different acylating agents. In liquid phase, the reaction produces mainly *p*-HAP using either Friedel-Crafts or solid acid catalysts; but the process is hampered because of environmental constraints and the decay of catalyst activity [6, 7]. In the gas phase, the phenol acylation on solid acids forms predominantly *o*-HAP, but the reported experimental *o*-HAP yields are still moderate, particularly because of significant formation of phenyl acetate [8, 9]. An analysis of the literature shows that the potential use of solid acids to obtain hydroxyacetophenones in gas or liquid phases via either phenol acylation or phenyl acetate rearrangement reactions is limited because of relatively low yields and rapid activity decay. The development of more selective and stable catalysts is therefore required to efficiently promote the synthesis of *o*-HAP. Taking this into account, our research group decided to perform a detailed study of the gas-phase acylation of phenol with acetic acid over different solid acids with the goal of relating the structural properties and the surface acid site density and strength of the solids to their ability to efficiently catalyze the phenol acylation reaction to yield *o*-HAP. Our studies also focused on the decay of catalyst activity.

TABLE 15.1
Sample Physical Properties and Acidity

Catalyst	Surface Area S_g (m²/g)	Pore diameter d_p (Å)	Si/Al	TPD of NH₃		IR of Pyridine	
				μmol/g	μmol/m²	B (area/g)	L (area/g)
HY	660	7.4	2.4	1380	2.1	310	465
HZSM-5	350	5.5	20	770	2.2	337	341
Al-MCM-41	925	30	18	340	0.4	32	135
HPA/MCM-41	505	29	—	352	0.7	—	—
SiO₂-Al₂O₃	560	45	11.3	1005	1.8	68	204

15.2.2 Sample Preparation and Characterization

The o-HAP synthesis was studied on 12-tungstophosfotic acid (HPA) supported on mesoporous MCM-41 silica (sample HPA[30%]/MCM-98; Al-MCM-41 (Si/Al = 18), zeolites HY (UOP-Y54) and HZSM-5 (ZeoCat PZ-2/54), and SiO_2-Al_2O_3 (Ketjen LA-LPV) catalysts. Sample preparation and characterization are detailed in Padró and Apesteguía [10]. Table 15.1 shows the physicochemical characteristics (surface area, pore diameter, chemical composition) and the acidity of the samples. Sample acid properties were probed by temperature-programmed desorption (TPD) of NH_3 preadsorbed at 373 K and by infrared (IR) spectroscopy of preadsorbed pyridine. The NH_3 surface densities for acid sites in Table 15.1 were obtained by deconvolution and integration of TPD traces (not shown here).

Sample HPA/MCM-41 showed a sharp NH_3 desorption peak at about 910 K, which accounts for the strong Brönsted acid sites present on this material. The evolved NH_3 from HY, HZSM-5, and SiO_2-Al_2O_3 gave rise to a peak at 483 to 493 K and a broad band between 573 and 773 K. In contrast, Al-MCM-41 did not exhibit the high-temperature NH_3 band, showing instead a single asymmetric broad band with a maximum around 482 to 496 K. On an areal basis, zeolites HY and HZM-5 exhibited the highest surface acid density (about 2.2 mmol/m²).

The density and nature of surface acid sites were determined from the IR spectra of adsorbed pyridine. The relative contributions of Lewis and Brönsted acid sites were obtained by deconvolution and integration of pyridine absorption bands appearing at around 1450 and 1540 cm⁻¹, respectively (Table 15.1). In agreement with the results obtained by TPD of NH_3, the amount of pyridine adsorbed on Al-MCM-41 after evacuation at 423 K, in particular on Brönsted sites, was clearly lower as compared with acid zeolites or SiO_2-Al_2O_3, reflecting the moderate acidic character of mesoporous Al-MCM-41 samples. The areal peak relationship between Lewis (L) and Brönsted (B) sites on Al-MCM-41 was L/B = 4.2, higher than on SiO_2-Al_2O_3 (L/B = 3). The L/B ratio on HY was 1.5, while zeolite HZSM-5 contained a similar concentration of Brönsted and Lewis acid sites.

15.2.3 Catalytic Results

The gas-phase acylation of phenol (P) with acetic acid (AA) was carried out in a fixed bed, continuous-flow reactor at 553 K and 101.3 kPa. Standard catalytic tests were conducted at a contact time (W / F_P^0) of 146 g·h/mol. Main products of phenol acylation with acetic acid were phenyl acetate (PA), o-HAP, and p-HAP; para-acetoxyacetophenone (p-AXAP) was detected in trace amounts. Phenol conversion (X_P, mol of phenol reacted/mol of phenol fed) was calculated as $X_P = \Sigma Y_i / (\Sigma Y_i + Y_P)$, where ΣY_i is the molar fraction of products formed from phenol, and Y_P is the outlet molar fraction of phenol. The selectivity to product i (S_i, mol of product i/mol of phenol reacted) was determined as: S_i (%) = $(Y_i/\Sigma Y_i)100$. Product yields (η_i, mol of product i/mol of phenol fed) were calculated as $\eta_i = S_i X_P$.

On all of the samples, o-HAP and PA were the predominant products. At similar phenol conversion levels, the initial o-HAP selectivity was between 67.1% (HZSM-5) and 39.1% (SiO$_2$-Al$_2$O$_3$), while the S_{p-HAP}^0 values were always lower than 8%. Figure 15.1 shows the evolution of the formation rate of o-HAP as a function of time onstream. By comparing the experimental data in Figure 15.1 at the beginning of the reaction, it is inferred that zeolites HZSM-5 and HY are clearly more active than the HPA/MCM-41 and SiO$_2$-Al$_2$O$_3$ samples for producing o-HAP from phenol, while Al-MCM-41 shows intermediate r_{o-HAP} values.

By determining the effect of contact time on the product distribution, we identified the primary and secondary reaction pathways involved in the synthesis of o-HAP from phenol and acetic acid [10]. Specifically, we proposed (Figure 15.2) that o-HAP is formed from phenol and AA via two parallel pathways:

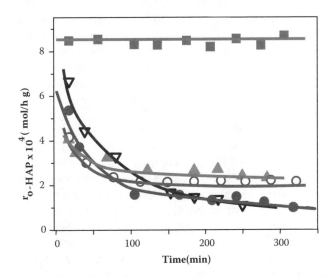

FIGURE 15.1 Formation rate of o-HAP as a function of time onstream on: HZSM-5 (■), HY (▼), HPA/MCM-41 (●), SiO$_2$-Al$_2$O$_3$ (O), Al-MCM-41 (▲) (553 K, 101.3 kPa total pressure, W / F_P^0 = 146 g·h/mol, P/AA = 1, N$_2$/[P + AA] = 45).

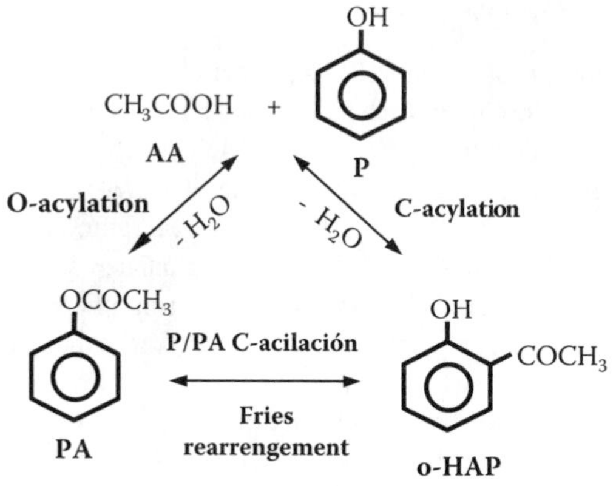

FIGURE 15.2 Synthesis of *o*-HAP by acylation of phenol with acetic acid.

Direct C-acylation of phenol
O-acylation of phenol forming the PA intermediate, which is consecutively
   transformed to *o*-HAP via intramolecular Fries rearrangement or intermolecular phenol/PA C-acylation

The relative rate of the different pathways involved in Figure 15.2 greatly
depends on the solid acid employed. In fact, as shown in Figure 15.1, the *o*-HAP
formation rate was higher on acid zeolites containing strong Brönsted and Lewis
acid sites (zeolites HY and HZSM-5) as compared with samples containing only
Brönsted acid sites (HPA/MCM-41) or exhibiting moderate acidity (SiO$_2$-Al$_2$O$_3$ and
Al-MCM-41). This result suggested that both strong Brönsted and Lewis sites are
required to produce efficiently *o*-HAP via both the direct C-acylation of phenol and
the acylation of phenyl acetate intermediate formed from O-acylation of phenol.

Figure 15.1 also shows that the *o*-HAP formation rate does not change with time
on HZSM-5, but rapidly decreases on the other samples, particularly on HPA/MCM-
41 and HY. Coke formation was determined by analyzing the samples after the catalytic
tests by temperature-programmed oxidation. The amount of carbon on the samples
ranged from 18.4% C on HY to 2.9% C on ZSM5. The %C formed on Al-MCM-41,
SiO$_2$-Al$_2$O$_3$, and HY increased with the sample acidity; i.e., %C was in the order Al-
MCM-41 < SiO$_2$-Al$_2$O$_3$ < HY. On the other hand, it was observed that catalyst deactivation increased with the amount of carbon on the sample, thereby suggesting that
the activity decay for the formation of *o*-HAP is caused by coke formation. Additional
studies were performed to determine the catalyst deactivation mechanism and to
ascertain the causes for the superior stability of zeolite HZSM-5 onstream.

Previous work [11] reported that formation of coke via the irreversible polymerization of highly reactive ketene is the main reason of the rapid deactivation observed
during the liquid-phase Fries rearrangement of PA on solid acids. Ketenes can be
formed via the conversion of PA to P according to Reaction 15.1; these are extremely
reactive and unstable compounds that dimerize to diketenes and polymerize very
quickly. However, we determined that ketenes are not present during the gas-phase

acylation of phenol with AA, because ketenes react rapidly with water formed in reaction (see Figure 15.2) to produce acetic acid [10].

$$ \text{(15.1)} $$

In an attempt to ascertain the nature of the species responsible for coke formation in the synthesis of o-HAP from acylation of P with AA, we studied the conversion of o-HAP with AA on zeolites HY and ZSM-5. The coinjection of o-HAP with AA on HY formed P, PA, and o-acetoxyacetophenone (o-AXAP), and a rapid activity decay was observed. In contrast, HZSM-5 did not produce o-AXAP and did not deactivate during o-HAP/AA conversion reactions. The observed HY deactivation was related therefore with the formation of o-AXAP. Neves et al. [12] reported that coke formed on MFI zeolites during the acylation of phenol with AA is mainly constituted by methylnaphthols, 2-methylchromone, and 4-methylcoumarine. Formation of 2m-cromone and 4m-coumarine can take place from o-HAP and AA via the initial formation of o-AXAP, as depicted in Reaction 15.2:

$$ \text{(15.2)} $$

The assumption that coke is formed essentially via Reaction 15.2 is also consistent with results showing that zeolite HZSM-5, which does not produce o-AXAP when cofeeding o-HAP and AA, does not deactivate. It seems therefore that the superior stability of zeolite HZSM-5 is due to a shape-selectivity effect that avoids formation of the coke precursor species. In other words, zeolite HZSM-5 does not deactivate because its narrow pore-size structure hinders the formation of o-AXAP, which is the key coke precursor in the gas-phase acylation of phenol with acetic acid.

The o-HAP yield on HZSM-5 can be improved by selecting proper reaction conditions. Figure 15.3 shows the evolution of initial o-HAP yield (η^0_{o-HAP}) as a function of contact time over zeolites HZSM-5 and HY. The initial o-HAP yield (η^0_{o-HAP}) increases with W/F^0_P on both zeolites, but formation of o-HAP is clearly favored on HZSM-5 at high W/F^0_P values. On the other hand, Figure 15.4 plots the evolution of η^0_{o-HAP} as a function of P_{AA} over HZSM-5 and Al-MCM-41 samples. By changing the reactant AA/P ratio from 0.5 to 4, the η^0_{o-HAP} is increased from 6.0 to 38.6% on HZSM-5. Figure 15.4 also confirms the superior activity of HZSM-5 to produce o-HAP as compared with Al-MCM-41.

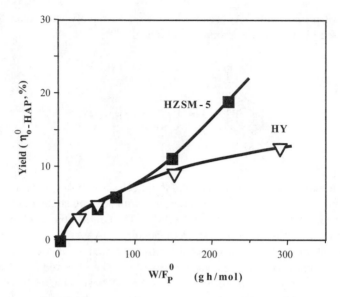

FIGURE 15.3 Initial *o*-HAP yield as a function of contact time (553 K, 101.3 kPa total pressure, P/AA = 1).

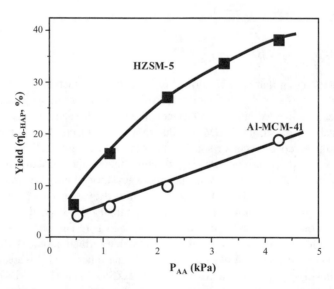

FIGURE 15.4 Initial *o*-HAP yield as a function of PAA (553 K, 101.3 kPa total pressure, PP = 1.10 kPa, W / F_P^0 = 146 g·h/mol).

In summary, our studies show that zeolite HZSM-5 is an active and stable catalyst for efficiently promoting the synthesis of *o*-HAP from phenol acylation in the gas phase. Zeolite HZSM-5 contains strong Lewis and Brönsted acid sites and effectively catalyzes the two main reaction pathways leading from phenol to *o*-HAP, i.e., the direct C-acylation of phenol and the O-acylation of phenol forming the PA intermediate, which is subsequently transformed via intermolecular phenol/PA C-acylation. In addition, on zeolite HZSM-5, the *o*-hydroxycetophenone yield remains stable onstream, and formation of coke is drastically suppressed. The superior stability of zeolite HZSM-5 is due to the fact that the microporous structure of this zeolite avoids the formation of bulky *o*-acetoxyacetophenone, which is the key intermediate for coke formation.

15.3 SUSTAINABLE CATALYTIC PROCESSES IN THE SYNTHESIS OF FINE CHEMICALS

15.3.1 INTRODUCTION

Menthol is a fine chemical of significant industrial interest because it is widely employed in pharmaceuticals, cosmetics, toothpastes, chewing gum, and cigarettes. The menthol molecule is a terpenoid containing three quiral centers that account for the four resulting enantiomers: menthol, isomenthol, neomenthol, and neoisomenthol. Because each individual enantiomer forms two optical isomers — the dextro(+) and levo(−) isomers — menthol comprises eight optically active isomers, but only (−)-menthol possesses the characteristic peppermint odor and exerts a unique cooling sensation on the skin and mucous membranes. Most of (−)-menthol is obtained from natural essential oils, but its production by synthesis has increased lately. In 1998, the production of synthetic menthol was 2500 tonnes (t), which represented about 20% of the world production of menthol [13]. Synthetic menthol is currently produced by two companies, Symrise (previously Haarmann & Reimer) and Takasago. The Haarmann & Reimer process [14, 15] uses thymol as a raw material to obtain racemic (±)-menthol, which is then resolved into pure (−)-menthol by a crystallization process. In the early 1980s, Takasago developed an asymmetric synthesis technology for producing (−)-menthol from myrcene [16]. The key to the Takasago process was the use of a chiral Rh BINAP catalyst to transform the diethylgeranylamine (obtained from myrcene) to the chiral 3R-citronellal enamine with more than 95% enantiomeric excess.

Considerable effort has been devoted to the production of (−)-menthol by synthetic or semisynthetic means from other more readily reliable raw materials. Our research group recently reported [17] for the first time the selective synthesis of menthols from citral in a one-step process that involves the initial hydrogenation of citral to citronellal, followed by the isomerization of citronellal to isopulegols, and the final hydrogenation of isopulegols to menthols. Production of menthols from citral is an attractive synthetic route because citral is a renewable raw material that is mainly obtained by distillation of essential oils, such as lemongrass oil, that contain ca. 70 to 80% citral. However, the citral conversion reaction network potentially involves a complex combination of serial and parallel reactions, as depicted in Figure 15.5. The direct synthesis of menthols from citral requires, therefore, the develop-

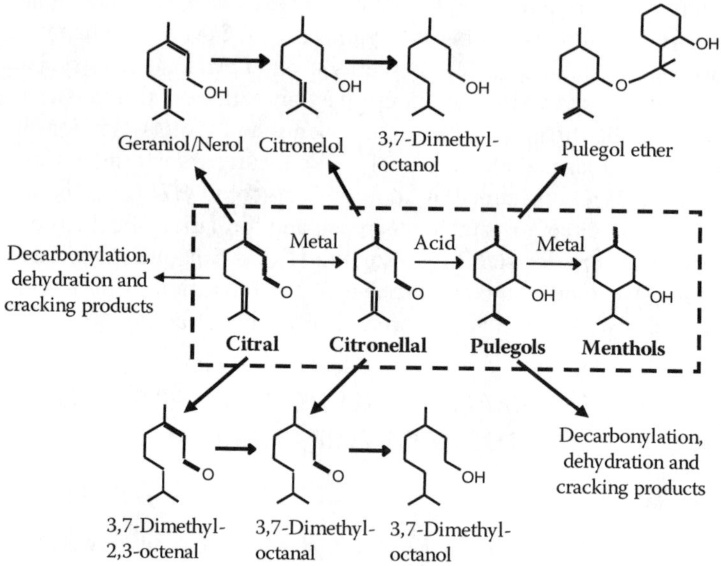

FIGURE 15.5 Reaction network for citral conversion reactions.

ment of highly selective bifunctional metal-acid catalysts. The main results obtained by our research group — with the aim of developing novel bifunctional catalysts to efficiently promote the liquid-phase synthesis of menthols from citral — are described here.

15.3.2 SAMPLE PREPARATION AND CHARACTERIZATION

Hydrogenation of citral to citronellal was studied on different metals (Pt, Pd, Ir, Ni, Co, and Cu) supported on a SiO_2 powder (Grace G62, 99.7%). Metals were supported by incipient-wetness impregnation at 303 K using metal nitrate solutions. Catalysts were characterized by X-ray diffraction (XRD), temperature-programmed reduction (TPR), and hydrogen chemisorption. The catalyst metal loadings and surface areas together with characterization results are shown in Table 15.2.

Isomerization of citronellal to pulegols was carried out on Al-MCM-41 (Si/Al = 10, S_g = 780 m²/g), zeolite Beta (Zeocat PB, Si/Al = 25, S_g = 630 m²/g), ZnO(25%)/SiO_2, and Cs-HPA ($Cs_{0.5}H_{2.5}PW_{12}O_{40}$, S_g = 130 m²/g). Sample preparation details have been reported elsewhere [17]. Acid site densities were determined by deconvolution and integration of TPD of NH_3, and the results are given in Table 15.3. On a weight basis, ZnO/SiO_2 exhibited the highest density of acid sites, probably reflecting the presence of chloride ions on the surface (sample was prepared using an aqueous solution of Cl_2Zn), but the TPD peak maximum appeared at relatively low temperature (about 523 K). The evolved NH_3 from zeolite Beta (HBEA) and Al-MCM-41 gave rise to a single asymmetric broad band with a maximum around 482 to 496 K. Cs-HPA desorbed NH_3 in a single peak centered at about 893 K, reflecting a superior acid site strength compared with the other

TABLE 15.2
Catalyst Characterization and Catalytic Data for Citronellal Hydrogenation

Catalyst	XRD	TPR T_{Max} (K)	S_g (m^2/g)	H_2 Chemisorption (cm^3/mol metal)	S_{Cit}^{max} [a/b] (%)	S_i [b] (%) Citronellal	G-N[c]	Others
Ni(12%)/SiO_2	NiO	665	250	0.53	99	97	0	3
Cu(12%)/SiO_2	CuO	523	218	0.08	53	53	41	6
Co(12%)/SiO_2	CoO	656	240	0.16	14	11	81	8
Pt(0.3%)/SiO_2	—	395	280	1.49	59	59	25	16
Pd(0.7%)/SiO_2	—	393	250	1.98	91	71	0	29
Ir(1%)/SiO_2	—	359	230	1.95	50	32	60	8

Note: T = 393 K, P = 1013 kPa, Wcat =1 g.

[a] Maxima selectivities to citronellal.

[b] Selectivities at 60 min.

[c] Selectivity to geraniol and nerol isomers.

TABLE 15.3
Characterization of Sample Acidity

Catalyst	TPD of NH_3 (μmol/g)	IR of Pyridine B (μmol /g)	L (μmol /g)	L/(L+B)
H-BEA	496	100	120	0.55
Al-MCM-41	110	20	36	0.64
ZnO/SiO_2	2200	0	50	1
Cs-HPA	37	—	—	—

samples. Table 15.3 also shows the nature of the surface acid sites as determined from IR spectra obtained after adsorption of pyridine at room temperature and evacuation at 423 K. As it is well known, Cs-HPA contains only Brönsted sites [18]. Sample ZnO/SiO_2 contained only Lewis sites, while the relative concentration of Brönsted and Lewis acid sites (L/B) on Al-MCM-41 and H-BEA were 1.8 and 1.2, respectively. Results of Table 15.2 and Table 15.3 reveal that zeolite H-BEA contains a higher density of stronger acid sites compared with A-MCM-41.

15.3.3 Catalytic Results

Figure 15.5 shows that the selective formation of menthol from citral requires bifunctional metal/acid catalysts with the ability of not only promoting coupled hydrogenation/isomerization reactions of the citral-to-menthols pathway, but also minimizing the parallel hydrogenation reactions of (a) citral to nerol/geraniol or 3,7-dimethyl-2,3-octenal and (b) citronellal to citronelol or 3,7-dimethyloctanal. In other words, from a kinetic point of view:

The formation rate of citronellal from citral must be much higher than the hydrogenation rates of citral to nerol/geraniol and 3,7-dimethyl-2,3-octenal

The formation rate of pulegols from citronellal must be much higher than the hydrogenation rates of citronellal to citronelol and 3,7-dimethyloctanal

Consequently, the individual steps involved in the reaction pathway leading to menthols from citral were studied separately to select the metallic and acid functions of the bifunctional catalyst.

The monometallic catalysts of Table 15.2 were tested for the liquid-phase hydrogenation of citral (T = 393 K, P = 1013 kPa, W_{cat} = 1 g) using isopropanol as solvent. Pd/SiO$_2$ and Ni/SiO$_2$ selectively hydrogenated the conjugated C=C bond of the citral molecule, initially giving more than 90% selectivity to citronellal (Table 15.2). This result showed that, on both catalysts, the citral hydrogenation to citronellal is clearly favored compared with parallel hydrogenations leading to nerol/geraniol and 3,7-dimethyl-2,3-octenal. The citronellal selectivity then decreased with reaction time because citronellal is, in turn, hydrogenated to citronelol or 3,7-dimethyloctanal. In contrast, the maximum selectivity to citronellal was never higher than 60% on the other catalysts, which formed significant amounts of nerol/geraniol isomers. Overall, the results in Table 15.2 are consistent with previous works on citral hydrogenation showing that Ni and Pd favor C=C bond hydrogenation [19, 20], while Co and Ir are more selective for C=O bond hydrogenation [20, 21].

The solid acids of Table 15.3 were tested in the liquid-phase isomerization of citronellal to isopulegols (T = 343 K, P_{N2} = 506.5 kPa, W = 0.200 g, citronellal:toluene (ml) = 2:150) using toluene as solvent. Isopulegol yields are shown in Figure 15.6 as a function of Wt / n^0_{Clal} , where W is the catalyst weight, t the reaction time, and n^0_{Clal} the initial moles of citronellal. The local slope of each product in Figure 15.6 gives its rate of formation at a specific value of reactant conversion and contact time. In all cases, pulegol isomers were the only products detected.

Figure 15.6 shows that the citronellal cyclization rate was clearly higher on zeolite Beta and Al-MCM-41 as compared with both ZnO/SiO$_2$ and Cs-HPA. The exact nature of the surface-active sites required for efficiently catalyzing the cyclization of citronellal to isopulegols is still debated. While several authors [22, 23] reported that the reaction is readily catalyzed on Lewis acids, others [24] correlated the cyclization activity on acid zeolites with accessible Brönsted acid sites. Chuah et al. [25] found that catalytic materials containing strong Lewis and weak Brönsted acidity show good activity and selectivity for cyclization of citronellal to isopulegol. The solid acids listed in Table 15.3 contain either Lewis (ZnO/SiO$_2$), Brönsted (Cs-HPA), or both Lewis and Brönsted acid sites (Al-MCM-41 and H-BEA). The superior activity showed by zeolite Beta and Al-MCM-41 samples for the formation of isopulegols are consistent with the assumption that Lewis/weak Brönsted dual sites are required to efficiently catalyze the citronellal cyclization.

Based on the above results, three bifunctional catalysts containing one of the metals most selective for hydrogenating citral to citronellal (Pd or Ni) and one of the solid acids more active for converting citronellal to isopulegols (zeolite Beta or Al-MCM-41) were prepared: Pd(1%)/H-BEA, Ni(3%)/H-BEA, and Ni(3%)/Al-MCM-41. These bifunctional catalysts were tested for the conversion of citral to

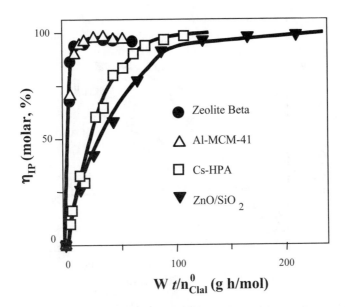

FIGURE 15.6 Cyclization of citronellal to isopulegols: isopulegol yield as a function of parameter Wt / n_{Clal}^{0} (343 K, 506.5 kPa nitrogen, W = 0.200 g, citronellal:toluene = 2:150 [ml]).

menthols (T = 343 K, P_T = 506.5 kPa, W = 1 g) using toluene as solvent. In all cases, citral and citronellal were totally converted after 5 h of reaction. Figure 15.7 shows the evolution of the yield of total menthols as a function of reaction time. It is observed that the menthol yield was only about 20% on Pd/H-BEA catalyst at the end of the catalytic test. This poor selectivity for menthol synthesis reflected the high activity of Pd/H-BEA to hydrogenate the C=C bond of citronellal, thereby forming considerable amounts of 3,7-dimethyloctanal. Pd/H-BEA also formed significant amounts of undesirable products (35%) via secondary decarbonylation and hydrogenolysis reactions. In contrast, Figure 15.7 shows that the menthol selectivity on Ni/H-BEA was 81% at the end of the catalytic test. None of the by-products formed from hydrogenation of citral or citronellal was detected on Ni/H-BEA, suggesting that this bifunctional catalyst satisfactorily combines the hydrogenation and isomerization functions needed to selectively promote the reaction pathway leading from citral to menthols. However, formation of secondary compounds, formed probably via decarboxylation and cracking reactions on the strong acid sites of zeolite Beta, was significant. The best catalyst was Ni/Al-MCM-41, which yielded ca. 90% menthol. The observed yield improvement for menthol on Ni/Al-MCM-41 is explained by considering that the moderate acid sites of Al-MCM-41 do not promote the formation of by-products via side cracking reactions. Table 15.4 shows the distribution of menthol isomers obtained at the end of catalytic runs. (±)-Neoisomenthol was never detected in the products. On Ni-based catalysts, the menthol mixture was composed of 70 to 73% of (±)-menthols, 15 to 20% of (±)-neomenthol, and 5 to 10% of (±)-isomenthol. On Pd/Beta the racemic (±)-menthol mixture represented only about 50% of total menthols.

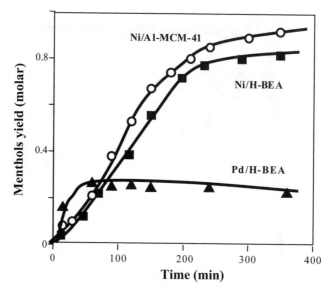

FIGURE 15.7 Menthol synthesis from citral: total menthol yields as a function of time (343 K, 506.5 kPa total pressure, W = 1 g, citral:toluene = 2:150 [ml]).

TABLE 15.4
Menthol Synthesis from Citral: Menthol Isomer Distribution

| Catalyst | Pressure (kPa) | Menthol Isomer Distribution (%) IR of Pyridine | | |
		(±)-Menthols	(±)-Neomenthol	(±)-Isomenthol
Pd/H-BEA	506.5	47.2	15.6	37.2
Ni/H-BEA	506.5	72.0	21.3	6.7
Ni/Al-MCM-41	506.5	72.3	20.2	7.5
Ni/Al-MCM-41	2026.0	71.1	20.0	8.9

Note: 343 K, W = 1 g, citral:toluene = 2:150 (ml).

Finally, an additional test was performed on Ni/Al-MCM-41 by increasing the hydrogen pressure to 2026 kPa. Results are presented in Figure 15.8 and Table 15.4. Figure 15.8 shows the evolution of product yields and citral conversion as a function of time. Citral was totally converted to citronellal on metallic Ni crystallites, but the concentration of citronellal remained very low because it was readily converted to pulegols on acid sites of mesoporous Al-MCM-41 support. Pulegols were then totally hydrogenated to menthols on metal Ni surface sites. The menthols yield reached 94% at the end of the test, showing the beneficial effect of increasing P_{H2}, probably because it diminishes the formation of undesirable products via secondary reactions. In contrast, menthol isomer distribution was not changed by increasing the hydrogen pressure (Table 15.4).

In summary, the results presented here show that the liquid-phase synthesis of menthols from citral was successfully achieved using proper bifunctional catalysts.

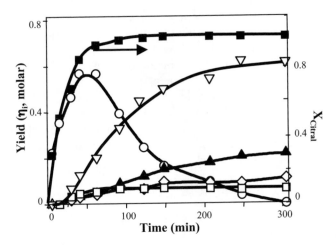

FIGURE 15.8 Menthol synthesis from citral on Ni(3%)/Al-MCM-41. Product yields and citral conversion as a function of time. Isopulegols (○), (±)-menthol (◻), (±)-neomenthol (▲), (±)-isomenthol (◇), others (□) (343 K, 2026 kPa total pressure, W = 1 g, citral:toluene = 2:150 [ml]).

In fact, at P_{H2} = 2026 kPa, Ni(3%)/Al-MCM-41 yields 94% menthols directly from citral and produces about 72% of racemic (±)-menthol into the menthol mixture.

15.4 BIODIESEL FROM LOW-VALUE RAW MATERIALS

15.4.1 INTRODUCTION

Interest in the use of renewable resources for fuel and feedstocks has increased lately due to the rising price of crude oil and concerns over production of carbon dioxide. Biodiesel is a nontoxic and biodegradable renewable fuel obtained from vegetable oils or animal fats that possesses physical and fuel properties similar to those of conventional oil-derived diesel fuel [26]. However, biodiesel has important advantages compared with petroleum diesel. For example, it is oxygenated, contains no sulfur, reduces unburnt and particulate matter in the exhaust, and does not cause a net increase of carbon dioxide in the atmosphere because it contains photosynthetic organic carbon [27, 28]. Commercial biodiesel is produced from renewable resources including rapeseed, sunflower, palm, or soybean oils, which are essentially edible in nature and contain almost 90 to 95% of fatty acid triglycerides [29]. These are converted to biodiesel by transesterification with short-chain alcohols, typically methanol or ethanol, using homogeneous basic catalysis [30].

For economic reasons, biodiesel has not been produced in Argentina on a large scale since December 2001. However, there are small-scale commercial units that produce biodiesel from low-value sources such as recycled frying oil, acid tallow, and brown greases. These inedible materials are usually very acid and contain free fatty acids that, in presence of a base and water, form soaps. Formation of soaps hampers the use of conventional base-catalyzed transesterification processes for efficient production of biodiesel. On the other hand, acid catalysts transform fatty acids and triglycerides into methyl esters [31], which react very slowly with methanol in acid media. Consequently, a process based only in acid catalysis cannot be carried out at a commercial scale. Thus, technological processes for obtaining biodiesel require several transforming stages when starting from very acid sources [32]. In Argentina, the research group directed by Querini (INCAPE, Santa Fe) has explored the use of a two-step process that consecutively employs acid and basic catalysts to efficiently obtain biodiesel from low-value acid raw materials [33]. The main aspects of these studies related to the processes developed for Argentine companies that manufacture biodiesel are presented.

15.4.2 Materials and Methods

Different raw materials of high acidity (A) were employed: beef tallow (A = 17), oil of Paraguayan coconut (A = 12), recycled vegetable oil (A = 35), and brown fat (A = 55). The acidity represents the amount of free acids, expressed as grams of oleic acid per 100 grams of material, and was determined by titration, dissolving the raw material in a mixture of toluene/ethanol and using NaOH and phenolphthalein as indicator. The amount of water in the samples was determined by the method of Dean Stark (ASTM D95). The biodiesel properties — total and free glycerin content, viscosity, flash point, pour point, cloud point, and sulfur and methanol contents — were determined following ASTM 6515 methods of analysis. The fatty acid ester composition was determined by gas chromatography.

The transesterification reaction for obtaining biodiesel from raw materials with acidity higher than 3 involved the following steps:

Reaction with methanol and acid catalysis, using sulfuric acid
Product separation
Reaction with methanol and alkaline catalysis
Separation and purification of biodiesel

Reactions were carried out in a batch reactor with reflux at temperatures between 333 and 343 K. In general, the reaction time was long enough to reach equilibrium conversions.

15.4.3 Catalytic Results

Biodiesel is an alkyl ester (R-COO-CH$_3$) mixture that is obtained by transesterification of triglycerides and free fatty acids (R-COOH) with short-chain alcohols such as methanol. The reaction scheme involved when the process uses methanol and is catalyzed by liquid bases is:

$$\text{Triglyceride} + 3\ CH_3OH \underset{}{\overset{NaOH}{\rightleftarrows}} \text{Glycerine} + 3\ R\text{-}COO\text{-}CH_3 \qquad (15.3)$$

$$\text{Triglyceride} + H_2O \underset{}{\overset{NaOH}{\rightleftarrows}} \text{Glycerine} + R\text{-}COO\text{-}Na \qquad (15.4)$$

$$R\text{-COOH} + NaOH \rightleftharpoons R\text{-COO-Na} + H_2O \qquad (15.5)$$

Strongly acidic materials contain a high concentration of free fatty acids that are transformed to soaps (R-COO-Na) and water via Reaction 15.5. Water formed in Reaction 15.5 promotes in turn the production of additional soaps via Reaction 15.4, which consumes the catalyst and reduces catalyst efficiency. Thus, for materials of high acidity, the scheme of Reactions 15.3 to 15.5 leads to low biodiesel yields. Besides, the presence of soap hinders biodiesel separation from the glycerine fraction [34]. Conversion of fatty acids via Reaction 15.5 can be avoided by pretreating the raw material with a strong acid, according to the following reaction scheme:

$$R\text{-COOH} + CH_3OH \xrightarrow{H_2SO_4} R\text{-COO-CH}_3 + H_2O \qquad (15.6)$$

$$Triglyceride + CH_3OH \xrightarrow{H_2SO_4} R\text{-COO-CH}_3 + glycerine \qquad (15.7)$$

Acid-catalyzed Reactions 15.6 and 15.7 convert fatty acids to biodiesel and water. Then, two phases are separated: a triglyceride-rich phase, and a methanol-rich phase containing most of the water formed through Reaction 15.6. Finally, the triglyceride-rich phase is transformed to biodiesel using basic catalysis according to Reaction 15.3.

To obtain high biodiesel yields when using strongly acid raw materials, it is required to efficiently convert the free fatty acids in a first step according to Reactions 15.6 and 15.7. Results obtained by Querini's research group [33] on this initial acid-catalyzed process for reducing the free fatty acids content for maximum biodiesel production are presented below.

The equilibrium constant (K) of fatty acid esterification is calculated using Equation 15.8:

$$K = \frac{\left(A_0 - A_f\right)^2 k(\rho, PM)}{A_0 \left(V_{MeOH} / V_{trig}\right)} \qquad (15.8)$$

where A_0 and A_f represent the initial and final acidity, respectively; k (ρ, PM) groups properties of the system; and V_{MeOH} and V_{trig} are the initial volume of methanol and triglyceride in the reaction, respectively.

Figure 15.9 shows the equilibrium constant values of the esterification of fatty acids obtained for several raw materials. Fatty acids of raw materials of Figure 15.9, excepting coconut, are all formed mainly by C18, followed by C16, and in smaller proportion by C14 and C20. The typical chain lengths of these fatty acids are similar then, and as a consequence their equilibrium constants are approximately the same. In contrast, the fatty acid distribution of coconut oil exhibits a predominant proportion of C12, followed by C10 and C8, which results in a smaller carbon chain length as compared with the other materials of Figure 15.9 and thus

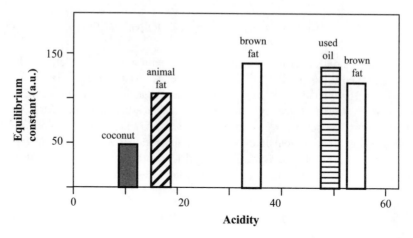

FIGURE 15.9 Esterification equilibrium constants determined for raw materials of different acidity.

a lower esterification equilibrium constant. Determination of K values is useful for selecting the methanol/raw-material ratio needed in each individual acid-catalyzed stage in order to diminish the raw material acidity to the values required for the final base-catalyzed step.

Figure 15.10 shows the evolution of the acidity as a function of time for a sample obtained by breakage of a vegetable oil emulsion. Figure 15.10 shows two different experiences carried out under the same experimental conditions using methanol (60%v) and sulfuric acid (0.17% H_2SO_4). Similar A vs. t curves were obtained, thereby indicating that the acid-catalyzed reaction can be satisfactorily reproduced.

Figure 15.11 shows the effect of H_2SO_4 concentration on the acidity decrease rate when using a chicken fat sample as raw material; sulfuric acid concentrations of 0.14 and 0.21% were used. The initial acidity of the sample ($A_0 = 65$) dropped

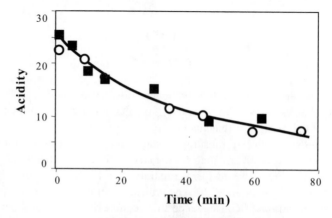

FIGURE 15.10 Evolution of acidity as a function of time for a degumming residue of sunflower oil (T = 333 K, 0.17% H_2SO_4, CH_3OH/raw material = 6).

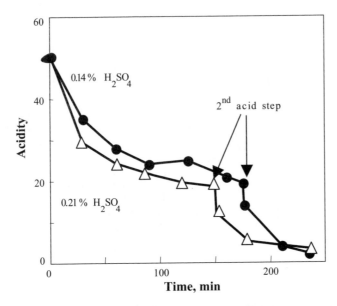

FIGURE 15.11 Effect of H_2SO_4 concentration on the acidity decrease rate (chicken fat, T = 333 K, CH_3OH/raw material = 6).

to about 50 after dilution with methanol. As expected, the activity decline was faster when using higher acid concentrations, but similar A values were obtained at the end of the catalytic runs, suggesting that the equilibrium was reached for both cases. After phase separation, a second acid-catalyzed step was carried out using the same acid concentrations as in the initial step. It is observed in Figure 15.11 that at the end of this second reaction stage the acidity of the triglyceride-rich phase was below 2, thereby indicating that this phase can now be efficiently converted to biodiesel via a final base-catalyzed step.

Esterification of fatty acids with ethanol 96% in the presence of sulfuric acid was also investigated. The use of ethanol 96%, less expensive than absolute ethanol, is an attractive alternative for the acid-catalyzed reaction because the presence of water does not promote the formation of soap, as explained above. Figure 15.12 shows results obtained when a beef tallow sample (A = 4.9) was reacted in acid media with ethanol 99.5% and 96%, respectively. It is observed that quasi-stable acidity values are reached in both cases after 60 min, suggesting that fatty acids conversion is limited by reaction equilibrium. Because quasi-equilibrium conditions are reached, the final A value was higher when using ethanol 99.5%, as predicted by Reaction 15.4. But it should be noted that even with ethanol 96%, the acidity of a raw material can be diminished by transforming free fatty acids into biodiesel.

In summary, low-value raw materials of high acidity can be efficiently transformed to biodiesel via a two-step reaction process involving a first acid-catalyzed stage for converting free fatty acids and a final base-catalyzed for converting triglycerides into biodiesel. Design of the initial acid-catalyzed process is based on

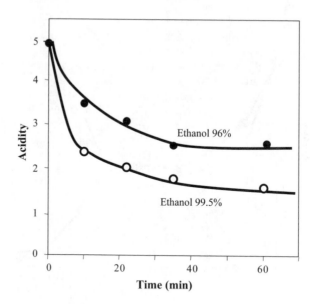

FIGURE 15.12 Esterification with ethanol: effect of the water content (beef tallow, 0.21% H_2SO_4, T = 333 K, C_2H_5OH/raw material = 6).

the determination of the equilibrium constant of fatty acid esterification, which can be conveniently used to select the alcohol/raw-material ratio and the number of required stages. Biodiesel that meets all the requirements of standards has been produced from brown fats of acidity higher than 50 and from emulsions derived from degumming of different types of vegetable oils.

REFERENCES

1. Commarieu, A., et al., Fries rearrangement in methane sulfonic acid, an environmental friendly acid, *J. Mol. Catal. A: Chemical*, 182, 137, 2002.
2. Uwaydah, I. et al., U.S. Patent 5,696,274, 1997.
3. Fritch, J., Fruchey, O., and Horlenko, T., U.S. Patent 4,954,652, 1990.
4. Vogt, A., Kouwenhoven, H., and Prins, R., Fries rearrangement over zeolitic catalysts, *Appl. Catal.*, 123, 37, 1995.
5. Jayat, F., Sabater Picot, M.J., and Guisnet, M., Solvent effects in liquid phase Fries rearrangement of phenyl acetate over a HBEA zeolite, *Catal. Lett.*, 41, 181, 1996.
6. Mueller, J. et al., U.S. Patent 4,508,924, 1985.
7. Freese, U., Hinrich, F., and Roessner, F., Acylation of aromatic compounds on H-Beta zeolites, *Catal. Today*, 49, 237, 1999.
8. Subba Rao, Y.V. et al., An improved acylation of phenol over modified ZSM-5 catalysts, *Appl. Catal. A: General*, 133, L1, 1995.
9. Jayat, F. et al., Acylation of phenol with acetic acid: effect of density and strength of acid sites on the properties of MFI metallosilicates, in *Stud. Surf. Sci. Catal.*, Vol. 105, Chon, H., Ihm, S.-K., and Uh, S., Eds., Elsevier, Amsterdam, 1997, p. 1149.

10. Padró, C.L. and Apesteguía, C.R., Gas-phase synthesis of hydroxyacetophenones by acylation of phenol with acetic acid, *J. Catal.*, 226, 308, 2004.

11. Heidekum, A., Harmer, M.A., and Hölderich, W.F., Highly selective Fries rearrangement over zeolites and Nation in silica composite catalysts: a comparison, *J. Catal.*, 176, 260, 1998.

12. Neves, I. et al., Acylation of phenol with acetic acid over HZSM-5 zeolite, reaction scheme, *J. Molec. Catal.*, 93, 169, 1994.

13. Clark, G.S., Menthol, *Perfumer Flavorist*, 23, 33, 1998.

14. Fleischer, J., Bauer, K., and Hopp, R., German Patent, DE 2,109,456, 1971.

15. Davis, J.C., *Chem. Eng.*, 11–12, 62, 1978.

16. Misono, M. and Nojiri, N. Recent progress in catalytic technology in Japan, *Appl. Catal.*, 64, 1, 1990.

17. Trasarti, A.F., Marchi, A.J., and Apesteguía, C.R., Highly selective synthesis of menthols from citral in a one-step process, *J. Catal.*, 224, 484, 2004.

18. Kozhevnikov, I.V., Catalysis by heteropoly acids and multicomponent polyoxometalates in liquid-phase reactions, *Chem. Rev.*, 98, 171, 1998.

19. Aramendía, M.A. et al., Selective liquid-phase hydrogenation of citral over supported palladium, *J. Catal.*, 172, 46, 1997.

20. Maki-Arvela, P. et al., Liquid phase hydrogenation of citral: suppression of side reactions, *Appl. Catal. A: General*, 237, 181, 2002.

21. Singh, U.K. and Vannice, M.A., Liquid-phase citral hydrogenation over SiO_2-supported group VIII metals, *J. Catal.*, 199, 73, 2001.

22. Ravasio, N. et al, Intramolecular ene reactions promoted by mixed cogels, in *Stud. Surf. Sci. Catal.*, Vol. 108, Blaser, H.U., Baiker, A., and Prins, R., Eds., Elsevier, Amsterdam, 1997, p. 625.

23. Milone, C. et al., Isomerisation of (+)citronellal over Zn(II) supported catalysts, *Appl. Catal. A: General*, 233, 151, 2002.

24. Fuentes, M. et al., Cyclization of citronellal to isopulegol by zeolite catalysis, *Appl. Catal.*, 47, 367, 1989.

25. Chuah, G.K. et al., Cyclisation of citronellal to isopulegol catalysed by hydrous zirconia and other solid acids, *J. Catal.*, 200, 352, 2001.

26. Dunn, R.O. and Knothe, G., Alternative diesel fuels from vegetable oils and animal fats, *J. Oleo. Sci.*, 50, 415, 2001.

27. Maa, F. and Hannab, M.A., Biodiesel production: a review, *Bioresource Technol.*, 70, 1, 1999.

28. Bondioli, P. et al., Storage stability of biodiesel, *JAOCS*, 72, 699, 1995.

29. Dmytryshyn, S.L. et al., Synthesis and characterization of vegetable oil derived esters: evaluation for their diesel additive properties, *Bioresource Technol.*, 92, 55, 2004.

30. Lang, K. et al., Preparation and characterization of biodiesel from various bio-oils, *Bioresource Technol.*, 80, 53, 200.

31. Freedman, B., Butterfield, R.O., and Pryde, E.H., Transesterification kinetics of soybean oil, *JAOCS*, 63, 1375, 1986.

32. Alcantara, R. et al., Catalytic production of biodiesel from soybean oil, used frying oil and tallow, *Biomass Bioenergy*, 18, 515, 2000.

33. Pisarello, M.L. et al., Biodiesel: producción a partir de materias primas de alta acidez, in *Proc. XIX Iberoamerican Symposium on Catalysis*, Mérida, México, 2004, p. 63.

34. Demirbas, A., Biodiesel fuels from vegetable oils via catalytic and non-catalytic supercritical alcohol transesterifications and other methods: a survey, *Energy Convers. Manage.*, 44, 2093, 2003.

16 Marketing Photovoltaic Technologies in Developing Countries

Gajanana Hegde and Chem V. Nayar

CONTENTS

16.1 INTRODUCTION

So-called developing countries are home to more than three-fourths of the world population while accounting for less than a third of the global commercial energy consumption. This is no surprise, given that an estimated 1.64 billion people worldwide live without access to grid electricity. Extension of the electricity grids to many of these locations is not a realistic option in the near future due to high costs, low energy demands, and the dispersed nature of the mainly rural communities. In the past, grid connections were regularly established in remote areas, even if economically not viable, primarily motivated by political priorities and a lack of alternatives. A World Bank study that included a variety of countries in the Asian and Latin American region concluded that an average US $10,000 per kilometer was being spent on grid extension.[1] Meeting the broad development needs in these nations places numerous competing demands on limited financial resources. Because electrification is just one of these demands, it is even more important that alternative approaches be researched, validated, and reviewed to further the efficient use of resources for sustainable energy supply.

In recent times, select renewable energy sources have matured to play an increasingly important role in meeting energy demand. Issues such as energy independence and mitigation of greenhouse gas emissions from energy generation are becoming important to the nations around the world. Coupled with these attributes, renewable energy technologies (RETs) are rapidly advancing into the mainstream of sustainable development initiatives after being at the periphery for the last couple of decades. Many of the RETs have proved to be economically viable options for electrification and have considerable potential to meet the needs of rural populations in a sustainable way.

The characteristic distributed nature of renewable energy sources requires local installation, operation, and maintenance capabilities, which implies a vital role for local entrepreneurs in the development of the market. Further, the modular nature of the technology allows sizing of systems to the present need of the end user for an initial investment that would be much less than the investment in major infrastructure such as grid extension for the predicted energy needs for the next couple of decades.

Despite the reasons above, it is only recently that significant market forces have come into play for the deployment of RE technologies. The market-development approach for renewable energy has undergone a paradigm shift. In the past, technical demonstration projects found favor with the donor agencies and were important markets for the industry. In this technology-driven paradigm, the focus was on assessment and demonstration of a technology and, to a lesser extent, the life-cycle cost. The new paradigm focuses on market assessment, policy and institutional issues, and demonstrations of sustainable business and social models.

In this chapter, we look at the factors that need consideration before embarking on a project to market photovoltaic (PV) systems, different types of implementation models that are currently being employed, their merits and demerits, and finally the financing of PV projects. Also discussed is a case study of a solar pumping initiative. An overview of the status of the PV market in South and Southeast Asian countries is also presented.

16.2 PV APPLICATIONS: PRESENT STATUS AND EMERGING TRENDS

PV systems that are being currently deployed can be broadly classified as off-grid domestic systems, off-grid nondomestic systems, and grid-connected systems (distributed and centralized). Off-grid systems also include hybrid energy systems when PV is combined with other renewable or fossil fuel sources.

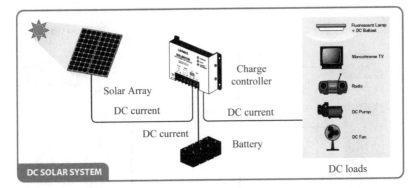

FIGURE 16.1 Solar home system (DC).

Off-grid domestic systems to provide electricity to households and villages that are not connected to the utility grid have been widely installed world-wide. Usually electricity is provided for lighting and other low-power loads. Figure 16.1 and Figure 16.2 show DC and AC solar-powered home systems, respectively. Systems ranging in size from a few watts to a few hundred watts have been installed. Solar home systems (SHS) for households (typically 20 to 150 W) and village power stations (typically 500 to 2500 W) are some of the types of systems that are being deployed. Solar home systems can displace or reduce the need for candles and kerosene in rural homes. An estimated 1.1 million solar home systems and solar lanterns have been installed in rural areas of developing countries.

Off-grid nondomestic systems provide power for a wide range of applications, such as telecommunications, water pumping, vaccine refrigeration, and navigational aids. These are remote-area applications where small amounts of electricity have a high value, thus making PV commercially cost competitive with other small generating sources.

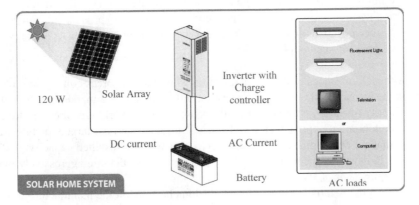

FIGURE 16.2 Solar home system (AC).

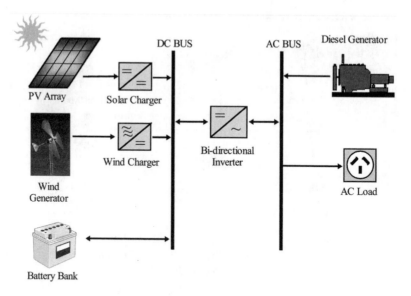

FIGURE 16.3 Hybrid energy system.

A combination of renewable energy sources — such as photovoltaic arrays, biogas generators, or wind turbines — with engine-driven generators and battery storage are generally classified as hybrid energy systems (Figure 16.3). Just as for PV off-grid systems, the potential market for hybrid systems is considered to be huge, in the GWe (gigawatt-electric) range, especially for pumping systems.

Grid-connected PV systems can be grid-connected-distributed or grid-connected-centralized systems. Distributed systems supply power to a building or other loads that are also connected to the utility grid. Typical systems are between 1 kW and 100 kW in size. Electricity is often fed back into the utility grid when the on-site generation exceeds the building loads. Grid-connected centralized systems have been installed as an alternative to conventional centralized power generation or to strengthen the utility distribution system. Experience with grid-connected systems in weak grid areas has been limited, and efforts are under way to develop inverter technology suitable for such applications.

Although instances of commercialization when market forces coming into play are on the increase (for example, in Kenya, a self-sustaining PV market has thrived), the majority of the PV systems currently being installed worldwide are still driven from the top-down approach (i.e., subsidized) through national targets or bilateral or multilateral development programs. Table 16.1 lists the installed capacity of PV systems in developing countries, mainly Asian countries, that are aggressively pursuing deployment of PV systems.

Among the Asian countries, India and China have led in the manufacture of PV equipment for many years, while Taiwan has recently emerged as a strong

TABLE 16.1
PV Systems Status in Asia and Africa

Country	Total Installed Capacity (MW)	Annual Sales (MW) and Key Targets
India	83	16 MW (2003); 280 MW of solar power by 2012 (includes solar thermal)
China (incl. Tibet)	58	10.5 MW (2003); 450 MW by 2010
Indonesia	28	—
Morocco	7.1	(projected 15 MW by 2010)
Thailand	6	250 MW by 2011, includes 300,000 SHS (36 MW) by end of 2005
Vietnam	5.4	—
Kenya	3.2 (1999)	>0.5 MW (around 20,000 SHS)
Nepal	2.7	0.7 MW (around 20,000 SHS)
Sri Lanka	40,000 SHS	1 MW (around 22,000 SHS)

manufacturing center with ambitious plans for future growth.[2] Thailand also appears set for a significant manufacturing growth in the next few years.

TATA BP Solar, a joint venture between the Tata Power Company and BP Solar, is the leading producer in India and accounted for nearly 50% of national production in 2003. The firm has a module-manufacturing capacity of as high as 38 MW and clocked a sales turnover of US $107.5 million in 2004–2005, with exports reaching US $77.26 million. WEBEL, Maharishi Solar Technology, Bharat Electronics, Bharat Heavy Electrical, Central Electronics, and Udhaya Semiconductors are the other significant Indian producers whose manufacturing capabilities include modules, cells, and wafers manufactured from ingots produced in-house.

Despite huge market potential, the manufacturing capacity of Chinese companies has been limited so far, but the situation is expected to change soon. Combined annual production in 2003 was about 8 MW in China, and over 96% of the output was crystalline silicon cells. Nevertheless, production capacity is expected to increase to over 100 MW in 2005. Baoding Yingli is currently the largest Chinese PV cell/module manufacturing firm. Wuxi Snitch Solar Power Company and Xi'an Jiayang are other significant players. Kyocera has established a PV-module manufacturing plant in Tianjin that uses cells imported from the Japanese parent company. Shandong Linuo and Shenzhen Clean Energy Company are both constructing cell/module production plants with capacities of 10 to 12 MW.

In Taiwan, Motech is rapidly emerging as an important international cell manufacturer, increasing production to 17 MW in 2003. The company's cells are supplied to module manufacturers worldwide. Planned further expansions are expected to take production capacity to beyond 50 MW by 2005. In Thailand, Solartron has established a 15-MW module manufacturing (encapsulation) facility, while Bangkok Solar has bought the Hungarian firm Dunasolar's 5-MW amorphous-silicon manufacturing plant.

As can be seen in Table 16.1, many countries have ambitious targets for PV system capacity, and growth rate of PV system deployment has been impressive in many countries and spectacular in some countries. Hence, there appears to be enormous emerging potential for marketing PV systems. In the next section, we look at various models that have been employed in PV programs worldwide. These models are indicative only; many times, what is actually implemented in the field may be a combination of more than one model.

16.3 MARKET DEVELOPMENT OF PV SYSTEMS: SUMMARY OF MODELS FOR IMPLEMENTATION

Studies looking into market introduction of products have identified distinctive stages of customers (Figure 16.4).[3] "Pioneer customers" make up for 2.5% of the market; they are willing to pay a high price for a new and attractive product or solution, even if it is not economical. The pioneer customers influence the "opinion leaders," and if a critical number of pioneer customers have developed, then the stage is set for the entry of opinion leaders. Pioneer customers could be the governments or educational institutions such as universities, but they could also include individual investors. Opinion leaders, who make up 13.5% of the market, include professionals such as engineers, doctors, lawyers, etc., who have enough free cash to spare. The "early majority" account for 34% of the market, and these investors expect the product to be technologically mature and economically viable. A significant 34% of the market is termed "late majority," and they are hard to catch. It is almost impossible to sell the product to the "hesitators," who make up 16% of the market.

PV systems, which are relatively new for many regions, have to pass through these stages before achieving an expanded market; however, many factors have proved to be advantageous for rapid implementation of PV projects. These are discussed in the next section.

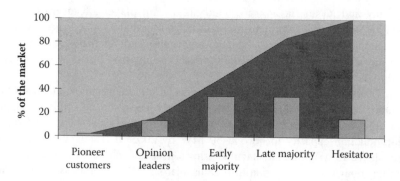

FIGURE 16.4 Market development of a new product.

16.3.1 CONSIDERATIONS IN SELECTING AN IMPLEMENTATION MODEL

An informed choice of implementation model tailored to local conditions would be crucial for developing a new market for PV products. Some of the factors that merit consideration before embarking on a PV project include:

- The electricity and energy service needs and expectations of the end users, the competing/conventional practices to cover the needs, and the expenditure for it.
- Economic activity and source of income (agriculture, services etc.) of the end users, the seasonal influences on income, and the willingness to pay for renewable energy services.
- Geographic location of the end users and the transport/communication infrastructure in the region.
- Experience of the end users with credit and the existence of microcredit organizations.
- Opportunities to enhance productive use of electricity/energy by the end users.
- Government policies toward rural electrification, existing electrification plans, and the potential role for PV.
- Role for private entities to generate and provide electricity.
- Role of utilities in electrification; their attitude, approach, image, and relationship with the customers. (Utilities are increasingly being restructured in many countries.)
- Cost of grid connection, the lead time to secure grid connection, and the current electricity fees for customers and subsidies within the energy and electricity sector. Government policies are only as good as the manner in which they are implemented. Hence, assessing a level playing field for different technology options is an important step (subsidies on fuel, PV component, grid extension; exemption of import duties; etc.).
- Ways of reaching the potential market; the distribution, installation, and servicing network for the hardware; and the collection of payments from the end users for systems sold on credit. It might be advantageous to use any existing infrastructure (e.g., agricultural cooperatives). Commercial practices suitable for PV implementation should be harmonized with prevailing practices and not disturb the often-fragile existing economies.[4] It is thus necessary to identify banks and other credit institutions with local presence, PV dealers/companies, other retail networks (who may not be dealing with PV presently, but who might have strong credibility with the user groups, for instance a rural retail shop for agricultural tools or irrigation equipment).
- The experience (success/failure) of the existing renewable-energy companies and the existing/ongoing electrification or renewable-energy programs and initiatives.

16.3.2 Cash Sales Model

In this model, the end user purchases the system by presenting full payment for the cost of the system to a PV supplier. The supplier may assume the responsibility of installation with an eye toward ensuring the long-term sustainability of the system, and the supplier's desire to preserve its reputation as a reliable vendor is also a factor. However, in many cases, the end user assumes the responsibility for installation, given that the initial cost of the system is often the biggest consideration in this transaction. The operation and maintenance of the system is the responsibility of the end user.

The criticism of this model is that it is prone to the initial-investment barrier, resulting in a small market, and it also tends to encourage the sale of smaller products, such as solar lanterns, or of cheaper, low-quality systems. The purchasing power of the end user might also have strong seasonal fluctuations, for instance, during post-harvest periods in agricultural societies. Further, unlike most other after-sales agreements, the PV supplier is expected to honor the warranty on its PV module for a lengthy period (as many as 20 years), enforcement of which might be difficult. Competition with cheap, low quality products is a problem, especially if the market is just starting, as there is no common knowledge within the market yet about good- and poor-quality brands. Insufficient attention to end-user training for operation and maintenance is yet another issue that can affect this model.

A factor in favor of this model is the limited number of players involved, which keeps the model simple and the transaction cost to a minimum. It is also self-sustaining, as it relies on market forces and is not dependent on external sources such as government or program support. Finally, this model can promote the growth of local infrastructure for installation, maintenance, after-sales services, and even some of the accessories or balance of systems that go with a PV system. Capital demand for the PV supplier is also among the lowest for this model.

16.3.3 Dealer Credit/End-User Credit Model

To expand its market, a PV supplier might choose to offer the PV system on credit, either with its own funds or from borrowed funds. Normally, these kinds of end-user credits are for short terms (mostly between 1 month and 1 year), involve high up-front payments (up to 75%), and relatively high interest rates (rates can be in the range of 15 to 25%). In some cases, the credit is through an informal arrangement involving the local representative/dealer of the PV supplier and the end-user client based on mutual trust. There is considerable experience in most countries in consumer-credit systems that are used to sell/buy consumer durables such as televisions and refrigerators. To minimize risks, the dealer usually employs a cautious approach to assess the creditworthiness of the client (the client must, for example, be an employee of a reputed firm or should have a support letter from a local credit organization such as a cooperative society).

The fact that one institution handles the sales, installation, and credit/recovery as well as the maintenance, training, and other after-sales services is seen as a major advantage of this model. Little involvement of the government or external agencies is also advantageous in some ways.

Criticism of this model is that it channels the expensive working capital of the PV supplier and excludes the poorest segment of the households due to high down payments, the short credit period involved, and more importantly, because of the strict credit track record that is usually required of the end user. PV companies oftentimes lack the skills and are not equipped to administer a credit scheme, as this requires special expertise and is time consuming. The operations of a particular supplier may also become geographically restricted in view of the infrastructure needed for the collection of the payments and possible retrieval of the collateral. The main risk lies with the PV company/dealer from nonpayment of the credit from the end user. This can be mitigated by thoroughly evaluating the creditworthiness of the client before the sale is made as well as by using the PV system as a collateral. Clear advice to the clients on the limitations of the system as well as sufficient attention to maintenance are some factors that can influence the success or failure of the system.

There are two variations of the dealer/user credit model, namely, end-user credit and hire purchase. Sometimes a sound financial institution, with rural outlets having credibility with the end users, is interested in financing PV credits. If the financial institution can implement the credit scheme, the PV supplier is relieved of the responsibility and risks of financing the project, and the valuable working capital remains available for the PV company. Although this can be advantageous, strong rural credit institutions are scarce, and such rural institutions often have economic development as their main objective and may have their focus on income-generating credits. Further, such institutions might be lacking in the knowledge of PV systems and might promote low-initial-cost systems, thereby compromising on maintenance costs.

In the case of the hire-purchase model, either the PV company or an intermediary credit institution offers the PV system on a hire-purchase basis. The client (lessee) makes periodic payments for a limited period, typically 2 or 3 years. The company (lessor) remains the owner of the system during the rental or lease term, and the ownership is handed over to the lessee at the end of the period. The installation and after-sales service is carried out by the PV company. The advantages of this model include the reduced initial down payment and prolonged repayment period. With this model, there is a good chance that higher-quality products are selected because of the long repayment period involved, which makes maintenance a higher priority.

Criticism of this model includes the citing of instances where end users may not have treated the systems with care, as the initial maintenance and ownership do not lie with them. In addition, PV companies are usually not capable of running a hire-purchase program, as it requires additional financial administration skills and can be time consuming.

16.3.4 FEE FOR SERVICE OR FEE FOR ENERGY

In this model, an energy service company (ESCO) invests in PV systems and sells an energy service to the end users, who might be in remote locations. The ESCO remains the owner of the hardware and is responsible for installation, maintenance, and repair/replacement of components. The end user pays a connection fee and makes a periodic payment (usually monthly). The end user pays as long as he receives the energy service and never becomes the owner of the system. However, most of the time, the end user owns the electrical loads (lamps, fans, and other appliances).

As the end user does not invest in a solar system but only has to make periodic payments for the energy delivered, a large segment of the population can choose to have access to electricity. Because long-term agreements will be in place, the quality of the installed systems will usually be high, and the maintenance will receive a professional approach.

Barriers and limitations to this model include the low rate of return and long payback period to the ESCO. The end user is not the owner of the system and hence may not treat the system with care. Also, the PV system should be theft and tamper proof, and monthly collection of the fees is time consuming and expensive. These high risks and high transaction costs may result in high monthly fees to the end user and may reduce affordability for the poor. End-user expectations are also often high, which might result in disappointments at certain times, such as when systems run out of energy purely due to weather conditions.

Once a particular model is selected for implementation and the role players are identified, the project developer has to determine the financial sources for the project on an urgent basis, for it is financing that makes or breaks most PV projects.

16.4 SOURCES OF FINANCING FOR PV-BASED RURAL ELECTRIFICATION

PV technologies, including solar home systems, are usually expensive relative to the average individual income in developing countries. Securing financing is therefore an important element of a PV project. Funds could be by way of loans, grants, equity investment, or other instruments. However, securing PV financing can be a long and winding road. Successful projects had persistent project developers who were able to demonstrate the profitability as well as the nonfinancial merits of the project.

Barriers exist at the program level (national level) as well as the project level. National governments can access and secure financing from large multilateral and bilateral development banks only by submitting to a lengthy and complicated process. Commercial lending and investment organizations may also still perceive PV lending as a high-risk area owing to a lack of familiarity with the technology or a lack of access to well-informed advice about PV system financing. The relatively small size of PV projects, especially by the standards of investment organizations, can also be a barrier. PV project developers may have to compete with a host of other important rural development projects, while the available funds could be limited. Development financiers also may have political agendas and technological preferences around which PV program developers will have to work.

To navigate these barriers, PV project developers should familiarize themselves with potential funding sources and initiate early conversations with them. Before arriving at a mix of financing options, a number of variables, such as return on investment (ROI) and the length, uncertainties, and risks of a given program, should be considered. The application process will usually require considerable dialogue and substantial paperwork; hence, a thoroughly worked out business plan that includes market analysis would be essential. The application time frame can vary widely from source to source, taking as long as a couple of years with the large multilateral development banks and as little as several months with commercial sources.[5]

16.4.1 OVERVIEW OF SOURCES

While international financing sources play an important role in financing PV projects, most funding is from the national development funds in the developing countries (80% or more) themselves. Commercial banks and investment firms at both the national and international level are also important sources for financing PV projects. Figure 16.5 presents an overview of important international sources for PV finance.

16.4.2 INTERNATIONAL CONCESSIONARY FINANCING

International concessionary finance is mainly provided by way of loans (some grant money may also be available) by multilateral development banks (MDB) such as those under the World Bank Group. However, most official development assistance (ODA) is accessible only to the national governments of the developing countries and may not be directly accessible to individual project planners. Prior to receiving any funding, the MDB application process — which can include compilation of information, application writing, and exchanges with the MDB — can take up to anywhere between 2 to 5 years!

Alternative international financing includes funds from private foundations (e.g., Shell Foundation, E & Co., etc.) as well as green investment mechanisms such as tradable certificates for CO_2 emissions. These are market-based efforts to externalize the environmental benefits of renewable energy and solar electrification.

A large pool of financing resources is available to developing countries through bilateral donor funds. Bilateral agencies, like MDBs, are more open to innovative or unproven technology than the commercial financiers. However, geographical restrictions for project location might apply, and sometimes the equipment suppliers or project managers must be sourced from the donor country. While MDB financing favors longer-term projects (5 to 10 years), bilateral financing favors projects with time-spans of just 2 to 3 years. However, in many cases, bilateral funding cannot ensure sustainability of the market development. If follow-up funding is not available, the situation can prove disruptive to market development.

16.4.3 NATIONAL AND COMMERCIAL FINANCING

Many developing countries provide their own development funds created through taxes or tariffs. Governments can also provide fiscal and financial incentives to offset the cost of development projects, such as tax and customs exemptions on equipment, accelerated depreciation benefits, etc. Subsidies are also provided by the local governments in the form of a refund of a portion of the cost of a PV system to end-users or to project developers. Sometimes, subsidies are provided direct by way of grant to PV system companies to assist them in marketing, selling, and maintaining PV systems. Although subsidies and incentives can assist market

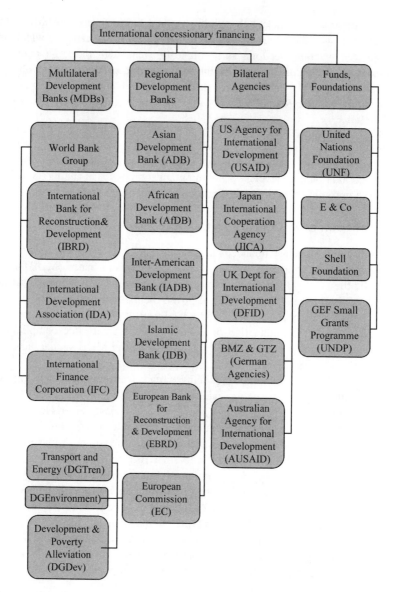

FIGURE 16.5 Sources of international concessionary financing.

development, if government fiscal policy changes (for instance, when a new government comes to power), the project developer may be faced with an unmanageable increase in the cost of equipment, or the end users may no longer be able to afford the planned PV systems. Thus subsidies have often been criticized for creating a false market. Moreover, while subsidies can assist in establishing a market where one could not exist before, their introduction into existing markets can create unsustainable market demands.

16.5 MARKETING SOLAR WATER PUMPS:
A CASE STUDY

As a case study, the pertinent issues of energy supply for a water-pumping operation in the agriculture sector are examined in this section. The case for solar-powered water pumping and experience with an implementation model involving an ESCO are presented.

16.5.1 ENERGY AND WATER USE IN AGRICULTURE

The problems affecting the electricity sector in the countries in the Asia Pacific region have been described by Padmanabhan and Sarkar.[6] A vicious cycle in energy and water use in agriculture begins with inefficient operation of the utilities in the public sector, resulting in poor voltage profiles, high distribution losses, and low load-diversity factor, culminating in remarkably low end-use efficiency of electricity and water (Figure 16.6). For instance, the agricultural sector in India consumes 27% of the electricity and 85% of available freshwater (irrigation efficiency 20 to 50%) while contributing only 5 to 10% of revenue. Although the operating costs are higher, diesel pumps are widely used to compensate for the unreliable power supply.

Problems in groundwater management can have a potentially huge implication for world carbon dioxide emissions. In the case of India, studies have projected an increase of 4.8 to 12% in emissions for each 1-m drop in groundwater levels. Solar-powered pumping, if implemented with sufficient attention to demand-side management (for instance including a drip-irrigation system with water-usage efficiency 90% or above), can potentially provide a viable alternative.

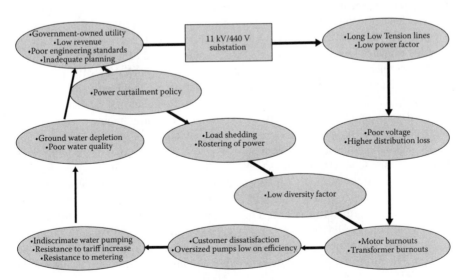

FIGURE 16.6 Vicious cycle of energy and water usage in the Indian agriculture sector.

16.5.2 Punjab PV Pumping Project

In 2000, the government of the northern Indian state of Punjab initiated an innovative program to install 1000 PV pumps in just 2 years (each pump with a PV array of 1800 Wp[*]). The state of Punjab, India's food basket bordering Pakistan, is known for its "green revolution" to accomplish massive increases in production of wheat, rice, etc., crops in the 1950s by implementing extensive canal irrigation systems as well as using groundwater sources. Farmers use a combination of diesel, electric (open as well as submersible, with high discharge capacity), and tractor pumps, usually in the capacity range of 5 to 10 hp (3.7 to 7.4 kW), with low irrigation efficiencies (<50%). Despite the fact that the state-run electricity board has been operating for several years with a negative rate of return of over 10%, the government has continued to supply electricity for pumping free of cost or with only a marginal cost recovery, mainly owing to political reasons (large vote bank, strong farming lobby). Thus the solar pumps chosen for the program, which were 1.8 kW (ca. 2 Hp) in capacity and cost over US $10,000 (2002 market price in India), had to compete with the larger pumps costing under US $500 that were receiving free electricity. In the absence of concessionary financing and subsidy from the government, there was no market. However, an implementation model integrating financial engineering that involved a leasing company drastically reduced the up-front price for the pumps for the farmers, thus creating a market.

16.5.3 Financial Engineering

The Ministry of Nonconventional Energy Sources (MNES) in India has been implementing a unique solar pumping program (launched in 1992–1993), and by 2002 some 4200 pumps had been installed under the program. The Indian Renewable Energy Development Agency (IREDA), the financial arm of the ministry, was offering soft loans for the purchase of pumps at an annual rate of interest of 2.5%, with a repayment period of 10 years as well as a moratorium of 2 years. Over and above this, the ministry also was providing a direct capital subsidy for the purchase of the pumps. Furthermore, if a profit-making company were to invest in a PV project, it would be eligible to claim an accelerated depreciation benefit on the cost of the project (i.e., a tax break taken in the first year that, in the normal course of events, would have been amortized over 10 years). For the Punjab pumping project, all of these elements of financial engineering were included.

A reputable company secured a soft loan from IREDA for a large number of pumps (at times, several hundred pumps), absorbed the accelerated depreciation benefit, obtained the available capital subsidy, and finally leased the pumping units to individual farmers upon payment of a one-time fee that is only a fraction of the true cost (costs to the farmers have ranged from US $900 to 1500 over the years). The Punjab Energy Development Agency (PEDA), which was the nodal agency for implementing the project, also pumped in additional grants to reduce the cost of the system. The farmer would only make an up-front payment, but theoretically would

---

[*] The rated power of PV panel expressed in watt peak, or Wp, is a measure of how much power a solar panel can produce under optimal conditions.

receive the ownership of the pump at the end of 10 years. As there is no further payment requirement from the farmer, there is little cause for default, and the only binding condition to the farmer was that the PV array and pump should be used for agriculture purposes during the lease period.

16.5.4 IMPLEMENTATION OF THE PROGRAM AND LESSONS LEARNED

Marketing efforts begin with newspaper advertisements. A preinstallation survey follows when the farmer expresses his interest by paying an initial deposit, and then a suitable site is mutually agreed upon between the energy service company (ESCO) and the farmer. Helped by an unreliable grid supply and a long waiting period for a new connection from grid, a total of 1000 pumps were successfully installed between October 2000 and March 2002.[7] Following this success, the project is now being repeated on a yearly basis, and neighboring states have picked up on the initiative and started their own programs. The success of this program highlights several novelties and winning factors:

The Punjab Energy Development Agency (PEDA) chose to award the contract to to a few leading domestic solar PV companies such as Tata BP Solar, Bharat Heavy Electricals, etc ., instead of contracting a single company. These companies, in turn, chose their own ESCOs to fulfill the contractual obligation with PEDA. These contracted ESCOs were contracted to provide both installation and maintenance service for 5 years after installation. Auroville Renewable Energy (AuroRE) was one of the chosen ESCOs, a company that also specialized in the manufacture and supply of balance-of-system (BOS) components (including accessories such as array-support structures, trackers, etc.) as well as installation. AuroRE, located in South India (over 1500 km from Punjab), was successfully able to identify key issues, train personnel, and implement the project through local representatives, which goes to prove that geographic proximity to the project location is not one of the biggest challenges to overcome if the chosen implementation model is right. AuroRE was awarded the Ashden Award (considered a green Oscar by many) for the year 2004. In presenting the award, the Ashden judges commended AuroRE for its integrated approach to supplying energy services, combining technical and business competence with a strong commitment to the greater use of sustainable energy, which sums up the requirements for a successful ESCO.

Rotomag, an indigenously and exclusively manufactured DC centrifugal pump of 2-hp capacity, was used under this program to replace more expensive (to buy and maintain) Grundfos pumps, which were earlier imported for the packaged systems. Two Axis tracking structures have been independently developed by various installation companies.

User training — including demonstration of the operation of the system at the field level, pasting of waterproof instruction sheets in the regional language on the array, etc. — were important elements of this program.

Maintenance problems often arise from equipment or devices that may not be part of the PV systems itself. For instance, a faulty foot valve may render the system inoperational. The motor brushes of the DC pump need replacement after 2 to 3 years. Surveys have indicated that these details have been attended to, and over 98% of the systems were still functioning satisfactorily after 1 year.

Surveys also indicated a high level of user satisfaction, and farmers were increasingly adopting efficient irrigation systems such as drip irrigation (>90% irrigation efficiency) to grow high-value plantation crops rather than field crops, which would generate higher income for the farmer.

Thus the Punjab solar-powered pumping program, relying heavily on subsidies, has been able to establish a market for solar pumping system where there was none before. However, in the opinion of the authors, the sustainability of this market is dependent upon how well the lessons of past mistakes with subsidies are integrated into this program. This requires that subsidies be tapered off gradually, that ESCO services be maintained, and that improved irrigation practices continue to be encouraged.

REFERENCES

1. Inversin, A.R., *Reducing the Cost of Grid Extension for Rural Electrification*, Joint UNDP/World Bank Energy Sector Management Assistance Programme (ESMAP), ESM227, World Bank, Washington, DC, February 2000, p. 48.
2. IEA, *Trends in Photovoltaic Applications Survey*, report of selected IEA countries between 1992 and 2003, Report IEA-PVPS T1-13:2004, International Energy Agency, Paris, 2004, p. 6.
3. Nordmann, T., in *Subsidies versus Rate-Based Incentives for Technology: Economical and Market Development of PV, the European Experience*, 3rd World Conference on Photovoltaic Energy Conversion, Osaka, 2003.
4. IEA, *Summary of Models for the Implementation of Solar Home Systems in Developing Countries*, Report IEA-PVPS T9-02:2003, International Energy Agency, Paris, 2003, pp. 33–38.
5. IEA, *Sources of Financing for PV-Based Rural Electrification in Developing Countries*, Report IEA-PVPS T9-08:2004, International Energy Agency, Paris, 2004.
6. Padmanabhan, S. and Sarkar, A., in *Electricity Demand Side Management in India: a Strategic and Policy Perspective*, International Conference on Distribution Reforms, New Delhi, 2001.
7. Akker, J.v.d. and Lamba, H., Thinking big: solar water pumping in the Punjab, *Refocus*, 3, 40–43, 2002.

Index

R

S